Hematology and Coagulation

Hematology and Coagulation

A Comprehensive Review for Board Preparation, Certification and Clinical Practice

Amer Wahed, MD

Assistant Professor of Pathology and Laboratory Medicine,
University of Texas Medical School at Houston

Amitava Dasgupta, PhD, DABCC

Professor of Pathology and Laboratory Medicine,
University of Texas Medical School at Houston

ELSEVIER

AMSTERDAM • BOSTON • HEIDELBERG • LONDON • NEW YORK • OXFORD
PARIS • SAN DIEGO • SAN FRANCISCO • SINGAPORE • SYDNEY • TOKYO

Elsevier
Radarweg 29, PO Box 211, 1000 AE Amsterdam, Netherlands
The Boulevard, Langford Lane, Kidlington, Oxford OX5 1GB, UK
225 Wyman Street, Waltham, MA 02451, USA

Notices
Knowledge and best practice in this field are constantly changing. As new research and experience broaden
our understanding, changes in research methods, professional practices, or medical treatment may become necessary.

Practitioners and researchers must always rely on their own experience and knowledge in evaluating and using any
information, methods, compounds, or experiments described herein. In using such information or methods they should be
mindful of their own safety and the safety of others, including parties for whom they have a professional responsibility.

To the fullest extent of the law, neither the Publisher nor the authors, contributors, or editors, assume any liability
for any injury and/or damage to persons or property as a matter of products liability, negligence or otherwise,
or from any use or operation of any methods, products, instructions, or ideas contained in the material herein.

Medical Disclaimer

Medicine is an ever-changing field. Standard safety precautions must be followed, but as new research and clinical
experience broaden our knowledge, changes in treatment and drug therapy may become necessary or appropriate.
Readers are advised to check the most current product information provided by the manufacturer of each drug to be
administered to verify the recommended dose, the method and duration of administrations, and contraindications.
It is the responsibility of the treating physician, relying on experience and knowledge of the patient, to determine
dosages and the best treatment for each individual patient. Neither the publisher nor the authors assume any liability
for any injury and/or damage to persons or property arising from this publication.

ISBN: 978-0-12-800241-4

British Library Cataloguing-in-Publication Data
A catalogue record for this book is available from the British Library

Library of Congress Cataloging-in-Publication Data
A catalog record for this book is available from the Library of Congress

For information on all Elsevier publications
visit our website at http://store.elsevier.com/

Typeset by MPS Limited, Chennai, India
www.adi-mps.com

Printed and bound in the United States of America

Dedication

Dedicated to my wife, Tania, to my sons, Arub, Ayman, and Abyan, and to my professors who taught me hematology and continue to do so—Professor Rashid, Professor Andy Nguyen, and Professor Margaret Uthman

Amer Wahed

Dedicated to my wife, Alice

Amitava Dasgupta

Contents

Preface

This is the second book in a series of books, designed for board review for Pathology residents. The first book in this series *Clinical Chemistry, Immunology and Laboratory Quality Control: A Comprehensive Review for Board Preparation, Certification and Clinical Practice* was published, also by Elsevier, in January 2014. The aim of this current book is to provide a strong foundation for students, residents, and fellows embarking on the journey of mastering hematology. It is expected that this will also act as a valuable resource for residents preparing for the clinical pathology board exam. Thus, this book should not be considered a conventional textbook of hematology. There are several such textbooks of hematology and most of them are excellent ones. We have deliberately refrained from including pictures, as conventional textbooks and the internet are excellent sources of such images. At the same time, this will allow the final cost of the book to remain reasonable. We have added a section, denoted as "Key Points" at the end of each chapter. We hope that this section will be a good resource for reviewing information, when time at hand is somewhat limited.

We are confident that medical students with a keen interest in hematology will find this book to be of benefit, as well as clinical hematologists who wish to have a better understanding of the diagnostic aspects of hematologic diseases.

We would like to thank our Department Chair, Dr. Robert Hunter, for encouraging us to write the second book in the series. We also thank our pathology residents who read various chapters of this book critically and made extremely helpful comments. In particular, we have to mention Dr. Alyaa Al-Ibraheemi, Dr. Andres Quesada, and Dr. Elizabeth Jacobi in this regard. If readers enjoy reading this book, our efforts will be duly rewarded.

Amer Wahed

Amitava Dasgupta

Houston, Texas

CHAPTER 1

Complete Blood Count and Peripheral Smear Examination

1.1 INTRODUCTION

A complete blood count (CBC) is one of the most common laboratory tests ordered by clinicians. Even for a routine health checkup of a healthy person, CBC is ordered to ensure there is no underlying disease when the individual may be asymptomatic. Tefferi *et al.* noted that at the Mayo Clinic in Rochester, Minnesota, approximately 10−20% of CBC results are reported as abnormal. Common abnormalities associated with abnormal CBC include anemia, thrombocytopenia, leukemia, polycythemia, thrombocytosis, and leukocytosis [1]. For CBC analysis, the specimen must be collected in an ethylene diamine tetraacetic acid (EDTA) tube (lavender or purple top).

CBC consists of numbers that are printed out from the hematology analyzer. In addition, the printout contains graphs and "flags." Flags are essentially messages provided by the analyzer to the interpreting person that certain abnormalities may be present. For example, an analyzer may flag that blasts are present. Therefore, a review of blood smear slides is required to verify the presence of blasts. To make a meaningful interpretation of the peripheral smear, a review of not only the CBC printout but also the patient's electronic medical records must be conducted. CBC parameters that are printed from an automated hematology analyzer are red blood cell (RBC)-related numbers, white blood cell (WBC)-related numbers, and platelet-related numbers (Box 1.1).

1.2 ANALYSIS OF VARIOUS PARAMETERS BY HEMATOLOGY ANALYZERS

Modern hematology analyzers are capable of counting as well as determining the size of various circulating blood cells in blood, including RBCs, WBCs, and platelets [2]. One such instrument, the Beckman−Coulter analyzer, generates an electrical pulse when a blood cell passes through the analyzer channel, which

CONTENTS

1

A. Wahed and A. Dasgupta: Hematology and Coagulation. DOI: http://dx.doi.org/10.1016/B978-0-12-800241-4.00001-2

BOX 1.1 VARIOUS PARAMETERS PRINTED BY A HEMATOLOGY ANALYZER FOLLOWING CBC ANALYSIS

Red Blood Cell (RBC)-Related Numbers
- RBC count
- Hemoglobin level
- Hematocrit
- Red cell differential width (RDW)
- Mean corpuscular volume (MCV)
- Mean corpuscular hemoglobin (MCH)
- Mean corpuscular hemoglobin concentration (MCHC)
- Reticulocyte count

White Blood Cell (WBC)-Related Numbers
- Total WBC count corrected
- Total WBC count uncorrected
- WBC differential
- Absolute count of each type of WBC

Platelet-Related Numbers
- Platelet count
- Mean platelet volume (MPV)
- Platelet differential width (PDW)

consists of a small aperture surrounded by electrodes. Each electrical pulse represents an individual cell, and pulse height indicates the cell volume. Modern hematology analyzers are also capable of multimodal assessment of cell size and cell count, thus providing additional information regarding various categories of WBCs, such as neutrophils, lymphocytes, monocytes, eosinophils, and basophils.

The following are examples of various channels in a hematology analyzer:

- Channel for red cells (and also platelets): This channel is capable of analyzing red blood cells and platelets.
- Channel for WBC and hemoglobin measurement: Lytic agents lyse red cells first before analysis.
- Channel for WBC differential count.
- Channel for reticulocyte count.
- Other channels: nucleated red blood cell (NRBC) channel, separate hemoglobin (Hb) channel, WBC/basophil channel, and immature granulocyte channel.

Different hematology analyzers may use different methods for counting, including the following (one analyzer may employ multiple methods):

- Impedance
- Conductivity measurements with high-frequency electromagnetic current (depends on the internal structure, including nuclear cytoplasmic ratio and nuclear density to granularity ratio)
- Light scatter
- Fluorescence-based methods.

1.2.1 RBC Count and Hemoglobin Measurement

Typically, one channel is used to detect RBCs and platelets. The detector is set such that any cell between 2 and 30 fL will be counted as a platelet and any cell between 40 and 250 fL will be counted as a red cell. If there are large platelets, these will be counted as red cells and will also result in a falsely low platelet count. Similarly, if there are fragmented red cells, these smaller red cells will be counted as platelets.

For hemoglobin measurement, the spectrophotometric method is used after the red cells are lysed. The principle of the method is oxidation of ferrous ion of hemoglobin by potassium ferricyanide into the ferric ion of methemoglobin, which is then converted into stable cyanomethemoglobin by potassium cyanide. However, sulfhemoglobin, if present, is not converted into cyanomethemoglobin under this reaction condition. Usually, dihydrogen potassium phosphate (added to lower pH and accelerate the reaction) is used in the reaction mixture. Nonionic detergents are also used to accelerate lysis and reduce turbidity. Finally, absorbance of light at 540 nm is measured, and the intensity of the signal corresponds to hemoglobin concentration.

Smokers have a higher than normal carboxyhemoglobin concentration because carboxyhemoglobin takes longer to convert into cyanomethemoglobin and also absorbs more light at 540 nm compared to cyanomethemoglobin. Thus, the hemoglobin value in smokers may be falsely elevated. Nordenberg *et al.* commented that cigarette smoking seems to cause a generalized upward shift of the hemoglobin distribution curve, which reduces the diagnostic value of detecting anemia in smokers using the hemoglobin value. The authors suggested that the minimum hemoglobin cutoff level for anemia should be adjusted for smokers [3]. Fetal hemoglobin may interfere with spectrophotometric measurement of carboxyhemoglobin, thus falsely indicating carbon monoxide poisoning in an infant [4].

Hyperlipidemia and hypergammaglobulinemia can also falsely elevate hemoglobin levels. In cold agglutinin disease, red cell agglutination usually takes place. In such situations, a clump of red cells may be counted as one red cell. Thus, the RBC count may be falsely low, and the mean corpuscular volume (MCV) will be falsely high. However, when the red cells are lysed, a true hemoglobin result will be available. Therefore, a clue to cold agglutinin disease is a disproportionate low RBC count compared to the hemoglobin level.

1.2.2 Hematocrit, Red Blood Cell Distribution Width, Mean Corpuscular Volume, Mean Corpuscular Hemoglobin, and Mean Corpuscular Hemoglobin Concentration

In the context of RBC parameters, measurement of MCV and red blood cell distribution width (RDW) is also performed. The hematocrit (Hct) value is calculated from the MCV and the RBC count using the following formula (normal Hct value approximately 45%):

$$Hct = MCV \times RBC\ count$$

RDW is a measure of the degree of variation of size of red cells—that is, it reflects the extent of anisocytosis. RDW is elevated in iron deficiency anemia, myelodysplastic syndrome, and macrocytic anemia secondary to vitamin B_{12} or folate deficiency. In contrast, RDW is usually normal or mildly elevated in thalassemia. The RBC histogram is plotted using volume as the x axis and percentage as the y axis. The RBC histogram has an ascending slope, a peak, and a descending slope. A perpendicular line drawn from the peak down to the x axis represents the MCV. Mean corpuscular hemoglobin (MCH) refers to the average amount of hemoglobin found in RBCs. Mean corpuscular hemoglobin concentration (MCHC) represents the concentration of hemoglobin in RBCs. Both MCH and MCHC are also calculated values (Box 1.2). In cold agglutinin disease, the RBC count is low, and therefore the Hct is also low; the MCHC is high. The laboratory scientist uses abnormally high MCHC as an indicator of possible cold agglutinin disease and warms the blood prior to repeating the CBC run on the analyzer. In hyperosmolar states, cells swell, causing increased MCV.

BOX 1.2 VARIOUS FORMULAE USED FOR CALCULATING DIFFERENT PARAMETERS IN CBC ANALYSIS

Hematocrit (Hct)

$$Hct = MCV \times RBC\ count$$

Mean Corpuscular Hemoglobin (MCH)

$$MCH = Hb/RBC\ count$$

Mean Corpuscular Hemoglobin Concentration (MCHC)

$$MCHC = Hb/Hct$$

Corrected Reticulocyte Count

$$Corrected\ reticulocyte\ count = reticulocyte\ count \times patient\ Hct/normal\ Hct$$

Reticulocyte Production Index (RPI)

$$RPI = [(\%\ reticulocyte \times hematocrit)/45] \times [1/correction\ factor]$$

Correction factor: 1 (hematocrit 40–45), 1.5 (hematocrit 35–39), 2.0 (hematocrit 25–34), 2.5 (hematocrit 15–24), or 3 (hematocrit <15)

MCH is decreased in patients with anemia caused by impaired hemoglobin synthesis. MCH may be falsely elevated in blood specimens with turbid plasma (usually caused by hyperlipidemia) or severe leukocytosis.

MCHC is decreased in microcytic anemias in which the decrease in hemoglobin mass exceeds the decrease in the size of the RBCs. It is increased in hereditary spherocytosis and in patients with hemoglobin variants, such as sickle cell disease and hemoglobin C disease.

1.2.3   Reticulocyte Count

Reticulocytes are immature red cells. They are named as such because they contain reticular material that is actually RNA. The RNA can be seen with special stains such as new methylene blue. Reticulocyte count is used to assess bone marrow response to anemia. It is important to use the corrected reticulocyte count when making such assessments. The formula is provided in Box 1.2. Reticulocytes may also be assessed using the reticulocyte production index (RPI), and the formula is also provided in Box 1.2.

1.2.4   WBC Count and Differential

The WBC histogram has three peaks. The first peak corresponds to lymphocytes, and the third peak corresponds to neutrophils, whereas the second peak corresponds to the remaining types of WBCs. When nucleated RBCs are present, these cells may be counted as WBCs, especially lymphocytes. The total WBC count may thus be falsely elevated. For an accurate WBC count, the analyzer must be run in the NRBC mode. This is referred to as the corrected WBC count. In some printouts, UWBC represents uncorrected WBC count, and WBC represents corrected WBC count. Also in some printouts, the corrected WBC count is denoted by the sign "&" before the WBC count. When significant myeloid precursors are present, the downward slope of the neutrophil peak might not touch the baseline.

1.2.5   Platelet Count, Mean Platelet Volume, and Platelet Differential Width

Pseudothrombocytopenia is an important issue. Causes of falsely low platelet count include the following:

- Traumatic venipuncture and activation of clotting
- A significant number of large platelets (platelets being counted as RBCs)
- EDTA-induced platelet clump
- EDTA-dependent platelet satellitism (platelets form a satellite around neutrophils).

The last two conditions are typically diagnosed when the slide is reviewed, although hematology analyzers are capable of flagging platelet clumps. Blood should be re-collected in citrate or heparin. If thrombocytopenia is due to peripheral destruction or consumption of platelets, then the bone marrow responds to the thrombocytopenia by releasing immature platelets, which are larger than normal. This increases the mean platelet volume (MPV) and also the platelet differential width (PDW). If thrombocytopenia is due to reduced production by the bone marrow, large platelets are not seen and thus MPV and PDW are not increased.

1.3 REVIEW OF PERIPHERAL SMEAR

A microscopic examination of appropriately prepared and well-stained blood smear slides by a pathologist or a knowledgeable laboratory professional is useful for clinical diagnosis. A blood smear analysis takes into account flagged automated hematology results and enables the determination of whether a manual differential count should be performed. Therefore, peripheral blood smear examination along with manual differential leukocyte count (if necessary) and CBC provide the complete hematological picture of the patient [5]. Review of the smear should start by ensuring that the name and accession number on the slide match those on the CBC printout. Sometimes naked eye examination of the slide may provide some important clues. For example, if the slide appears blue, there is a possibility of underlying paraproteinemia or myeloma. When paraproteins are present in significant amounts in blood, they are stained blue by the Wright–Giemsa stain. Sometimes tumor emboli are visible as clumps on the slide. Cryoglobulinemia may appear as "blobs" on the slide.

The slide should be scanned at first at low power to assess overall cellularity (especially of white cells), to find an appropriate area where red cell morphology is best assessed (under higher power), and to check for platelet clumps. Red cell morphology is typically best assessed where the cells are evenly distributed, and this area is away from the tail, toward the body. Rouleaux formation and red cell agglutination can also be appreciated under low power. With experience, blasts can also be picked up on low power. It is then best to assess each cell line under higher power.

Red cells are assessed for size, shape, anisocytosis, central pallor, and red cell inclusions, if any. WBCs should be checked for reactive (toxic) changes, left shift, the presence of blasts, and the degree of segmentation as well as the presence of dysplasia. Platelets should be checked for clumps, size, and adequacy of granules.

1.3.1 Red Cell Variations and Inclusions

Normal red cells are normocytic normochromic. This means that the average size of a red cell in an adult is the size of the nucleus of a mature lymphocyte. Only one-third of the central portion of the red cell has central pallor. Increased pallor means that the red cell is hypochromic. Thus, red cells may be the following:

- Normocytic normochromic
- Microcytic hypochromic
- Macrocytic.

Variation in shape refers to poikilocytosis. Examples of the types of cell found in poikilocytosis are sickle cells, target cells, ovalocytes, elliptocytes, stomatocytes, echinocytes (Burr cells), acanthocytes, schistocytes, spherocytes, dacryocytes (teardrop red cells), and bite cells. Each of these poikilocytes is associated with one or more underlying clinical conditions and is discussed in Chapter 3. There are also various red cell inclusions that can be observed during peripheral blood smear examination. These are listed in Box 1.3.

1.3.2 WBC Morphology

Reactive changes are predominantly appreciated in neutrophils and lymphocytes. Reactive neutrophils have prominent azurophilic granules, cytoplasmic vacuoles, and Dohle bodies (blue cytoplasmic bodies). Dohle bodies are named after the German pathologist, Karl Gottfried Paul Dohle, and these bodies represent rough endoplasmic reticulum. Dohle bodies in reactive neutrophils are typically seen at the periphery of the cell. Another condition in which "Dohle-like" bodies may be seen is May–Hegglin anomaly, in which Dohle-like bodies are randomly distributed throughout the cell. They are devoid of organelles. Rather, they are thought to consist of a mutant form of the nonmuscle myosin heavy-chain protein.

Reactive lymphocytes (also known as Downey cells) may be of three types:

- A larger than usual lymphocyte with abundant cytoplasm that appears to surround (or hug) the red cells (Downey type II cell) is the most common type of reactive lymphocyte

---

BOX 1.3 VARIOUS RED CELL INCLUSIONS

- Howell–Jolly bodies
- Pappenheimer bodies
- Cabot rings
- Basophilic stippling (punctate basophilia)
- Heinz bodies
- Hemoglobin C crystals
- Malarial parasite
- Nucleated RBCs

- A small lymphocyte with nuclear membrane irregularity (Downey type I cell)
- A larger lymphocyte with blue cytoplasm and nucleoli (Downey type III cell).

When there is neutrophilic leukocytosis, a pathologist may observe the presence of immature myeloid precursors in the peripheral blood. This is referred to as left shift. Rare blasts may also be present. Occasionally, patients may present with very high WBC count and left shift mimicking leukemia, although the process is reactive. This is referred to as a leukemoid reaction. Neutrophilic leukocytosis without reactive changes and with basophilia with or without eosinophilia is suspicious for chronic myelogenous leukemia. Numerous blasts may indicate an acute leukemic process.

When the WBC count is high, smudge cells may be observed, which are distorted white cells produced as an artifact during the process of making the slide. The presence of smudge cells with a high lymphocyte count should raise suspicion for chronic lymphocytic leukemia. Pretreatment (prior to making the slide) with albumin ensures that smudge cells are reduced.

Dysplasia is probably the most difficult feature to establish from the peripheral blood. It is most often assessed in the neutrophils. Normal mature neutrophils have two to five nuclear segments and have fine granules in the cytoplasm. Hypogranulation is a feature of dysplasia. Hyposegmentation and hypersegmentation, if present, may represent dysplasia. A bilobed polymorphonuclear leukocyte (PMN) with hypogranulation is referred to as a "pseudo-Pelger–Huët" cell. This cell is considered to be dysplastic. Hypersegmented PMN is a PMN with more than five segments. This can be seen in megaloblastic anemia as well as myelodysplastic syndrome. It may also be inherited, without any clinical significance. Here, the majority (>75%) of the neutrophils are hypersegmented.

Benign disorders of WBCs, such as May–Hegglin anomaly and Alder–Reilly and Chediak–Higashi diseases, may all be diagnosed from the peripheral smear. Rarely, cells such as hairy cells representing hairy cell leukemia may be seen. Patients with lymphoma may have lymphoma cells circulating in the peripheral blood. These will appear as atypical lymphocytes—that is, lymphoid cells that are neither mature nor reactive in appearance. The presence of a Barr body, which is a nuclear appendage, denotes the inactivated X chromosome and implies female sex of the patient.

1.3.3   Platelets

The normal platelet count is 150,000–450,000/μL. Thrombocytopenia is defined as platelet count below the 2.5th lower percentile of the normal platelet count distribution. Results of the third U.S. National Health and Nutrition Examination Survey support the traditional value of platelet count below 150,000/μL as the definition of thrombocytopenia, but adoption of a cutoff value below 100,000 may be more practical. Thrombocytopenia is a common hematological finding with variable clinical expression or may reflect a life-threatening disorder such as thrombotic microangiopathy [6]. Thrombocytopenia is also a common hematological abnormality found in newborns [7]. Thrombocytopenias can be broadly categorized as follows:

- Thrombocytopenias due to decreased production—congenital and acquired (any cause of bone marrow failure)
- Thrombocytopenia due to increased destruction—for example, idiopathic thrombocytopenia purpura
- Thrombocytopenias due to increased consumption—for example, disseminated intravascular coagulation (DIC), thrombotic thrombocytopenia purpura (TTP), and hemolytic uremic syndrome (HUS)
- Thrombocytopenias due to sequestration—sequestration in hemangiomas (Kasabach–Merritt syndrome).

Congenital thrombocytopenias can be broadly divided into three groups: thrombocytopenia with small platelets, thrombocytopenia with normal-sized platelets, and thrombocytopenia with large platelets.

Thrombocytopenia with small platelets can be associated with Wiskott—Aldrich syndrome, X-linked thrombocytopenia, or inherited macrothrombocytes.

Thrombocytopenia with normal-sized platelets can be due to Fanconi's anemia, thrombocytopenia with absent radii (TAR syndrome), amegakaryocytic thrombocytopenia, or Quebec platelet disorder.

Thrombocytopenia with large platelets may be associated with Bernard—Soulier syndrome, May—Hegglin anomaly, Sebastian syndrome, Epstein syndrome, Fechtner syndrome, or gray platelet syndrome. Normally, the α granules of the platelets are stained by the Wright—Giemsa stain. The δ granules are not stained. The absence of the α granules will result in gray-appearing platelets (i.e., gray platelet syndrome). The platelets are dysfunctional.

Thrombocytosis is defined as platelet counts higher than 450,000/μL. Thrombocytosis may be due to reactive thrombocytosis (associated with infection, inflammation, neoplasms, or iron deficiency); rebound thrombocytosis following thrombocytopenia; redistributional (e.g., post-splenectomy), myeloproliferative disorders such as essential thrombocythemia; or familial thrombocytosis. Platelet disorders are discussed in greater detail in Chapter 5.

1.4 SPECIAL SITUATIONS WITH CBC AND PERIPHERAL SMEAR EXAMINATION

There are special situations involving CBC and peripheral blood smear review. Pancytopenia is an important hematological finding in which all three major cells present in blood (RBCs, WBCs, and platelets) are decreased in number. Pancytopenia may not be a disease entity but, rather, a triad of findings that may result from a number of diseases primarily or secondarily involving bone marrow. The severity of pancytopenia determines the course of therapy [8]. Important causes of pancytopenia include the following:

- Bone marrow failure: Any cause of bone marrow failure may result in pancytopenia. Examples include aplastic anemia, bone marrow fibrosis, leukemias, metastatic diseases, and granulomas.
- Vitamin B_{12} or folate deficiency
- Myelodysplastic syndrome
- Autoimmune destruction of cells
- Hypersplenism.

1.4.1 Splenic Atrophy or Postsplenectomy

The absence of spleen is characterized by the presence of Howell–Jolly bodies (in RBCs), acanthocytes, and target cells. There may be transient thrombocytosis and leukocytosis as well.

1.4.2 Microangiopathic Hemolysis

There are three important causes of microangiopathic hemolysis:

- TTP (thrombotic thrombocytopenia purpura)
- HUS (hemolytic uremic syndrome)
- DIC (disseminated intravascular coagulation).

All are characterized by low platelets and the presence of schistocytes in the peripheral smear. Typically, there are numerous schistocytes in TTP and HUS, in contrast to DIC, in which there are lower numbers. In DIC, the coagulation profile (e.g., prothrombin time (PT) and partial thromboplastin time (PTT)) are abnormal. In TTP and HUS, the coagulation profile is typically normal. TTP is a medical emergency and requires urgent therapeutic plasma exchange (TPE).

1.4.3 Leukoerythroblastic Blood Picture

This term refers to the presence of red cell precursors (i.e., nucleated red cells) in the peripheral blood as well as WBC precursors (i.e., left shift with blasts). This may be seen in patients with significant hemolysis or hemorrhage. In such situations, clinicians should also search for teardrop red cells. Leukoerythroblastic blood picture with teardrop red cells may be due to a bone marrow infiltrative process. Anemias due to such bone marrow infiltrative processes are known as myelophthisic anemias. This infiltration may be due to many causes, such as fibrosis, infiltration by tumor, or even leukemia.

1.4.4 Parasites, Microorganisms, and Nonhematopoietic Cells in the Peripheral Blood

Several species of parasites, including malaria parasites, and microorganisms, may be seen in the peripheral blood (Box 1.4).

Occasionally, nonhematopoietic cells may be seen in the peripheral blood. These include the following:

- Epithelial cells
- Fat cells
- Endothelial cells

BOX 1.4 PARASITES AND MICROORGANISMS THAT MAY BE PRESENT IN THE PERIPHERAL BLOOD SMEAR

Parasites
- Malaria: Malarial parasites (see Chapter 3)
- Toxoplasmosis: Trophozoite forms are present in monocytes and may also be seen free from ruptured monocytes.
- Babesiosis: Trophozoites are seen within red cells. The trophozoites are ring form, similar to that of *Plasmodium falciparum*. They have one to three chromatin dots. Sometimes they are pear shaped. The pointed ends of four parasites may come into contact and assume the shape of a Maltese cross, which helps in their identification.
- Trypanosomiasis: Flagellated parasites of *Trypanosoma brucei* or *cruzi* may be seen in the peripheral smear.

Parasitic Microfilaria
- Bancroftian filariasis: Caused by human parasitic roundworm *Wuchereria bancrofti*. The microfilaria form may be seen in the peripheral blood.

- Loiasis: Caused by parasitic worm *Loa loa*. Again, the microfilaria may be seen in the peripheral blood.

Bacteria
- Bartonellosis: Rod-shaped coccobacilli may be seen within the red cells due to infection with bacteria of genus *Bartonella*.
- Borreliosis: *Borrelia* spirochetes may be seen in the peripheral blood due to Lyme borreliosis (Lyme disease) infection.

Fungi
- Individuals with impaired immunity and individuals with indwelling catheters may show fungi in their peripheral blood smear. *Candida*, *Histoplasma capsulatum*, and *Cryptococcus neoformans* may be seen within neutrophils.

- Malignant cells—for example, neuroblastoma cells, rhabdomyosarcoma cells, and medulloblastoma cells. These cells may resemble lymphoblasts. The presence of carcinoma cells is referred to as carcinocythemia. This is most often observed with carcinoma of the lung and breast. Melanoma cells and Reed–Sternberg cells have been described in the peripheral blood.

1.4.5 Buffy Coat Preparation

Buffy coat films are sometimes made to concentrate nucleated cells (i.e., white cells). This is done to search for low-frequency abnormal cells or bacteria or other microorganisms.

KEY POINTS

- For hemoglobin measurement, the red cells should be lysed first. Then hemoglobin (and methemoglobin and carboxyhemoglobin) is converted to cyanomethemoglobin (sulfhemoglobin is not converted). The absorbance of light at 540 nm is then measured.
- Smokers have higher than usual carboxyhemoglobin. Carboxyhemoglobin takes longer to be converted to cyanmethemoglobin. Carboxyhemoglobin absorbs more light at 540 nm than does cyanomethemoglobin, giving rise to a falsely higher hemoglobin level.
- Hyperlipidemia, hypergammaglobulinemia, cryoglobulinemia, fat droplets from hyperalimentation, and leukocytosis (>50,000) can falsely elevate hemoglobin levels.
- In cold agglutinin disease, red cell agglutination is observed. In such situations, a clump of red cells may be counted as one red cell. Thus, the RBC count will be falsely low, and the MCV will be falsely elevated. However, when the red cells are lysed, a true hemoglobin result should be observed. Thus, a clue to cold agglutinin disease is disproportionate low RBC count compared to hemoglobin level. In cold agglutinin disease, MCHC should also be high. The lab uses abnormally high MCHC as an indicator of possible cold agglutinin disease and warms the blood prior to repeating the CBC run on the analyzer.
- In patients with severe hyperglycemia (glucose > 600 mg/dL), osmotic swelling of RBCs may spuriously elevate the MCV.
- MCH is decreased in patients with anemia caused by impaired hemoglobin synthesis. The MCH may be falsely elevated in blood specimens with turbid plasma (usually caused by hyperlipidemia) or severe leukocytosis.

- RDW is elevated in iron deficiency anemia, myelodysplastic syndromes, and macrocytic anemia secondary to vitamin B_{12} or folate deficiency. In contrast, RDW is usually normal or only mildly elevated in thalassemia.
- MCHC is decreased in microcytic anemias in which the decrease in hemoglobin mass exceeds the decrease in the size of the RBC. It is increased in hereditary spherocytosis and in patients with hemoglobin variants, such as sickle cell disease and hemoglobin C disease
- Corrected reticulocyte count is as follows: Reticulocyte count × patient Hct/normal Hct.
- Pseudothrombocytopenia may be due to traumatic venipuncture and activation of clotting, a significant number of large platelets (platelets being counted as RBCs), EDTA-induced platelet clump, or EDTA-dependent platelet satellitism (in which platelets form a satellite ring (rosetting) around neutrophils).
- Features of reactive neutrophils: Prominent granules, cytoplasmic vacuoles, and Dohle bodies (blue cytoplasmic bodies). Dohle bodies represent rough endoplasmic reticulum. Dohle bodies in reactive neutrophils are typically seen at the periphery of the cell. Another condition in which "Dohle-like" bodies may be seen is May−Hegglin anomaly, in which they are randomly distributed throughout the cell. They are devoid of organelles. Rather, they are thought to consist of a mutant form of the nonmuscle myosin heavy chain protein.
- Reactive lymphocytes (also known as Downey cells) may be of three types: A larger than usual lymphocyte with abundant cytoplasm appearing to be "hugging" the red cells (Downey type II cell) is the most common type of reactive lymphocyte. The other two types are a small lymphocyte with nuclear membrane irregularity (Downey type I cell) and a larger lymphocyte with blue cytoplasm and nucleoli (Downey type III cell).
- Important causes of pancytopenia are bone marrow failure, vitamin B_{12} or folate deficiency, myelodysplastic syndrome, autoimmune destruction, and hypersplenism.
- The absence of spleen is characterized by the presence of Howell−Jolly bodies (in RBCs), acanthocytes, and target cells. There may be transient thrombocytoses and leukocytosis as well.
- There are three important causes of microangiopathic hemolysis: TTP, HUS, and DIC. All are characterized by low platelets and the presence of schistocytes in the peripheral smear. In DIC, the coagulation profile (e.g., PT and PTT) is abnormal. In TTP and HUS, the coagulation profile is typically normal. TTP is a medical emergency and requires urgent TPE.

- Leukoerythroblastic blood picture: This term refers to the presence of red cell precursors (i.e., nucleated red cells) in the peripheral blood as well as WBC precursors (i.e., left shift with blasts). This may be seen in patients with significant hemolysis or hemorrhage. In such situations, we should also search for teardrop red cells. Leukoerythroblastic blood picture with teardrop red cells may be due to a bone marrow infiltrative process. This infiltration may be due to many causes, such as fibrosis, infiltration by tumor, or even leukemia.

References

[1] Tefferi A, Hanson CA, Inwards DJ. How to interpret and pursue and abnormal complete blood count in adults. Mayo Clin Proc 2005;80:923−36.

[2] Gulati GL, Hyun BH. The automated CBC: a current perspective. Hematol Oncol Clin North Am 1994;8:593−603.

[3] Nordenberg D, Yip R, Binkin NJ. The effect of cigarette smoking on hemoglobin levels and anemia screening. JAMA 1990;264:1556−9.

[4] Mehrotra S, Edmonds M, Lim RK. False elevation of carboxyhemoglobin: a case report. Pediatr Emerg Care 2011;27:138−40.

[5] Gulati G, Song J, Florea AD, Ging J. Purpose and criteria for blood smear scan, blood smear examination and blood smear review. Ann Lab Med 2013;33:1−7.

[6] Stasi R. How to approach thrombocytopenia. Hematology Am Soc Hematol Educ Program 2012;2012:191−7.

[7] Gunnink SF, Vlug R, Fijnvandraat K, van der Bom JG, et al. Neonatal thrombocytopenia: etiology, management and outcome. Expert Rev Hematol 2014;7:387−95.

[8] Gayathri BN, Rao KS. Pancytopenia: a clinico-hematological study. J Lab Physicians 2011;3:15−20.

CHAPTER 2

Bone Marrow Examination and Interpretation

2.1 INTRODUCTION

Complete blood count (CBC), examination of peripheral blood smear, and other routine laboratory tests may not provide enough information for unambiguous diagnosis of hematological or nonhematological disease in certain patients. For these patients, direct microscopic examination of the bone marrow is required for a proper diagnosis. The bone marrow, which is disseminated within the intertrabecular and medullary spaces of bone, is a complex organ with dynamic hematopoietic and immunological functions. The role of bone marrow in hematopoiesis was first described by Neumann in 1868; since then, methods for bone marrow procedures have undergone many improvements. Following the development of newer techniques and equipment, bone marrow aspiration and bone marrow biopsy have become important medical procedures for diagnosis of hematological malignancies and other diseases and also for follow-up evaluation of patients undergoing chemotherapy, bone marrow transplantation, and other forms of therapy [1]. Bone marrow trephine biopsy should be carried out by a trained health care professional, and bone marrow aspirate should be collected during the same procedure. Because a diagnostic specimen is a small representation of the total marrow, it is important that material be adequate and representative of the entire marrow. The specimen must also be of high technical quality. Cytochemical analysis and various other diagnostic procedures can be performed on the liquid bone marrow aspirate, and bone marrow biopsy material can be stained using immunoperoxidase and other stains. The recent development of bone marrow biopsy needles with specially sharpened cutting edges and core-securing devices has reduced the discomfort of the procedure and improved the quality of the specimen obtained [2]. Today, bone marrow examination is considered an important and effective way to diagnose and evaluate primary hematological and metastatic neoplasm as well as nonhematological disorders [3]. Common indications for performing bone marrow examination are listed in Box 2.1.

CONTENTS

15

A. Wahed and A. Dasgupta: Hematology and Coagulation. DOI: http://dx.doi.org/10.1016/B978-0-12-800241-4.00002-4

BOX 2.1 COMMON INDICATIONS FOR BONE MARROW EXAMINATION

Diagnosis of Diseases
- Acute or chronic unexplained anemia including hypoplastic or aplastic anemia
- Differentiating megaloblastic anemia from normoblastic maturation
- Unexplained leukopenia
- Unexplained thrombocytopenia, pancytopenia
- Myelodysplastic syndrome
- Myeloproliferative disease
- Plasma cell dyscrasia
- Hodgkin and non-Hodgkin lymphoma
- Suspected leukemia
- Disseminated granulomatous disease

- Primary amyloidosis
- Metabolic bone disease
- Suspected multiple myeloma
- Suspected storage diseases (e.g., Gaucher's disease)
- Fever of unknown origin
- Confirmation of normal marrow in a potential allogeneic donor

Follow-Up of Medical Treatment
- Chemotherapy/bone marrow transplant follow-up
- Treatment of isolated cytopenia

2.2 FUNDAMENTALS OF BONE MARROW EXAMINATION

Prior to a bone marrow examination, the relevant history of the patient, CBC, and the report from the peripheral blood smear examination must be reviewed [4]. During a routine bone marrow examination, slides obtained from the aspirate, slides from the clot sections, slides from the trephine biopsy, touch preparation slides obtained from the trephine biopsy, and iron strains must be carefully examined for proper interpretation of results. Occasionally, examination of a well-prepared aspirate slide, core biopsy specimen, and iron strain by a well-trained professional may be adequate for arriving at a diagnosis [1]. However, additional tests, such as flow cytometry and cytogenetics studies, may be needed in other cases. Additional steps that may be performed during bone marrow examination are listed in Box 2.2.

The aspirate slides are typically used to assess morphology by performing a differential count and thus obtaining the myeloid:erythroid (M:E) ratio. If the aspirate lacks particulates or is unsatisfactory, morphology may be assessed from the touch prep slides.

The architecture of the bone marrow is best assessed from the trephine biopsy slides. Infiltrates (e.g., granulomas, lymphomatous infiltrates, and metastatic tumors), if any, and their distribution can also be assessed from the biopsy slides. The cellularity of the bone marrow is usually assessed from the biopsy slides. In addition, reticulin or collagen fibrosis is also assessed from the

> **BOX 2.2 ADDITIONAL STEPS THAT MAY BE PERFORMED AS PART OF A BONE MARROW EXAMINATION**
>
> - Immunophenotyping by flow cytometry (performed on the aspirate specimen)
> - Immunophenotyping by immunohistochemistry (performed on the biopsy or clot section slides)
> - Special stains—for example, acid fast bacilli (AFB), Grocott's methenamine silver (GMS), reticulin, trichrome,
> Wright—Giemsa stain, Prussian blue stain
> - Cytogenetic studies
> - Molecular studies—for example, polymerase chain reaction (PCR), fluorescence *in situ* hybridization (FISH)
> - Electron microscopy

biopsy slides. Bone marrow stroma and the bone itself are assessed from the biopsy slides. In the absence of a good trephine biopsy specimen, the slides from the clot section may be used as an alternate means of assessment.

2.2.1 Dry Tap

Causes of dry tap while performing a bone marrow procedure include the following:

- Faulty technique
- Packed marrow (e.g., with leukemia)
- Fibrotic marrow (e.g., myelofibrosis)
- Hairy cell leukemia.

In cases of dry tap, one must improvise to obtain the greatest possible amount of information. One way of achieving this is to obtain two trephine biopsies and to submit the first for flow cytometry and the other for cytogenetic studies. Good touch preps from the second biopsy should provide adequate morphological and architectural information.

2.2.2 Granulopoiesis

Granulopoiesis involves maturation of myeloblasts into mature polymorphonuclear neutrophils, basophils, and eosinophils. The steps include the transformation of myeloblasts to promyelocytes to myelocytes to metamyelocytes to bands to mature granulocytes.

Myeloblasts are large cells with a high nuclear to cytoplasmic (N:C) ratio, moderately blue cytoplasm (less blue than the cytoplasm of an erythroblast), and prominent nucleoli. Promyelocytes are larger cells compared to myeloblasts and have prominent nucleoli, a Golgi hof, and granules. These granules are primary granules and appear reddish-purple. The promyelocytes of the three granulocytic lineages cannot be differentiated by routine light microscopy. Myelocytes no longer have nucleoli but continue to have granules. However, these granules are secondary and specific granules. Thus, cells of the three granulocytic lineages can now be distinguished. Myeloblasts, promyelocytes, and myelocytes are all capable of cell division. Metamyelocytes have indented nuclei and cannot undergo cell division. The nucleus of bands is "U" shaped. Granulopoiesis in a normal marrow is seen adjacent to the bony trabecular surface (as a layer two or three cells thick) and to arterioles.

The following is one approach to the accurate identification of the various stages of granulopoietic cells:

- If the cell has nucleoli, then it is either a blast or a promyelocyte. If there is a Golgi hof and many granules are present, it is a promyelocyte; otherwise, it is a blast (high N:C ratio). Erythroblasts may resemble myeloblasts, but the cytoplasm of a myeloblast is moderately blue and may have granules. The cytoplasm of an erythroblast is deep blue and does not have granules.
- If the nucleus is indented but not U shaped, it is a metamyelocyte. If there are no nucleoli and no nuclear indentation is present, then it is a myelocyte. Cells with a U-shaped nucleus without proper nuclear segmentation represent a band.

2.2.3   Erythropoiesis

The stages of erythropoiesis are proerythroblast to basophilic normoblast (early normoblast) to polychromatic normoblast (intermediate normoblast) to orthochromatic normoblast (i.e., late normoblast) to reticulocyte to mature red cell. The proerythroblast is a large cell with deep blue cytoplasm, high N:C ratio, and prominent nucleoli. Subsequent to proerythroblasts, nucleoli are no longer seen. In the three normoblasts, the chromatin in the nuclei becomes progressively clumped. In the late normoblast stage, the chromatin is dark, dense, and clumped, ready to be extruded. Once it is extruded, the cell is known as a reticulocyte. In the three normoblasts, the cytoplasm will change color from blue (basophilic normoblast) to gray to gray-orange (late normoblast).

Erythropoiesis can be seen as erythropoietic islands, where cells of the erythropoietic series are seen surrounding a macrophage. The macrophage is referred to as the nurse cell, providing nourishment to the maturing cells. The macrophage also engulfs defective erythroblasts as well as the extruded nuclei from late normoblasts. Erythropoiesis typically occurs close to the sinusoids in the bone marrow.

2.2.4 Monopoiesis, Megakaryopoiesis, Thrombopoiesis, and Other Cells in Bone Marrow

The steps involved in monopoiesis are monoblasts to promonocytes to mature monocytes. The latter are present in very small numbers in a normal bone marrow. Megakaryopoiesis is the process by which mature megakaryocytes develop from hematopoietic stem cell [5]. Thrombopoiesis is the generation of platelets from megakaryocytes. Megakaryocytes are highly specialized large nuclear cells (50−100 μm in diameter) that differentiate to produce platelets. Megakaryocytes mature through endomitosis, in which the nuclear DNA content increases in multiples of two without nuclear or cytoplasmic division. Thus, in the bone marrow, megakaryocytes are 2 N, 4 N, 8 N, 16 N, and 32 N (32 N DNA or 16 copies of normal complement of DNA). The dominant ploidy category is 16 N. The development and formation of megakaryocytes are regulated by a multitude of cytokines, principally Tpo (thrombopoietin), which is produced in the liver and marrow stroma [6]. Megakaryocytes can be classified into three stages of maturation:

- Stage I: These have strong basophilic cytoplasm and a high N:C ratio.
- Stage II: Less basophilic cytoplasm with some azurophilic granules and a lower N:C ratio.
- Stage III: Plentiful cytoplasm; cytoplasm is weakly basophilic; the cytoplasm has abundant azurophilic granules. The cytoplasm of the cell margin is agranular. This is the mature megakaryocyte.

Megakaryocytes may engulf other hematopoietic cells; this is known as emperipolesis. Megakaryocytes are typically found close to the sinusoids. On average, a high power field should demonstrate two to four megakaryocytes. Megakaryocytes do not cluster. Clustering of more than three megakaryocytes may be observed in a regenerating marrow, following chemotherapy and bone marrow transplantation or in pathological states (e.g., myelodysplasia). Various other cells are also present in bone marrow, including mast cells and lymphocytes; these are listed in Box 2.3.

2.3 BONE MARROW EXAMINATION FINDINGS AND BONE MARROW FAILURE

Bone marrow findings can be discussed under the following categories:

- Leukemias
- Lymphomas
- Bone marrow failure
- Disorders of erythropoiesis, granulopoiesis, or thrombopoiesis

BOX 2.3 VARIOUS OTHER CELLS FOUND IN BONE MARROW

- Mast cells: These are seen occasionally and have deep purple granules in their cytoplasm. The granules do not obscure the nucleus, and the nucleus is not lobulated. These features help to distinguish mast cells from basophils.
- Lymphocytes: These are a small population of bone marrow cells (approximately 10%) in which T cells outnumber B cells. Bone marrow of children and recovering marrow exhibit cells that resemble lymphoblasts. These are actually B lymphocyte precursors and are known as hematogones.

- Plasma cells: These represent typically 1% or less of a normal bone marrow.
- Fat cells.
- Osteoblasts: These may resemble plasma cells. However, osteoblasts are larger, and the Golgi hof is separate from the nucleus.
- Osteoclasts: These are multinucleated giant cells. They may resemble megakaryocytes. The nuclei of osteoclasts are separate, and the azurophilic granules are coarser than those of megakaryocytes.

- Infections
- Granulomatous changes
- Storage disorders
- Metabolic bone diseases
- Metastatic tumors
- Miscellaneous: hemophagocytic syndrome, necrosis/infarction, serous atrophy, bone marrow fibrosis, reactive lymphoid surrogate, and amyloidosis.

Bone marrow findings in leukemias and lymphomas are discussed in the sections on various leukemias and lymphomas.

There are also various scenarios for bone marrow failure, including the following:

- In aplastic anemia, trephine biopsy is crucial to document bone marrow hypocellularity. Hematopoietic cells are markedly reduced and fat cells are increased. Lymphocytes are increased and reactive lymphoid aggregates may be seen. Other cells that may be increased are plasma cells, mast cells, and macrophages. Erythroid cells may show features of dysplasia.
- Anticancer and immunosuppressive agents may also cause bone marrow hypoplasia. Dyserythropoiesis may be seen along with megaloblastoid changes. In the early stages of chemotherapy administration, there may be interstitial edema. Subsequently, serous

atrophy (also known as gelatinous transformation) may become evident.

- With radiation damage, there may be necrosis of bone marrow with bone necrosis. Cells next to bony trabeculae are particularly vulnerable to radiation. Hematopoietic cells may be permanently replaced by fat or fibrous tissue.

Mirzai et al. commented that bone marrow examination had the highest diagnostic value with respect to suspicion of leukemia, multiple myeloma, myeloproliferative disorders, and lymphoma and was least helpful in the diagnosis of infection and storage disorders [7].

2.3.1 Disorders of Erythropoiesis, Granulopoiesis, and Thrombopoiesis

Most of the anemias, such as iron deficiency anemia, megaloblastic anemia, and hemolytic anemia, will show features of erythroid hyperplasia and dyserythropoiesis in the bone marrow. In iron deficiency, the erythroblasts are smaller than normal, whereas in megaloblastic anemia they are larger than normal. In megaloblastic anemia, granulopoiesis is also increased with the presence of large forms. Giant myelocytes and metamyelocytes are usually present. With increased intramedullary destruction, an increased number of macrophages may be seen. The increase in cell turnover may be so marked in cases with congenital dyserythropoietic anemia type II that pseudo-Gaucher cells may be apparent. Patients with sickle cell anemia may show evidence of bone marrow necrosis.

Agranulocytosis typically occurs as an idiosyncratic reaction to drugs or chemicals. Marked reductions in mature neutrophils are seen. Bone marrow recovery occurs earlier when myeloid precursors such as promyelocytes and myelocytes are present compared to when they are absent.

In Kostmann's syndrome and agranulocytosis associated with thymoma, myeloid differentiation shows apparent arrest at the promyelocyte stage. In agranulocytosis with superimposed sepsis, promyelocytes may be predominant, mimicking acute promyelocytic leukemia (APL). However, these promyelocytes possess a Golgi hof, and Auer rods are absent.

When thrombocytopenia is due to increased destruction or consumption and the process is sustained, megakaryocytes are increased in the bone marrow. There is also a reduction in the average size of megakaryocytes. In myelodysplastic syndrome (MDS), features of megakaryocytic dysplasia are evident. In reactive thrombocytosis, megakaryocytes are increased in number, with an increase in average size as well as increased variation in size.

2.3.2 Infections

Bone marrow examination is indicated for diagnosis of fever of unknown origin. In general, mycobacterial bone marrow infection is the most common diagnosis established by bone marrow examination performed for fever of unknown origin. In a study of 24 patients, Lin *et al.* reported that 11 patients (46%) had infection due to *Mycobacterium tuberculosis* and 10 patients showed nontuberculosis mycobacteria such as *Mycobacterium avium* or *Mycobacterium kansasii*. In addition, 9 patients were positive for HIV [8]. Bone marrow abnormalities are frequently observed in all stages of patients infected with HIV. The most common abnormality is dysplasia affecting one or more cell lines; erythroid dysplasia is the most common type of dysplasia, observed in more than 50% of patients. Moreover, abnormal granulocytic and megakaryocytic development is encountered in one-third of all HIV-infected patients. Plasma cells are also strikingly increased in the bone marrow [9]. Various characteristics of bone marrow findings observed in infected patients include the following:

- Bacterial infections lead to bone marrow hypercellularity due to granulocytic hyperplasia. Granulopoiesis may be left shifted—that is, immature cells predominate. With chronic infection, an increase in plasma cells is noted. Plasma cell satellitosis (macrophage surrounded by plasma cells) may be evident as well as secondary hemophagocytic syndrome.
- Viral infections cause an increase in bone marrow lymphocytes, plasma cells, and macrophages. Hemophagocytosis may also be evident. In infections with cytomegalovirus (CMV) or human herpesvirus-6 (HHV-6), intranuclear inclusions may be seen. In post-transplant patients, CMV or HHV-6 infection can cause bone marrow hypoplasia. Chronic hepatitis B and C infection can result in the presence of reactive lymphoid aggregates.
- Fungal infections in bone marrow are typically seen in immunocompromised individuals. Usually, the organisms are within macrophages or within necrotic tissue. Rarely, they may be found within megakaryocytes. Granulomas may also be present.
- The presence of parasites is an unusual finding in the bone marrow. Bone marrow examination may be done when leishmaniasis is being considered. In visceral leishmaniasis, the organisms are seen within the macrophages. There may be granuloma formation. This may be confused with *Histoplasma capsulatum*. However, *H. capsulatum* is Grocott's methenamine silver (GMS) positive, whereas *Leishmania* is not. Also, with the Giemsa stain, the kinetoplast of *Leishmania* is stained, giving a characteristic double-dot appearance.

2.3.3   Granulomatous Changes

Causes of bone marrow granulomas include the following:

- Infections—for example, tuberculosis, fungal infections, brucellosis
- Sarcoidosis
- Malignancy: Hodgkin and non-Hodgkin lymphoma, metastatic disease
- Drugs
- Lipogranulomas: These are characterized by focal aggregates of macrophages with lipid vacuoles, where plasma cells, lymphocytes, and eosinophils are associated with them. Rarely, giant cells may also be seen. Lipogranuloma is a benign disorder.

2.3.4   Storage Disorders

Lysosomal storage disorders (caused by enzyme deficiency and transmitted in an autosomal recessive manner) may be evident from bone marrow examinations. Clinically, these disorders may result in pancytopenia. Partially degraded lipids accumulate in macrophages of liver, spleen, bone marrow, etc. This results in organomegaly and cytopenias. Storage diseases that show characteristic bone marrow abnormalities include the following:

- Gaucher's disease, which is due to lack of the enzyme glucocerebrosidase. "Gaucher" cells are seen in the bone marrow. These are macrophages with a "wrinkled cigarette paper" appearance of the cytoplasm. The cells are periodic acid–Schiff (PAS) positive.
- Indistinguishable from Gaucher cells are pseudo-Gaucher cells, which are seen in conditions with high cell membrane turnover. Examples of such states are chronic myelogenous leukemia (CML), hemoglobinopathies, and myeloma. Gaucher cells are mimicked in *Mycobacterium avium intracellulare* (MAI) infection, in which macrophages are packed with organisms and negatively staining organisms mimic striations of Gaucher cells.
- In Niemann–Pick disease, which is caused by a lack of the enzyme sphingomyelinase, foamy macrophages with bubbly cytoplasm are seen. These macrophages are weakly PAS positive and oil red O and Sudan black positive. Sea blue histiocytes may also be seen in Niemann–Pick disease.
- Sea blue histiocytes are histiocytes with sea blue cytoplasm. They are macrophages containing ceroid. With hematoxylin and eosin (H&E) staining, they appear yellow-brown; with Giemsa staining, they appear bright blue-green and are PAS positive. They are seen in high turnover states—CML, idiopathic thrombopoietic purpura (ITP), and sickle cell disease (SCD). They may also be seen in lipidoses and hyperlipidemia.

2.3.5 Metabolic Bone Diseases

Examples of metabolic bone diseases that may be evident in trephine biopsies include the following:

- Osteopetrosis (Albers–Schonberg or marble bone disease): There is a functional defect in osteoclasts with resultant abnormal accumulation of dense bone (with persistence of cartilaginous cores). It may be transmitted as autosomal recessive (severe form) or autosomal dominant (may be asymptomatic). In the autosomal recessive form, there may be reduced hematopoiesis (myelophthisic anemia) with extramedullary hematopoiesis.
- Osteomalacia: This disorder is due to defective mineralization of bone as a result of vitamin D deficiency. There is hyperosteoidosis, which is increased unmineralized osteoid.
- Renal osteodystrophy: This condition is seen in chronic renal insufficiency. There is deficiency of active vitamin D, which results in hyperosteoidosis. There is also secondary hyperparathyroidism with resultant increased osteoclast activity. This leads to irregular scalloping of bony trabeculae and peritrabecular fibrosis. Ultimately, there may be osteitis fibrosa cystica.
- Paget's disease (osteitis deformans): This disease is due to disordered bone remodeling, characterized by thickened trabeculae and irregular cement lines giving rise to a "mosaic" pattern. Both osteoblasts and osteoclasts are prominent.

2.3.6 Metastatic Tumors

Common metastatic tumors to the bone marrow include the following:

- Adults: Breast, prostate, lung, and gastrointestinal tract carcinoma
- Pediatric: Neuroblastoma

2.3.7 Hemophagocytic Syndrome

Hemophagocytic syndrome occurs when there is increased hemophagocytosis (phagocytosis of nucleated cells) by macrophages with resultant cytopenias. This may be triggered by bacterial or viral infections. Lymphomas, especially T cell lymphomas, are also associated with hemophagocytic syndrome. Familial hemophagocytosis is transmitted as autosomal recessive and manifests in early childhood.

2.3.8 Bone Marrow Necrosis/Infarction

Bone marrow necrosis is noted more often in trephine biopsies than in aspirate slides. The aspirate slides may show amorphous debris with the presence of karyorrhectic nuclear material. Bone marrow necrosis may also be accompanied by necrosis of the adjacent bone. Bone marrow necrosis may be seen in leukemias, lymphomas, metastatic disease, SCD, and infections.

2.3.9 Serous Atrophy

This is also known as gelatinous transformation of the bone marrow. Here, there occurs loss of fat cells and hematopoietic cells with replacement by an increased amount of ground substance. H&E stain shows amorphous pink material. Causes of serous atrophy include cachexia due to chronic debilitating illnesses such as malignancy, renal failure, high-dose radiation, and overwhelming infections.

2.3.10 Bone Marrow Fibrosis

This may be due to increased reticulin or reticulin and collagen in the bone marrow. With marked collagen fibrosis, there may also be osteosclerosis. Collagen deposition is more uncommon than reticulin deposition and causes significantly greater abnormality. Increased reticulin and subsequently collagen deposition may be seen in the chronic myeloproliferative neoplasm. Increased reticulin without collagen deposition is seen in HIV infection and hairy cell leukemia. The peripheral blood changes seen in bone marrow fibrosis are the presence of teardrop cells with leukoerythroblastic blood picture. Attempts at bone marrow aspirate may yield a dry tap.

2.3.11 Reactive Lymphoid Aggregate

Reactive or benign lymphoid aggregates are seen in the bone marrow, especially with increasing age. Causes include infection, inflammation, and autoimmune diseases. Increased frequency of benign lymphoid aggregate has also been documented with Castleman's disease. Benign aggregates are usually low in number and should not be located in the paratrabecular area. They should be well circumscribed and have a heterogeneous population of cells. These cells should be small, mature lymphocytes with some plasma cells, macrophages, and occasional eosinophils and mast cells. With immunohistochemistry, the lymphocytes are typically predominantly T cells, or they may have a central core of T cells surrounded by a rim of B cells or a mixed distribution of B and T cells.

2.3.12 Bone Marrow Infiltration in Lymphoproliferative Disorders

Bone marrow infiltration is a common finding in lymphoproliferative disorders. The particular pattern of infiltration is best assessed from the trephine

BOX 2.4 PATTERNS OF LYMPHOID INFILTRATION OF BONE MARROW IN LYMPHOPROLIFERATIVE DISORDERS

- Nodular lymphoid aggregates: Paratrabecular or nonparatrabecular
- Interstitial (individual neoplastic cells interspersed between hematopoietic cells)
- Intrasinusoidal (this pattern alone is quite rare; usually seen in combination with a second pattern)
- Random focal (irregular, randomly distributed foci of neoplastic cells)
- Diffuse (bone marrow elements replaced by neoplastic cells)

biopsy. The various patterns include nodular (which in turn may be paratrabecular and nonparatrabecular), interstitial (individual neoplastic cells interspersed between hematopoietic cells), intrasinusoidal (this pattern alone is quite rare; it is usually seen in combination with a second pattern), random focal (irregular, randomly distributed foci of neoplastic cells), and diffuse (bone marrow elements are replaced by neoplastic cells). With B cell lymphomas, bone marrow infiltration is more common in low-grade lymphomas. Bone marrow infiltration is more often seen in B cell lymphomas than in T cell lymphomas. Paratrabecular aggregates are a classical feature of follicular lymphoma. Diffuse involvement is more often seen in T cell lymphomas. Patterns of lymphoid infiltration of bone marrow in lymphoproliferative disorders are listed in Box 2.4.

2.3.13 Amyloidosis

In amyloidosis, there is extracellular deposition of amorphous eosinophilic material. Amyloidosis can be seen in a number of conditions. In amyloidosis seen with immunocyte dyscrasias, immunoglobulin light chain is the main protein content. When amyloidosis occurs due to an underlying reactive process, serum amyloid protein A is the major component. The presence of amyloid may be confirmed with Congo red staining. This demonstrates apple green birefringence under polarized light.

KEY POINTS

- Causes of dry tap while performing a bone marrow procedure include faulty technique, packed marrow (e.g., with leukemia), fibrotic marrow (e.g., myelofibrosis), and hairy cell leukemia.
- A myeloblast is a large cell with a high N:C ratio, moderately blue cytoplasm, and prominent nucleoli. A promyelocyte is a larger cell than a myeloblast, with prominent nucleoli, a Golgi hof, and granules.

These granules are primary granules and appear reddish-purple. The promyelocytes of the three granulocytic lineages cannot be distinguished by routine light microscopy. Myelocytes no longer have nucleoli but continue to have granules. However, these granules are secondary and specific granules. Thus, cells of the three granulocytic lineages can now be distinguished. Myeloblasts, promyelocytes, and myelocytes are all capable of cell division. Metamyelocytes have indented nuclei and cannot undergo cell division. The nucleus of bands is U shaped.

- A proerythroblast is a large cell with deep blue cytoplasm, high N:C ratio, and prominent nucleoli. Subsequent to proerythroblasts, nucleoli are no longer seen. In the three normoblasts, the chromatin in the nuclei becomes progressively clumped. In the late normoblast stage, the chromatin is dark, dense, clumped, and ready to be extruded. Once it is extruded, the cell is known as a reticulocyte. In the three normoblasts, the cytoplasm will change color from blue (basophilic normoblast) to gray to gray-orange (late normoblast).

- Megakaryocytes mature through endomitosis. Here, the nuclear DNA content increases in multiples of two. There is no nuclear or cytoplasmic division. Thus, in the bone marrow there exists megakaryocytes, which are 2 N, 4 N, 8 N, 16 N, and 32 N. The dominant ploidy category is 16 N.

- Megakaryocytes do not cluster. Clustering of more than three megakaryocytes is seen in regenerating marrow, following chemotherapy and bone marrow transplantation, and in pathological states (e.g., myelodysplasia).

- In aplastic anemia, trephine biopsy is crucial to document bone marrow hypocellularity. Hematopoietic cells are markedly reduced and fat cells are increased. Lymphocytes are increased, and reactive lymphoid aggregates may be seen. Other cells that may be increased are plasma cells, mast cells, and macrophages.

- Anticancer and immunosuppressive agents may also cause bone marrow hypoplasia. Dyserythropoiesis may be seen along with megaloblastoid changes. In the early stages of chemotherapy administration, there may be interstitial edema. Subsequently, serous atrophy (gelatinous transformation) may become evident.

- With radiation damage, there may be necrosis of bone marrow with bone necrosis. Cells next to bony trabeculae are particularly vulnerable to radiation. Hematopoietic cells may be permanently replaced by fat or fibrous tissue.

- In megaloblastic anemia, erythroids are larger than normal. In megaloblastic anemia, granulopoiesis is also increased with the presence of large forms. Giant myelocytes and metamyelocytes are usually present. With increased intramedullary destruction, an increased number of macrophages may be seen.

- Causes of bone marrow granulomas include infections (e.g., tuberculosis, fungal infections, and brucellosis), sarcoidosis, malignancy (Hodgkin and non-Hodgkin lymphoma and metastatic disease), drugs and lipogranuloma (these are characterized by focal aggregate of macrophages with lipid vacuoles; plasma cells, lymphocytes, and eosinophils are associated with them; rarely, giant cells may also be seen).
- In Gaucher's disease, which is caused by a lack of the enzyme glucocerebrosidase, "Gaucher" cells are seen in the bone marrow. These are macrophages with a "wrinkled cigarette paper" appearance of their cytoplasm. The cells are PAS positive.
- Indistinguishable from Gaucher cells are pseudo-Gaucher cells, which are seen in conditions with high cell membrane turnover. Examples of such states are CML, hemoglobinopathies, and myeloma. Gaucher cells are mimicked in MAI infection, in which macrophages are packed with organisms and negatively staining organisms mimic striations of Gaucher cells.
- Common metastatic tumors to the bone marrow include breast, prostate, lung, and gastrointestinal carcinoma (for adults) and neuroblastoma (for children).
- Bone marrow necrosis may be seen in leukemias, lymphomas, metastatic disease, SCD, and infections.
- Serous atrophy is also known as gelatinous transformation of the bone marrow. Here, there occurs loss of fat cells and hematopoietic cells with replacement by an increased amount of ground substance. H&E shows amorphous pink material. Causes of serous atrophy include cachexia due to chronic debilitating illnesses such as malignancy, renal failure, high-dose radiation, and overwhelming infections.
- Reactive or benign lymphoid aggregates are seen in the bone marrow, especially with increasing age. Causes include infection (most often viral, such as HIV and Epstein–Barr virus), inflammation, and autoimmune diseases. Increased frequency of benign lymphoid aggregate has also been documented with Castleman's disease. Benign aggregates are usually low in number. They should not be located in the paratrabecular area. They should be well circumscribed and have a heterogeneous population of cells. These cells should be small, mature lymphocytes with some plasma cells, macrophages, and occasional eosinophils and mast cells. With immunohistochemistry, the lymphocytes are typically predominantly T cells, or they may have a central core of T cells surrounded by a rim of B cells or a mixed distribution of B and T cells.

- The various patterns of bone marrow involvement in lymphoproliferative disorders are nodular (paratrabecular and nonparatrabecular), interstitial (individual neoplastic cells interspersed between hematopoietic cells), intrasinusoidal (this pattern alone is quite rare; usually seen in combination with a second pattern), random focal (irregular, randomly distributed foci of neoplastic cells), and diffuse (bone marrow elements are replaced by neoplastic cells). With B cell lymphomas, bone marrow infiltration is more common in low-grade lymphomas. Bone marrow infiltration is more often seen in B cell lymphomas than in T cell lymphomas. Paratrabecular aggregates are a classical feature of follicular lymphoma. Diffuse involvement is more often seen in T cell lymphomas.

References

[1] Oudat RI. Bone marrow: indications and procedure set-up. Jordan Med J 2010;44:88−94.

[2] Riley RS, Hogan TF, Pavot DR, Forysthe R, et al. A pathologist's perspective on bone marrow aspiration and biopsy: I. Performing a bone marrow examination J Clin Lab Anal 2004;18:70−90.

[3] Hyun BH, Stevenson AJ, Hanau CA. Fundamentals of bone marrow examination. Hematol Oncol Clin North Am 1994;8:651−63.

[4] Hyun BH, Gulati GL, Ashton JK. Bone marrow examination: techniques and interpretation. Hematol Oncol Clin North Am 1988;2:513−23.

[5] Yu M, Cantor AB. Megakaryopoiesis and thrombopoiesis: an update on cytokines and linkage surface markers. Methods Mol Biol 2012;788:291−303.

[6] Italiano JE, Shivdasani RA. Megakaryocytes and beyond: the birth of platelets. J Thromb Haemost 2003;1:1174−82.

[7] Mirzai AZ, Hosseini N, Sadeghipour A. Indications and diagnostic utility of bone marrow examination in different bone marrow disorders in Iran. Lab Hematol 2009;15:38 44.

[8] Lin SH, Lai CC, Huang SH, Hung CC, et al. Mycobacterial bone marrow infections at medical centre in Taiwan 2001−2009. Epidemiol Infect 2014:1524−32.

[9] Tripathi AK, Misra R, Kalra P, Gupta N, et al. Bone marrow abnormalities in HIV disease. J Assoc Physicians India 2005;53:705−10.

CHAPTER 3

Red Blood Cell Disorders

3.1 INTRODUCTION

Anemia is defined as reduction in the concentration of hemoglobin, taking into account the age and sex of the individual. Anemia is the most common blood disorder, affecting millions of people worldwide, and it is relatively more common among the elderly population. During the past decade, anemia has emerged as a risk factor with a variety of adverse outcomes in aging adults, including hospitalization, disability, and mortality. Because anemia is a multifactorial condition, and due to the increased comorbidity in older adults, it is difficult to establish whether anemia is a marker of disease burden or a mediator in a pathway leading to adverse effects. The World Health Organization (WHO) criteria for anemia are hemoglobin levels less than 13 g/dL in men and less than 12 g/dL in women. Endres *et al.* reported that among 6880 individuals (2905 men and 3975 women) aged 65−95 years, the prevalence of mild anemia (hemoglobin 10 g/dL or higher) was 6.1% for women and 8.1% for men. The authors further observed that in the typical elderly population without severe comorbidity, mild anemia was associated with greater mortality in men but not in women [1]. Anemia is very common among older adults age 85 years or older (26.1% men, 20.1% female), but in adults 75−84 years old, the prevalence was 15.7% in men and 10.3% in women. In contrast, in adults 50−64 years old, the prevalence of anemia was 4.4% in men and 6.8% in women [2]. Ania *et al.* reported that the prevalence of anemia was lower among men than among women younger than age 55 years, but it became more frequent in men older than age 55 years, reaching a prevalence of 44.4% among men 85 years of age or older [3].

CONTENTS

A. Wahed and A. Dasgupta: Hematology and Coagulation. DOI: http://dx.doi.org/10.1016/B978-0-12-800241-4.00003-6

3.2 ANEMIA: MORPHOLOGICAL AND ETIOLOGICAL CLASSIFICATION

Hemoglobin level is highest at birth, ranging from 16 to 20 g/dL, and then declines, with the lowest hemoglobin level observed at 3−6 months, when values of 9−11 g/dL are considered normal. Men have slightly higher levels of hemoglobin, and this is thought to be due to the stimulatory effect of androgens on the bone marrow. The WHO definitions of hemoglobin less than 12 g/dL in nonpregnant women 15 years of age or older, less than 11 g/dL in pregnant woman, and less than 13 g/dL in men 15 years of age or older are widely used for diagnosis of anemia worldwide. Anemia can be classified on the basis of morphology and etiology. Morphological classification of anemia includes the following:

- Normocytic normochromic
- Microcytic hypochromic
- Macrocytic.

Etiologic classification of anemia includes the following:

- Anemia due to blood loss
- Anemia due to deficiency of hematopoietic factors
- Anemia due to bone marrow failure
- Anemia due to increased red cell breakdown.

Examples of etiological classification of anemia are listed in Table 3.1.

3.3 COMMON CAUSES OF ANEMIA

Causes of normocytic normochromic anemia include anemia due to acute blood loss or bone marrow failure, or anemia of chronic disease. Causes of

Table 3.1 Etiological Examples of Anemia	
Etiological Cause	**Specific Examples**
Anemia due to blood loss	Gastrointestinal blood loss, menorrhagia
Anemia due to deficiency of hematopoietic factors	Iron deficiency, folate deficiency, vitamin B_{12} deficiency, erythropoietin deficiency
Anemia due to bone marrow failure	Aplastic anemia, bone marrow infiltration (e.g., metastatic cancer, bone marrow fibrosis), myelodysplasia, bone marrow toxicity (alcohol abuse, chemotherapy)
Anemia due to increased red cell breakdown	Inherited or acquired defects (see Box 3.1)

microcytic hypochromic anemia include iron deficiency, anemia of chronic disease, hemoglobinopathies, thalassemias, sideroblastic anemia, and lead poisoning. Causes of macrocytic anemia include megaloblastic macrocytic anemia due to vitamin B_{12} deficiency, folate deficiency, and normoblastic macrocytic anemia related to hypothyroidism, chronic liver disease, alcohol, or pregnancy.

3.3.1 Anemia Due to Blood Loss and Iron Deficiency Anemia

Blood loss may be acute or chronic. In acute blood loss, replacement of plasma occurs within 24 hours and thus there is hemodilution. Anemia is normocytic normochromic. The maximum decline in hematocrit is usually observed within 3 days, whereas maximum reticulocytosis is seen in 10 days. There is also transient leukocytosis and thrombocytosis. With chronic blood loss, iron deficiency may occur, which results in microcytic hypochromic anemia.

Iron deficiency, which may occur due to reduced dietary intake, malabsorption, chronic blood loss, or parasitic infestations, is a common cause of anemia. Iron deficiency anemia occurs in 2–5% of adult men and postmenopausal women in developed countries. Menstrual blood loss is the most common cause of iron deficiency anemia in premenopausal woman, and blood loss from the gastrointestinal tract is the most common cause in adult men and postmenopausal woman. Asymptomatic colonic and gastric carcinoma may also present with iron deficiency anemia. Celiac disease resulting in malabsorption and the use of nonsteroidal anti-inflammatory drugs (NSAIDs) with resultant chronic blood loss may also cause iron deficiency anemia [4]. Common causes of iron deficiency anemia are listed in Table 3.2.

Iron deficiency results in microcytic hypochromic anemia, and mean corpuscular volume (MCV) is usually less than 80 fL. In iron deficiency, increased anisocytosis is observed; as a result, red blood cell distribution width (RDW) values are high. Anisocytosis precedes microcytosis and hypochromasia and is significant. This is in contrast to thalassemia, in which microcytic hypochromic anemia is observed with normal or near normal RDW. Examination of the peripheral blood smear may also show the presence of pencil cells and occasional or rare target cells. Patients with iron deficiency anemia may also have thrombocytosis. Iron studies show low serum iron levels, low serum ferritin (<12 ng/mL), and low serum transferrin saturation. Serum transferrin iron binding capacity is increased. Free erythrocyte protoporphyrin or zinc protoporphyrin and soluble transferrin receptor levels are increased. Castel *et al.* observed that serum ferritin is the best single laboratory test for the diagnosis of iron deficiency anemia and that a serum ferritin value less than

Table 3.2 Common Causes of Iron Deficiency Anemia

Cause	Commonly Encountered Conditions	Rarely Encountered Conditions
Gastrointestinal tract (GI) blood loss	Gastric ulceration	Esophagitis
	Gastric carcinoma	Esophageal carcinoma
	Colon cancer	Small bowl tumor
	GI bleeding due to chronic use of aspirin/NSAIDs	Gastric antral vascular ectasia
Non-GI bleeding	Menstruation	Hematuria
	Frequent blood donation	Epistaxis
Malabsorption	Coeliac disease	Gut resection
	Gastrectomy	Gastric bypass surgery
	Helicobacter pylori colony formation	Bacterial overgrowth

20 ng/mL (20 μg/L) is highly indicative of iron deficiency anemia, whereas a serum ferritin value greater than 100 ng/mL (100 μg/L) usually excludes the possibility of iron deficiency anemia. However, ferritin values between 20 and 100 ng/mL may be inconclusive, and for these patients, transferrin:log (ferritin) ratio may be useful for diagnosis of iron deficiency anemia [5]. Ferritin is also an acute phase reactant and thus in states of inflammation, values of serum ferritin may be misleading. It is important to note that iron deficiency anemia may mask the presence of β-thalassemia trait. This is because HbA_2 levels are falsely low in the setting of iron deficiency.

3.3.2 Lead Poisoning

Lead is a metal that has been associated with human civilization for more than 6000 years. However, it is also a toxic metal, with its most deleterious effects on the hematopoietic, nervous, and reproductive systems as well as the urinary tract. Hyperactivity, anorexia, decreased play activity, low intelligence quotient, and poor performance in school have been observed in children with high lead levels. Lead is capable of crossing the placenta during pregnancy and has been associated with intrauterine death, prematurity, and low birth weight. The US Centers for Disease Control and Prevention defines an elevated blood level of lead as any value exceeding 10 μg/dL [6]. In lead poisoning, two important enzymes involved in heme synthesis are inhibited: δ-aminolevulinic (5-aminolevulinic) acid dehydratase and ferrochelatase. Inhibition of 5-aminolevulinic acid dehydratase prevents the formation of porphobilinogen; as a result, accumulation of 5-aminolevulinic acid is observed, which is also seen in acute intermittent porphyria. By also inhibiting mitochondrial ferrochelatase, lead prevents the incorporation of iron into protoporphyrin IX, which then forms a metal chelate with zinc; as a result, an elevated level of zinc protoporphyrin in serum may be seen, indicating lead poisoning. Lead poisoning may result in hypochromic anemia and punctate basophilia in the peripheral blood, with the presence of ringed sideroblasts in the bone marrow. The punctate basophilia (basophilic stippling of erythrocytes) occur due to inhibition of the enzyme 5′-pyrimidine nucleotidase by lead. This enzyme is responsible for degradation of the RNA material in reticulocytes.

3.3.3 Anemia of Chronic Disease

Anemia of chronic disease is seen in the setting of chronic infection, inflammation, or malignancy. This is characterized by the following:

- Low serum iron, reduced transferrin saturation, reduced iron binding capacity, and normal or raised serum ferritin.
- It has been hypothesized that cytokines from inflammatory cells cause reduced iron release from reticuloendothelial (RE) cells (RE blockade). There is also reduced red cell survival and inadequate erythropoietin response.

3.3.4 Sideroblastic Anemia

Sideroblastic anemia is anemia due to defective utilization of iron. It is characterized by the presence of ringed sideroblasts in the bone marrow. The iron is present in the mitochondria. Sideroblastic anemia can be congenital or acquired. There are several forms of congenital sideroblastic anemia:

- X-linked sideroblastic anemia due to mutation of the 5-aminolevulinic acid synthase gene (*ALAS2*) or mutation in the *ABC7* gene that encodes a half-type ATP binding cassette (ABC) transporter
- Autosomal recessive sideroblastic anemia
- Mitochondrial DNA defect (e.g., Pearson syndrome, in which there occurs sideroblastic anemia, neutropenia, thrombocytopenia, exocrine pancreatic dysfunction, and hepatic dysfunction)
- Wolfram syndrome, also known as DIDMOAD (diabetes insipidus, diabetes mellitus, optic atrophy, and deafness), also accompanied by sideroblastic or megaloblastic anemia. Patients usually present with diabetes mellitus followed by optic atrophy in the first decade of life followed by cranial diabetes insipidus and sensorineural deafness in the second decade, which later lead to further complications, including multiple neurological abnormalities followed by premature death [7]. This syndrome is due to mutation in the *WFS1* (Wolfram syndrome 1) gene.

Congenital sideroblastic anemia usually exhibits dimorphic red cells (normocytic normochromic and microcytic hypochromic) or microcytic hypochromic red cells. This is in contrast to acquired sideroblastic anemia, which shows normocytic or macrocytic red cells. Acquired sideroblastic anemia may be primary or secondary. Primary acquired sideroblastic anemia is actually myelodysplasia. Secondary acquired sideroblastic anemia is most often due to drugs (e.g., antituberculous drugs), alcohol, and lead poisoning.

3.3.5   Megaloblastic Anemia

Megaloblastic anemias are a group of disorders characterized by peripheral blood cytopenia due to ineffective hematopoiesis in the bone marrow. The most common cause is folate or vitamin B_{12} deficiency or both. Vitamin B_{12} or folate deficiency may result due to poor nutrition, malabsorption, and drugs (e.g., methotrexate and hydroxyurea). Pernicious anemia is an important cause of vitamin B_{12} deficiency and is due to autoimmune destruction of the parietal cells. The parietal cells normally secrete intrinsic factor, which is required for the absorption of vitamin B_{12}. Antiparietal cell antibody and anti-intrinsic factor antibodies are found in patients with pernicious anemia. In megaloblastic anemia, macrocytic red cells are observed in the peripheral blood, and these are classically oval macrocytes. Hypersegmented polymorphonuclear leukocytes may be seen. Megaloblastic anemia is a cause of pancytopenia. The bone marrow shows erythroid hyperplasia with large erythroid precursors. This is known as megaloblastoid change. The myeloid precursors may also be large. Giant myelocytes and giant metamyelocytes may also be observed. Nuclear cytoplasmic dyssynchrony may also be seen. Overt features of dysplasia are seen in megaloblastic anemia. As such, this condition remains a differential diagnosis of myelodysplasia. Both conditions are associated with peripheral blood cytopenias, bone marrow hyperplasia, and dysplasia.

3.3.6   Bone Marrow Failure

Bone marrow failure may be due to bone marrow aplasia or related to bone marrow infiltration (myelophthisic anemia). Aplastic anemia is an example of potentially fatal bone marrow failure-related anemia. Aplastic anemia may be constitutional (inherited) or acquired (Box 3.1). Patients with aplastic anemia present with pancytopenia and reticulocytopenia. The bone marrow cellularity is markedly decreased. In the peripheral blood and bone marrow, the lymphocytes and plasma cells are intact.

Fanconi's anemia is an autosomal recessive disease characterized by bone marrow failure and represents two-thirds of all constitutional aplastic anemias. This disorder is due to a DNA repair defect. Cells derived from patients

BOX 3.1 CAUSES OF APLASTIC ANEMIA

Constitutional (Inherited) Causes
- Fanconi's anemia
- Dyskeratosis congenita
- Shwachman—Diamond syndrome
- Diamond—Blackfan anemia (constitutional red cell disorder)

Acquired Causes
- Idiopathic, drugs, toxins, radiation, infections, paroxysmal nocturnal hemoglobinuria
- Transient erythroblastopenia of childhood (acquired red cell disorder), human parvovirus infection that causes acquired red cell aplasia

with this disorder display hypersensitivity to DNA cross-linking agents, resulting in increased numbers of chromosomal abnormalities, including translocations and radial chromosomes. Aplasia develops by mid-childhood and may progress to acute myelogenous leukemia (AML). These patients are also susceptible to developmental abnormalities (short stature, skeletal defects, etc.), renal defects, mental retardation, and hypogonadism. They have a markedly increased risk of squamous cell carcinoma of the head and neck and genitourinary tract [8].

Dyskeratosis congenita (also known as Zinsser—Engman—Cole syndrome) is a rare progressive bone marrow failure syndrome characterized by skin hyperpigmentation, nail dystrophy, and oral leukoplakia and aplastic anemia. The chromosomal pattern is normal in this disorder, but patients show evidence of telomerase dysfunction, ribosome deficiency, and protein synthesis disorder. Early mortality in these patients may be related to bone marrow failure, infections, malignancy, or fatal pulmonary complications.

Shwachman—Diamond syndrome is a rare genetic disorder typically characterized by exocrine pancreatic dysfunction with malabsorption, malnutrition, growth failure, hematological abnormalities with cytopenia, susceptibility to myelodysplasia syndrome, and acute myelogenous leukemia. In almost all affected children, persistent or intermittent neutropenia is seen.

Diamond—Blackfan anemia is a disease with pure congenital red cell aplasia (inherited erythroblastopenia). Usually, white blood cell and platelet counts are normal in these patients (in contrast to Shwachman—Diamond syndrome, in which neutropenia is common). Skeletal defects are often seen in these patients, but renal or chromosomal abnormalities are not observed.

Acquired bone marrow failure causing aplastic anemia may be drug induced. Drugs such as chemotherapeutic agents and benzene cause dose-dependent bone marrow failure. Drugs such as chloramphenicol cause dose-independent, idiosyncratic bone marrow failure. Ionizing radiation can cause bone marrow failure as an early complication. Late complications of such radiation exposure include myelofibrosis and leukemia.

Viral hepatitis is a rare but recognized cause of aplastic anemia. It typically occurs 6 weeks after onset of symptoms. This complication is not related to the severity of hepatitis. Parvovirus B19-induced aplastic crisis tends to occur in individuals who have a background of chronic hemolysis. The red cells are selectively involved. Giant erythroblasts are seen. Nuclear inclusions may be seen. Immunostains are available for confirmation. The condition is self-limiting.

Aplastic anemia is characterized by pancytopenia with low corrected reticulocyte count. Anemia results in weakness and fatigue. Leukopenia predisposes to infections, and thrombocytopenia results in bleeding.

Other causes of pancytopenia include acute leukemia, myelophthisic anemia, megaloblastic anemia, myelodysplastic syndrome, and hypersplenism.

Severe aplastic anemia is characterized by any three of the following for 2 weeks:

- $<40,000$ reticulocytes/μL
- $<10,000$ platelets/μL
- <500 granulocytes/μL
- $>80\%$ of bone marrow nonmyeloid.

3.3.7 Congenital Dyserythropoietic Anemia

This disorder is characterized by anemia at an early age with multinucleation of erythroid precursors and internuclear bridging between erythroid precursors in the bone marrow. There are three types (I–III), of which type II, transmitted as autosomal recessive, is most common. The Ham acidified serum test (checks the fragility of erythrocytes when placed in mild acid), which is usually used for diagnosis of paroxysmal nocturnal hemoglobinuria (PNH), is also positive in patients with type II congenital dyserythropoietic anemia.

3.4 HEMOLYTIC ANEMIA

Hemolysis is the destruction or removal of red blood cells from circulation before their normal life span of 120 days. Although hemolysis can be an asymptomatic condition, patients commonly present with anemia when erythrocytosis cannot overcome the pace of erythrocyte destruction. Patients

BOX 3.2 UNDERLYING CAUSES OF HEMOLYTIC ANEMIA

Corpuscular Defects (All Inherited, Except PNH)
- Membrane defects (e.g., hereditary spherocytosis)
- Enzyme defects (G6PD deficiency, pyruvate kinase deficiency)
- Hemoglobinopathies and thalassemias
- PNH

Extracorpuscular Defects (Acquired)
- Immune mediated (hemolytic disease of newborn, mismatch transfusion, autoimmune hemolytic anemia)
- Non-immune mediated (infections, march hemoglobinuria, karate-induced hemoglobinuria, microangiopathic hemolysis)

may complain of dyspnea or fatigue due to anemia, and they may also present with dark urine and jaundice [9]. Hemolytic anemia is clinically characterized anemia, jaundice, increased reticulocyte count, splenomegaly, leg ulcers, increased serum concentration of lactate dehydrogenase, and decreased serum haptoglobin if intravascular hemolysis is present. Hemolytic anemia may be due to corpuscular defects or extracorpuscular defects (Box 3.2).

3.4.1 Hemolytic Anemia Due to Corpuscular Defects

This category includes hemolytic anemias due to membrane defects, enzyme defects, hemoglobinopathies, thalassemias, and PNH. All hemolytic anemias in this category are inherited, with the exception of PNH. Hemoglobinopathies and thalassemias are discussed in Chapter 4.

3.4.2 Hemolytic Anemia Due to Membrane Defects

Various hemolytic anemias are related to erythrocyte membrane defects. These anemias are summarized in this section.

3.4.2.1 Hereditary Spherocytosis

Spectrin is a major component of the normal red blood cell (RBC) cytoskeleton and consists of two intercoiled non-identical filamentous subunits that form heterodimers. The chain heads of each dimer pair bind with opposite subunit heads of another dimer pair to form tetramers. The tail of spectrin tetramers binds with a protein cluster of short actin protofilaments, and this interaction is markedly enhanced by protein 4.1. Spectrin−actin−protein 4.1 assembly is secured to the overlying lipid bilayer by ankyrin.

Hereditary spherocytosis is the most common of the hereditary hemolytic anemias among people of northern European descent. In the United States, the incidence is approximately 1 per 5000 people. It is transmitted as autosomal dominant, but in 25% of cases, it is due to a spontaneous mutation. Four abnormalities in RBC membrane proteins have been identified in this disorder: spectrin deficiency alone, combined spectrin and ankyrin deficiency, band 3 deficiency, and protein 4.1 defects. Spectrin deficiency is the most common cause of this disorder.

Spectrin deficiency leads to loss of erythrocyte surface area, which produces spherical RBCs. Spherocytic RBCs are rapidly removed from the circulation by the spleen. Patients with hereditary spherocytosis develop splenomegaly. Spectrin deficiency is due to impaired synthesis. In some cases, there are quantitative or qualitative deficiencies of other proteins that integrate spectrin into the cell membrane. In the absence of these binding proteins, free spectrin is degraded, leading to spectrin deficiency. The degree of spectrin deficiency is reported to correlate with the extent of spherocytosis, the degree of abnormality on osmotic fragility test results, and the severity of hemolysis. Most cases of hereditary spherocytosis are heterozygous because homozygous states are lethal.

A peripheral smear from these patients will demonstrate spherocytes with polychromasia (increased reticulocytes). Spherocytes are also seen in autoimmune hemolytic anemia. It is logical to perform a direct antiglobulin test (DAT) to rule out autoimmune hemolytic anemia. The confirmatory test for hereditary spherocytosis is the osmotic fragility test. The principle of this test is that spherocytes will easily lyse when placed in hypotonic solution. If the standard osmotic fragility test is negative, incubated osmotic fragility tests can be performed to identify more subtle cases. A normal osmotic fragility test does not necessarily exclude the diagnosis of hereditary spherocytosis. Such normal tests may be seen in 10−20% of cases. The test may be normal in the presence of iron deficiency, obstructive jaundice, and increased reticulocytes. Patients with hereditary elliptocytosis and hemolysis may yield a false-positive result. A relatively newer test is eosin-5-maleimide binding by flow cytometry.

3.4.2.2 Hereditary Elliptocytosis

This disorder is characterized by greater than 20% elliptocytes in the peripheral blood. The condition is transmitted as autosomal dominant. Clinical manifestations range from an asymptomatic carrier state to severe hemolytic anemia. Members of the same family may exhibit different clinical courses, and an individual's frequency and severity of hemolysis may change with time. Most patients with hereditary elliptocytosis or its variants lead healthy lives. There are three major categories: the common form, the spherocytic form, and the stomatocytic form. Hereditary elliptocytosis and its related disorders are caused by mutations that disrupt the RBC cytoskeleton; the most frequent is defective spectrin dimer−dimer interaction due to an abnormal α chain of spectrin. Quantitative deficiencies of protein 4.1 are associated with the spherocytic form of hereditary elliptocytosis.

3.4.2.3 Hereditary Pyropoikilocytosis

This is a severe form of congenital hemolytic anemia that is clinically similar to, and now considered a subtype of, homozygous hereditary elliptocytosis. RBCs show increased susceptibility to thermal injury. Mode of inheritance and exact defects are not fully understood, and similarly to hereditary ellipto-cytosis, a defective spectrin dimer−dimer interaction due to an abnormal α-chain of spectrin is speculated to be the cause of this disorder.

The peripheral smear is striking and pathognomonic, and it shows bizarre forms, anisocytosis, fragments, micropoikilocytosis, microspherocytes, and budding red cells. The MCV may be as low as 25−55 fL.

3.4.2.4 Hereditary Stomatocytosis

This disease is due to abnormal RBC membrane cation permeability and con-sequent defective cellular hydration. It is transmitted as autosomal dominant. Hereditary stomatocytosis represents a spectrum of disorders with variable phenotypic expression. At one end of the spectrum is hydrocytosis, which results when erythrocytes are swollen with water, and at the other end of the spectrum is xerocytosis, which results when cells are dehydrated. Active Na/K ATPase pumps are incapable of counterbalancing flow of cations across the cell membrane. The net influx of sodium brings with it water, resulting in a swollen cell that is susceptible to osmotic and mechanical lysis. The hallmark is a stomatocyte—a red cell with slitlike or "fish-mouth" area of central pal-lor. Greater than 35% stomatocytes and evidence of hemolysis are required to establish the diagnosis.

3.4.2.5 Rh Null Disease

In Rh null disease, an individual lacks all Rh determinants. Stomatocytes and spherocytes are found in the peripheral smear. It is clinically characterized by a mild compensated hemolytic anemia, which requires no therapy.

3.4.2.6 Hemolytic Anemia Due to Enzyme Defects

Hemolytic anemia due to enzyme defects may be related to deficiencies of enzymes in the glycolytic pathway or defective enzymes in the HMP shunt (hexose monophosphate shunt or pentose phosphate pathway). The mature RBC is totally dependent on glucose as a source of energy. Glucose usually (90%) is catabolized to pyruvate and lactate in the major anaerobic glycolytic pathway (Embden−Myerhof pathway). Because mitochondria are absent in RBCs, they depend entirely on anaerobic glycolysis to provide energy.

The glycolytic pathway generates adenosine triphosphate (ATP), which is required for the functioning of the Na/K ATPase pump. The proper functioning of the Na/K ATPase pump ensures that sodium and water are constantly pumped out of the red cell, allowing it to maintain its biconcave shape.

The pentose phosphate pathway produces nicotinamide adenine diphosphate (NADPH), which maintains the reduced state of glutathione. If glutathione is not in its reduced state, disulfide bond (S—S) formation takes place, with subsequent denaturation and precipitation of globin as Heinz bodies. Heinz bodies are removed by the reticuloendothelial cells as the red cells pass through the spleen. The damaged red cells have the appearance of bite or blister cells. Heinz bodies are not seen in the peripheral smear with routine Wright—Giemsa stain. However, bite cells are apparent in the peripheral smear. The presence of bite cells should be a clue to order special stains for Heinz bodies and evaluate for enzymes of the pentose phosphate pathway. Hereditary deficiencies of some of the glycolytic enzymes have been documented, and several cause a hereditary nonspherocytic hemolytic anemia (HNSHA), whereas others cause multisystem disease. Most are rare, and pyruvate kinase deficiency is the most common, occurring in 90% of affected patients.

3.4.2.7 Pyruvate Kinase Deficiency

Pyruvate kinase deficiency is transmitted as autosomal recessive. The condition is detected in infancy or childhood due to anemia, jaundice, splenomegaly, and gallstones. The severity of the condition is widely variable, even among patients with the same level of deficiency. Fluorescent screening tests are available for pyruvate kinase deficiency. Diagnosis is confirmed by enzyme assay, and DNA analysis by polymerase chain reaction (PCR) or single-strand conformation polymorphism are available to confirm the diagnosis and to identify the carrier state necessary.

3.4.2.8 Glucose-6-Phosphate Dehydrogenase Deficiency

Glucose-6-phosphate dehydrogenase (G6PD) deficiency is the most common enzyme deficiency in the pentose phosphate pathway, affecting more than 400 million people worldwide. The gene for the G6PD enzyme is located on the X chromosome, and 200 different mutations have been reported, resulting in a variety of clinical conditions. Clinical manifestations of G6PD deficiency include neonatal jaundice and acute hemolytic anemia arising from oxidative stress on RBCs induced by some medications, an infection, or ingestion of fava beans. However, many individuals with this disorder remain asymptomatic throughout their lives and may not be even aware of it. Drug-induced hemolytic anemia in patients with G6PD deficiency was first described with the antimalaria drug primaquine, but many other drugs are capable of causing hemolytic anemia in these patients [10]. The highest prevalence of this disorder is in Sub-Saharan Africa, followed by North Africa and the Middle East. Interestingly, this disorder protects individuals from infection by the malaria parasite.

Normally, two isotypes of G6PD—A and B—can be differentiated based on electrophoretic mobility, and the B isoform is the most common type of enzyme found in all population groups. However, the A isoform, found in 20% of black men in the United States, migrates more rapidly on electrophoretic gels than does the B isoform and has similar enzyme activity to the B isoform. Eleven percent of US black men have the G6PD variant (G6PD A$^-$). It has the same electrophoretic mobility as isoform A, but it is unstable, resulting in enzyme loss and, ultimately, enzyme deficiency. Older RBCs have only 5−15% enzyme, and G6PD A$^-$ is the most clinically significant type of abnormal G6PD among US blacks. Various G6PD variants predominate in other racial groups—for example, G6PD−Mediterranean in Sicilians, Greeks, Sephardi Jews, and Arabs; and G6PD−Canton in the Asian population. Five classes of G6PD deficiency exist based on enzyme activity levels as recommended by WHO (Table 3.3).

The two most common mutations involved in G6PD deficiency are G6PD A$^-$ and G6PD−Mediterranean [11]. In G6PD A$^-$ deficiency, the hemolytic anemia is self-limited because the young RBCs produced in response to hemolysis have nearly normal enzyme and are less affected by favism. In contrast, favism is associated with G6PD−Mediterranean. Patients with G6PD develop acute hemolysis with formation of Heinz bodies (denatured globin) and bite cells (eccentrocytes). The Heinz bodies may be demonstrated by special stains such as crystal violet. Further testing includes fluorescent screening tests for G6PD deficiency as well as G6PD enzyme activity assay [12].

Table 3.3 Classification of Glucose-6-Phosphate Dehydrogenase (G6PD) Deficiency

Class	Residual Enzyme Activity (%)	Clinical Manifestation	Mutation Type
I	≤1	Chronic hemolytic anemia—very rare condition. Total absence of G6PD deficiency, which is very unusual, is incompatible with life.	G6PD−Buenos Aires G6PD−Durham
II	<10	Severe enzyme deficiency causing acute hemolytic anemia induced by certain drugs or ingestion of kava beans.	G6PD−Mediterranean G6PD−Santamaria
III	10−60	Moderate to mild enzyme deficiency that may cause occasional acute hemolytic anemia.	G6PD A$^-$ G6PD−Canton
IV	60−90	Very mild to moderate enzyme deficiency. These individuals are asymptomatic. This occurs rarely.	G6PD−Orissa G6PD−Montalbano
V	>110	Increased enzyme activity; individuals are asymptomatic. Also rarely observed.	Not known

3.4.2.9 *Paroxysmal Nocturnal Hemoglobinuria*

This is the only acquired disorder among the intrinsic red cell disorders causing hemolytic anemia due to spontaneous mutation in a pluripotent stem cell of the *PIG-A* (phosphatidylinositol glycan anchor biosynthesis, class A) gene. The *PIG-A* gene encodes for GPI protein (glycophosphatidylinositol anchored protein). GPI functions as a cell membrane "anchor" and serves as attachment for approximately 20 cell surface proteins. Among the proteins that are thus absent on the cell surface are cell surface proteins that regulate complement, including the following:

- CD55/decay-accelerating factor (DAF): CD55 inhibits association of C4b and C2 and promotes dissociation of C4bC2a (C3 convertase) complex
- Homologous restriction factor (HRF)
- CD59/membrane inhibitor of reactive lysis (MIRL): CD59 (MIRL) and HRF prevent formation of membrane attack complex and cell lysis.

PNH is thus a clonal disorder and affects all hematopoietic cells where there is increased susceptibility to complement-mediated lysis. Nucleated hematopoietic cells are able to endocytose membrane attack complex, which causes cell lysis, allowing them to resist complement-mediated lysis. Non-nucleated cells such as red cells are unable to resist lysis. PNH is clinically characterized by the following:

- Hemolytic anemia: This may take the form of chronic intravascular hemolysis. There may be paroxysmal episodes of acute hemolysis. It is postulated that sleep causes relative acidosis, and under acidotic conditions there occurs red cell lysis. This results in hemoglobinuria, and when the patient wakes up in the morning and passes urine, the urine is dark due to hemoglobinuria (thus the name). The urine will clear during the day.
- Thrombosis: Complement activity creates a pro-thrombotic state, and thrombosis is a major cause of morbidity/mortality. Clinically significant thrombosis is seen in one-third of patients. The thrombosis is most often venous and includes Budd–Chiari syndrome (hepatic vein thrombosis).
- Cytopenias/bone marrow failure: Patients may develop neutropenia, thrombocytopenia, and aplastic anemia. There is an increased incidence of transformation to AML/myelodysplastic syndrome (MDS).

The screening test for PNH is the sucrose lysis test, in which the patient's red cells are added to a solution of isotonic sucrose and serum. Sucrose aggregates globin onto red cells, and serum is a source of complement. The test is positive if greater than 5% of red cells are lysed. Lysis makes the supernatant red.

Confirmatory tests for PNH include Ham's acidified serum test and flow cytometry. Characteristics of Ham's acidified serum test include the following:

- The basis of this test is that under acidified conditions, a patient's own complement destroys his or her red cells. The patient's red cells are placed in three test tubes, to which are added the patient's serum. In two tubes, acid is added. One of these tubes is heated to 56°C to destroy complement. Thus, there is one tube in which there is no acid, but it contains the patient's red cells and serum. In another tube are the patient's red cells, acidified serum, but no complement. The third tube has all three components required for hemolysis. All are incubated for 1 hour. Hemolysis present only in the third tube, but not in the first two tubes, constitutes a positive test.
- A positive test is greater than 1% lysis in acidified serum (acid and complement are both present).
- This test may also be positive in congenital dyserythropoietic anemia, type II.

Ham's acidified serum test is very sensitive but may not be specific. In flow cytometry, expression of GPI-anchored proteins CD55 and CD59 is analyzed on hematopoietic cells using monoclonal antibodies. This test is highly specific because there is no other condition in which RBCs are a mosaic of normal and GPI-linked protein-deficient cells.

3.4.3 Hemolytic Anemias Due to Extracorpuscular Defects

Extracorpuscular causes of hemolytic anemia can be classified as immune and non-immune. Examples of non-immune causes are malaria and red cell destruction due to trauma (e.g., march or karate hemoglobinuria) or very strenuous exercise. Immune hemolytic anemia is due to increased destruction of RBCs by antibody against antigens on RBCs. The following are examples of immune hemolytic anemias:

- Hemolytic disease of the newborn
- Hemolytic transfusion reactions
- Autoimmune hemolytic anemias: Patients make antibodies to antigens on their own RBCs. Autoimmune hemolytic anemias are classified as warm autoimmune hemolytic anemia (WAHA), cold hemagglutinin disease (CHAD), and paroxysmal cold hemoglobinuria (PCH).
- Drug-induced immune hemolytic anemia caused by antibody directed against a red cell membrane–drug complex (e.g., penicillin), immune complex deposition on red cell surface (e.g., quinine and rifampin), or true autoimmune hemolytic anemia (e.g., due to methyldopa).

In WAHA, IgG antibodies are formed against the patient's own red cell antigens. These antibodies are often formed against the broad Rh antigens. Hemolysis is classically extravascular. The IgG-coated red cells are destroyed by splenic macrophages. Splenomegaly is thus a feature. In CHAD, IgM antibodies are formed against red cell antigens. Agglutination of red cells can occur at low temperatures, and complement activation may result in intravascular hemolysis. Antibodies are often directed against I antigens. There are several subtypes of this condition:

- Acute postinfectious: Acute, self-limited, younger patients
- Chronic idiopathic: Insidious, older patients
- Cold agglutinin disease (CAD): Insidious, elderly women, associated with lymphoproliferative malignancy.

In PCH, antibodies are typically formed against P antigens. The antibody binds to red cells at low temperature and, when warmed, activates complement causing hemolysis. These antibodies are also known as biphasic antibodies or Donath−Landsteiner antibodies. PCH is a rare condition and sometimes seen in children following viral infections with sudden onset of hemolysis.

Causes of WAHA and CHAD may be idiopathic, drug induced (more often with WAHA), related to infection (more common with CHAD), or lymphoproliferative disorders. Laboratory findings in WAHA include positive DAT to IgG or IgG and C3, in addition to the presence of spherocytes in peripheral smear. Osmotic fragility test may be positive. In CHAD, DAT is only positive for C3. Red cell agglutination on peripheral smear may also be observed.

3.5   RED CELL POIKILOCYTOSIS

Variation in shape of red cells refers to poikilocytosis. The following are examples of cells associated with poikilocytosis:

- Sickle cell (drepanocyte): Under low oxygen tension (6−8 hemoglobin S (HbS)), molecules condense to form a tubular structure. This is called a tactoid. This distorts the red cell to form boat-shaped cells or sickle-shaped cells. Sickle cells cause occlusion of small vessels and can cause infarctions. Circulating red cells are effectively decreased, resulting in impaired oxygen delivery to tissue. When oxygen tension is improved, sickle cells resume the shape of normal red cells. Repeated sickling and unsickling makes the red cell membrane fragile and results in hemolysis.
- Target cell (codocyte): These cells are formed as a consequence of the presence of redundant membrane in relation to volume of cytoplasm. Excess cell membrane lipid occurs in cholestatic liver disease and lecithin−cholesterol acyltransferase deficiency, resulting in target cells.

Target cell formation due to a reduction of cytoplasm is seen in thalassemia and hemoglobinopathies.

- Ovalocyte: This is a red cell whose long axis is less than twice as long as its short axis. Ovalocytes, especially macro ovalocytes, are seen in folate and vitamin B_{12} deficiency.
- Elliptocyte: This is a red cell whose long axis is more than twice as long as its short axis. Significant elliptocytosis (>20%) can be seen in hereditary elliptocytosis. Other causes include thalassemia, HbS trait, and hemoglobin C (HbC) trait. Causes of a lesser degree of elliptocytosis include cirrhosis, iron deficiency anemia, megaloblastic anemia, and myelophthisic anemia.
- Stomatocyte: This is a red cell that has a slitlike area of pallor. These cells may be seen in hereditary stomatocytosis. Small numbers of stomatocytes may be seen in acute alcohol intake, cirrhosis, obstructive liver disease, and Rh null disease.
- Echinocyte or burr cell: This is a red cell with 10−30 short blunt spicules. These cells may be observed in storage artifact, liver and kidney disease, or pyruvate kinase deficiency.
- Acanthocyte: The term acanthocyte is derived from the Greek *acantha*, meaning "thorn." Acanthocytes are dense contracted RBCs with multiple thorny projections that vary in width, length, and surface distribution. This is presumably due to an increase in relative sphingomyelin content of RBC membrane, resulting in a rigid wall. Acanthocytes must be distinguished from echinocytes (Greek *echinos*, "urchin"). Echinocytes, also known as burr cells, have multiple small projections that are distributed uniformly on the red cell surface. Acanthocytes are red cells with 2−20 unequal, irregular spicules. Occasional acanthocytes may be seen in postsplenectomy, hemolytic anemia due to pyruvate kinase deficiency, microangiopathic hemolytic anemia, autoimmune hemolytic anemia, renal disease, and thalassemias, and with McLeod phenotype. If the majority of RBCs are acanthocytes, this may be due to abetalipoproteinemia.
  - In abetalipoproteinemia, there is an absence of apolipoprotein B resulting in an inability to transport triglycerides in the blood. Clinically, it presents in infancy with steatorrhea, progressive development of acanthocytosis, ataxic neuropathy, and an atypical form of retinitis pigmentosa.
  - Acanthocytes are also found with McLeod blood group, an X-linked disorder in which red cell Kx antigen, the precursor substance for the Kell blood group system, is absent. The McLeod phenotype may be associated with chronic granulomatous disease because of the proximity of the genetic loci for these two disorders.
- Keratocyte: This is a red cell with a pair(s) of spicules. These cells are seen in microangiopathic hemolytic anemia and renal disease.

- Schistocyte: This is a fragmented red cell. Increased numbers suggest microangiopathic hemolytic anemia. True schistocytes do not have central pallor.
- Spherocyte: This is a red cell without central pallor. These cells are observed in autoimmune hemolytic anemia and hereditary spherocytosis.
- Dacryocyte (teardrop red cell): If dacryocytes are present in significant numbers, this may imply conditions with bone marrow infiltration. Other causes include megaloblastic anemia, hemolytic anemias, and hypersplenism.
- Bite cell: These cells are seen in conditions with Heinz body formation, such as G6PD deficiency. They are usually accompanied by blister cells. They are associated with oxidative stress to red cells. Bite cells can also be seen in normal individuals receiving large amounts of aromatic drugs (or their metabolites).
- Blister cell: This is an RBC with vacuoles or very thin areas at the periphery of the cell. Causes are similar to those for bite cells.

3.6   RED CELL INCLUSIONS

Several red cell inclusions have been described, including the following:

- Howell–Jolly bodies: These are usually single peripheral bodies within red cells representing DNA material. These bodies may be seen in post-splenectomy, megaloblastic anemia, severe hemolysis, and myelophthisic anemia.
- Pappenheimer bodies: These are smaller than Howell–Jolly bodies and are multiple, often stacked like cannon balls. They represent iron material. These are found within the mitochondria.
- Cabot rings: These are mitotic spindle remnants seen as ring-shaped or as figure-eight inclusions.
- Basophilic stippling: These represent RNA material that could be fine or physiological or coarse but all are pathological in nature. Causes include lead poisoning, hemolytic anemia, and pyrimidine 5'-nucleotidase deficiency. Pyrimidine 5'-nucleotidase is the enzyme responsible for degradation of the RNA material.
- Heinz bodies: These are denatured globin. They require supra vital staining for their demonstration. They can be seen in G6PD deficiency. Heinz bodies are cleared by splenic macrophages. The damaged cells are known as bite cells.
- HbC crystals: These are an *in vitro* phenomenon. They are seen in individuals with HbC hemoglobinopathy.
- Malarial parasites
- Nucleated RBCs

Table 3.4 Diagnostic Points for Malaria Parasite Identification

Plasmodium falciparum	Plasmodium vivax	Plasmodium malariae	Plasmodium ovale
▪ Red cells not enlarged ▪ Rings appear fine/ delicate, several per cell ▪ Some rings with two chromatin dots ▪ Presence of marginal/ appliqué forms ▪ Unusual to see developing forms ▪ Crescent-shaped gametocytes ▪ Maurer's dots may be present	▪ Red cells containing parasites usually enlarged ▪ Schüffner's dots frequently present ▪ Mature ring forms large and coarse ▪ Developing forms frequently present	▪ Ring forms have squarish appearance ▪ Band forms characteristic ▪ Mature schizonts may have typical "daisy head" appearance with ≤10 merozoites ▪ Red cells not enlarged ▪ Chromatin dot may be on inner surface of the ring	▪ Red cells enlarged ▪ Comet forms common ▪ Rings large and coarse ▪ Schüffner's dots prominent when present ▪ Mature schizonts similar to those of P. malariae but larger and coarser

3.6.1   Malaria Parasites

Malaria is an RBC disorder caused by *Plasmodium* infection and, as expected, anemia is a common manifestation of this disease, which is responsible for substantial morbidity and mortality of infected individuals. There are four species of malarial parasite: *Plasmodium falciparum*, *P. vivax*, *P. ovale*, and *P. malariae*. Acute falciparum malaria results in increased removal from circulation of parasitized and, to a greater extent, nonparasitized RBCs through a combination of splenic filtration, schizont rupture, macrophage phagocytosis, complement-mediated hemolysis, and increased free radical damage. Outside of Africa, *P. falciparum* invariably coexists with other *Plasmodium* species, the most important being *P. vivax* [13]. It is important to differentiate *P. falciparum* from others because *P. falciparum* infection is very serious and may cause death within a short period of time. Red cell exchange is one of the therapeutic modalities for *P. falciparum* infection. Clinicians may also want to determine the parasite load (the percentage of red cells affected). Diagnostic points for identification of various malaria parasites are summarized in Table 3.4.

KEY POINTS

- Causes of normocytic normochromic anemia include anemia due to blood loss (acute), bone marrow failure, and anemia of chronic disease.
- Causes of microcytic hypochromic anemia include iron deficiency, anemia of chronic disease, hemoglobinopathies, thalassemia, sideroblastic anemia, and lead poisoning.

- Causes of macrocytic anemia include the following: megaloblastic macrocytic—vitamin B_{12} and/or folate deficiency; normoblastic macrocytic—hypothyroidism, chronic liver disease, alcohol, and pregnancy.
- Iron deficiency results in microcytic hypochromic anemia with anisocytosis. Anisocytosis precedes microcytosis and hypochromasia. This is in contrast to thalassemia, in which microcytic hypochromic anemia with normal or near normal RDW is observed. Examination of the peripheral smear may also show the presence of pencil cells and occasional or rare target cells. Patients with iron deficiency anemia may also have thrombocytosis. Iron studies show low serum iron levels, low serum ferritin, and low serum transferrin saturation. Serum transferrin iron binding capacity is increased. Free erythrocyte protoporphyrin or zinc protoporphyrin and soluble transferrin receptor levels are increased. It is important to note that iron deficiency anemia may mask the presence of the β-thalassemia trait because HbA2 levels are falsely low in the setting of iron deficiency.
- Sideroblastic anemia is characterized by the presence of ringed sideroblasts in the bone marrow. The iron is present in the mitochondria. Sideroblastic anemia can be congenital or acquired.
- In lead poisoning, two important enzymes of heme synthesis are inhibited: δ-aminolevulinic acid dehydratase and ferrochelatase. The punctuate basophilia occurs due to inhibition of the enzyme 5′pyrimidine nucleotidase by lead.
- Megaloblastic anemia is due to folate or vitamin B_{12} deficiency. Macrocytic red cells are seen in the peripheral blood, and these are classically oval macrocytes. Hypersegmented PMNs may be seen. Megaloblastic anemia is a cause of pancytopenia. The bone marrow shows erythroid hyperplasia with large erythroid precursors. This is known as megaloblastoid changes. The myeloid precursors may also be large. Giant myelocytes and giant metamyelocytes may be seen. Nuclear cytoplasmic dyssynchrony may also be observed.
- Aplastic anemia: The causes of aplastic anemia include constitutional (e.g., Fanconi's anemia, dyskeratosis congenita, Shwachman—Diamond syndrome, and Diamond—Blackfan anemia); and acquired—idiopathic, drugs, toxins, radiation, infections, PNH, transient erythroblastopenia of childhood (acquired red cell disorder), and human parvovirus infection, which causes acquired red cell aplasia.
- Congenital dyserythropoietic anemia is characterized by anemia at an early age with multinucleation of erythroid precursors and internuclear bridging between erythroid precursors in the bone marrow.
- Classification of hemolytic anemia: Corpuscular defects (membrane defects, enzyme defects, hemoglobinopathies and thalassemia, and

PNH) and extracorpuscular defects (these may be non-immune or immune in nature).

- Hereditary spherocytosis is the most common of the hereditary hemolytic anemias among people of northern European descent. It is transmitted as autosomal dominant; however, in 25% of cases, it is due to a spontaneous mutation. Spectrin deficiency is the most common defect. The confirmatory test for hereditary spherocytosis is the osmotic fragility test.
- Hereditary stomatocytosis is due to abnormal RBC membrane cation permeability and consequent defective cellular hydration. It is transmitted as autosomal dominant.
- Hereditary elliptocytosis is characterized by greater than 20% elliptocytes in the peripheral blood. The condition is transmitted as autosomal dominant. Clinical manifestations range from an asymptomatic carrier state to severe hemolytic anemia.
- Hereditary pyropoikilocytosis is a severe form of congenital hemolytic anemia, and it is clinically similar to, and now considered a subtype of, homozygous hereditary elliptocytosis. RBCs show increased susceptibility to thermal injury.
- Hemolytic anemia caused by enzyme defects may be due to deficiencies of the glycolytic pathway or disorders of HMP shunt (pentose phosphate pathway).
- Pyruvate kinase deficiency is transmitted as autosomal recessive and is the most common enzyme deficiency of the glycolytic pathway.
- G6PD deficiency is the most common enzyme deficiency in the pentose phosphate pathway. Normally, two isotypes of G6PD—A and B—can be differentiated based on electrophoretic mobility. B isoform is the most common type of enzyme found in all population groups. A isoform, found in 20% of black men in the United States, migrates more rapidly on electrophoretic gels than does the B isoform. It has similar enzyme activity to that of the B isoform. Eleven percent of US black men have G6PD variant (G6PD A$^-$). It has the same electrophoretic mobility as the A isoform but is unstable, resulting in enzyme loss and ultimate enzyme deficiency. Older RBCs have only 5–15% enzyme. G6PD A$^-$ is the most clinically significant type of abnormal G6PD among US blacks.
- PNH is the only acquired disorder among the intrinsic red cell disorders causing hemolytic anemia. There occurs spontaneous mutation in a pluripotent stem cell of the *PIG-A* gene. The *PIG-A* gene encodes for GPI protein. GPI functions as a cell membrane "anchor" and serves as attachment for approximately 20 cell surface proteins. Among the proteins that are thus lacking on the cell surface are cell surface proteins that regulate complement, and these include CD55/DAF, HRF,

and CD59. PNH is clinically characterized by hemolytic anemia, thrombosis, and cytopenias/bone marrow failure. There is an increased incidence of transformation to AML/MDS. The screening test for PNH is the sucrose lysis test while confirmatory tests are Ham's acidified serum test and flow cytometry.

- WAHA: IgG antibodies are formed against a patient's own red cell antigens. These antibodies are often formed against broad Rh antigens. Hemolysis is classically extravascular. The IgG-coated red cells are destroyed by splenic macrophages. Splenomegaly is thus a feature.
- CHAD: IgM antibodies are formed against red cell antigens. Agglutination of red cells can occur at low temperatures, and complement activation may result in intravascular hemolysis. Antibodies are often directed against I antigens.
- PCH: Antibodies are typically formed against P antigens. The antibody binds to red cells at low temperature and, when warmed, activates complement causing hemolysis. These antibodies are also known as biphasic antibodies or Donath−Landsteiner antibodies. PCH is a rare condition and is sometimes seen in children following viral infections with sudden onset of hemolysis.
- Examples of cells seen in association with poikilocytosis are sickle cells (drepanocytes), target cells (codocytes), ovalocytes (macro ovalocytes are seen in folate and vitamin B_{12} deficiency), elliptocytes (when a red cell's long axis is more than twice the size of its short axis; can be seen in hereditary elliptocytosis), stomatocytes, echinocytes or burr cells (these are red cells with 10−30 short blunt spicules; seen in storage artifact, liver and kidney disease, and pyruvate kinase deficiency), acanthocytes (these are red cells with 2−20 unequal, irregular spicules), schistocytes, spherocytes, dacryocytes (teardrop red cells), bite cells, and blister cells.
- Red cell inclusions include Howell−Jolly bodies (which represent DNA material), Pappenheimer bodies (iron), Cabot rings (mitotic spindle remnants), basophilic stippling (represents RNA material; seen in lead poisoning, hemolytic anemia, and pyrimidine 5′-nucleotidase deficiency), Heinz bodies (denatured globin seen in G6PD deficiency; Heinz bodies are cleared by splenic macrophages; the damaged cells are known as bite cells), HbC crystals, nucleated RBCs, and malarial parasite.
- Diagnostic points of *P. falciparum*: Red cells are not enlarged; rings appear fine/delicate, with several per cell; some rings have two chromatin dots; the presence of marginal/appliqué forms; unusual to see developing forms; crescent-shaped gametocytes and Maurer's dots may be present.

References

[1] Endres HG, Wedding U, Pittrow D, Thiem U, et al. Prevalence of anemia in elderly patients in primary care: impact on 5-year mortality risk and difference between men and women. Curr Med Res Opin 2009;25:1143−58.

[2] Patel KV. Epidemiology of anemia in older adults. Semin Hematol 2008;45:210−17.

[3] Ania BJ, Suman VJ, Fairbanks VF, Melton III LJ. Prevalence of anemia in medical practice: community versus referral patients. Mayo Clin Proc 1994;69:730−5.

[4] Goddard AF, James MV, McIntyre AS, Scott BB, et al. Guidelines for the management of iron deficiency anemia. Gut 2011;60:1309−16.

[5] Castel R, Tax MG, Droogendijk J, Leers MP, et al. The transferrin/log(ferritin) ratio: a new tool for the diagnosis of iron deficiency anemia. Clin Chem Lab Med 2012;50:1343−9.

[6] Papanikolaou NC, Hatzidaki EG, Belivanis S, Tzanakakis GN, et al. Lead toxicity update: a brief review. Med Sci Monit 2005;11:RA329−36.

[7] Barrett TG, Bundey SE. Wolfram (DIDMOAD) syndrome. J Med Genet 1997;34:838−41.

[8] Kupfer GM. Fanconi anemia: a signal transduction and DNA repair pathway. Yale J Biol Med 2013;86:491−7.

[9] Dhaliwai G, Cornett PA, Tierney LM. Hemolytic anemia. Am Fam Physicians 2004; 69:2599−606.

[10] Beutler E. G6PD deficiency. Blood 1994;84:3613−36.

[11] Frank JE. Diagnosis and management of G6PD deficiency. Am Fam Physicians 2005;72: 1277−82.

[12] Minucci A, Giardina B, Zuppi C, Capoluongo E. Glucose 6-phosphate dehydrogenase— Laboratory assay: how, when and why? IUBMB Life 2009;61:27−34.

[13] Fouglas NM, Lampah DA, Kenangalem E, Simpson JA, et al. Major burden of severe anemia from non-falciparum malaria species in southern Papua: a hospital-based surveillance study. PLoS Med 2013;10:e1001575.

CHAPTER 4

Hemoglobinopathies and Thalassemias

4.1 INTRODUCTION

Hemoglobinopathies, inherited disorders of hemoglobin, are the most common monogenic diseases in the world. Hemoglobinopathies are a group of autosomal recessive disorders that can be broadly categorized into two major groups: thalassemias and structural variants of hemoglobin. However, hereditary persistence of fetal hemoglobin (Hb F), a relatively benign condition, can also be considered as hemoglobinopathy. The wide variation in clinical manifestations of these disorders is attributable to both genetic and environmental factors. Interestingly, α-thalassemia is very prevalent in endemic regions of malaria, and it protects against the severe form of *Plasmodium falciparum* infection [1]. Some hemoglobinopathies may be fatal if not treated; fortunately, hematopoietic stem cell transplantation, which is the only established cure, is becoming increasingly safe and cost-effective [2].

4.2 HEMOGLOBIN STRUCTURE AND SYNTHESIS

Hemoglobin, the oxygen-carrying pigment of erythrocytes, consists of a heme portion (iron-containing chelate) and four globin chains. Six distinct species of normal hemoglobin are found in humans—three in normal adults and three in fetal life. The globulins associated with the hemoglobin molecule (both embryonic stage and after birth) include alpha (α) chain, beta (β) chain, gamma (γ) chain, delta (δ) chain, epsilon (ϵ) chain, and zeta (ζ) chain. In the embryonic stage, hemoglobin Grower and hemoglobin Portland are found, but these are replaced by Hb F (two α chains and two γ chains) in fetal life. Interestingly, Hb F has higher oxygen affinity than adult hemoglobin and is capable of transporting oxygen in peripheral tissues in the hypoxic fetal environment. In the third trimester, genes responsible for β- and γ-globulin synthesis are activated; as a result, adult hemoglobin such as hemoglobin A (Hb A: two α chains and two β chains) and hemoglobin A_2 (Hb A_2: two α chains and

CONTENTS

A. Wahed and A. Dasgupta: Hematology and Coagulation. DOI: http://dx.doi.org/10.1016/B978-0-12-800241-4.00004-8

Table 4.1 Embryonic, Fetal, and Adult Hemoglobins

Period of Life	Hemoglobin Species	Globulin Chains	% Present in Adult
Embryonic	Gower-1	Two ζ, two ε	
	Gower-2	Two α, two ε	
	Portland-1	Two ζ, two γ	
	Portland-2	Two ζ, two β	
Fetal	Hemoglobin F	Two α, two γ	
Adult	Hemoglobin A	Two α, two β	92–95
	Hemoglobin A_2	Two α, two δ	<3.5
	Hemoglobin F	Two α, two γ	<1

two δ chains) may also be found in neonates, but Hb F is still the major component. Newborn babies and infants up to 6 months old do not depend on Hb A synthesis, although the switch from Hb F to Hb A occurs at approximately 3 months of age. Therefore, disorders due to β-chain defects such as sickle cell disease tend to manifest clinically after 6 months of age, although diseases due to α-chain defect are manifested *in utero* or following birth. Embryonic, fetal, and adult hemoglobins are summarized in Table 4.1. The different types of naturally occurring embryonic, fetal, and adult hemoglobins vary in their tetramer–dimer subunit interface strength (stability) in the liganded (carboxyhemoglobin or oxyhemoglobin) state [3].

The normal hemoglobin (Hb A) in adults contains two α chains and two β chains. Each α chain contains 141 amino acids, and each β chain contains 146 amino acids. Hb A_2 contains two α chains and two δ chains. The gene for the α chain is located in chromosome 16 (two genes in each chromosome, for a total of four genes), whereas genes for β (one gene in each chromosome, for a total of two genes), γ, and δ chains are located on chromosome 11. Adults have mostly Hb A and small amounts of Hb A_2 (<3.5%) and Hb F (<1%). A small amount of fetal hemoglobin persists in adults due to a small clone of cells called F cells. When hemoglobin is circulating with erythrocytes, glycosylation of the globin chains may occur. These are referred to as X1c (with X being any hemoglobin; e.g., Hb A1c). When the hemoglobin molecule is aging, glutathione is bound to cysteine at the 93rd position of the β chain. This is Hb AIII or Hb A1d. Just like Hb A1c and Hb A1d, Hb C1c, Hb C1d, Hb S1c, and Hb S1d may also exist in circulation.

Heme is synthesized in a complex manner involving enzymes in both mitochondrion and cytosol. In the first step, glycine and succinyl CoA combine in mitochondria to form δ-aminolevulinic acid, which is transported into cytoplasm and converted into porphobilinogen by the action of the enzyme aminolevulinic acid dehydratase. Next, porphobilinogen is converted into coproporphyrinogen III through several steps involving multiple enzymes.

Then coproporphyrinogen III is transported into mitochondria and converted into protoporphyrinogen III by the coproporphyrinogen III oxidase enzyme. Protoporphyrinogen III is then converted into protoporphyrin IX by the protoporphyrinogen III oxidase enzyme, and protoporphyrin IX is converted into heme by ferrochelatase enzyme. Finally, heme is transported into cytosol and combines with globulin to form the hemoglobin molecule.

4.3 INTRODUCTION TO HEMOGLOBINOPATHIES

Hemoglobinopathies can be classified into three major categories:

- Quantitative disorders of hemoglobin synthesis: Production of structurally normal but decreased amounts of globin chains (thalassemia syndrome)
- Disorder (qualitative) in hemoglobin structure: Production of structurally abnormal globulin chains such as hemoglobin S, C, O, or E. Sickle cell syndrome is the most common example of such disease.
- Failure to switch globin chain synthesis after birth—for example, hereditary persistence of Hb F, which is a relatively benign condition. It may coexist with thalassemia or sickle cell disease and will result in decreased severity of such diseases (protective effect).

Hemoglobinopathies are transmitted in an autosomal recessive manner. Therefore, carriers who have one affected chromosome and one normal chromosome are usually healthy or slightly anemic. When both parents are carriers, their children have a 25% chance of being normal, a 25% chance of being affected by the disease, and a 50% chance of being a carrier. Hemoglobinopathies are caused by inherent mutation of genes coded for globin synthesis. Mutations may disrupt gene expression causing less production of α- or β-chain globin (thalassemias) or point mutation of the gene in the coding region (exons), which may cause the production of defective globin that results in the formation of abnormal hemoglobin (hemoglobin variants) [4].

It has been estimated that approximately 5% of the world's population are carriers of hemoglobin disorders. Moreover, hemoglobinopathies affect approximately 370,000 newborns each year. The hemoglobin variants of most clinical significance are Hb S, C, and E. In West Africa, approximately 25% of individuals are heterozygous for the Hb S gene, which is related to sickle cell diseases. In addition, high frequencies of Hb S gene alleles are also found in people who live in the Caribbean, south and central Africa, the Mediterranean region, the Arabian Peninsula, and East India. Hb C is found mostly in people who live in or originate from West Africa. Hb E is widely distributed between East India and Southeast Asia, with the highest prevalence in Thailand, Laos, and Cambodia; however, it may be sporadically observed in areas of China and Indonesia. Thalassemia syndrome is not due to a structural defect in the globin chain but, rather, to lack of sufficient synthesis of the globin chain, and it is also a genetically inherited disease. Thalassemic syndrome can be categorized as α-thalassemia and β-thalassemia. In general, β-thalassemia is observed in the Mediterranean, Arabian Peninsula, Turkey, Iran, West and Central Africa, India, and other Southeast Asian countries, whereas α-thalassemia is commonly observed in areas of Africa, the Mediterranean, the Middle East, and throughout Southeast Asia [5]. Of more than 1000 hemoglobinopathies reported, most are asymptotic. However, in other cases, significant clinical disorders can be noted, including the following:

- Thalassemias (both α and β)
- Sickling disorders (Hb SS, Hb SC, Hb SD, and Hb SO)
- Cyanosis (e.g., Hb Kansas)
- Hemolytic anemias (e.g., Hb H)
- Erythrocytosis (e.g., Hb Malmo)

4.3.1 α-Thalassemia

There are two genetic loci for the α gene resulting in four genes (alleles) for α hemoglobin (α/α, α/α) on chromosome 16. Two alleles are inherited from each parent. α-Thalassemia occurs when there is a defect or deletion in one or more of four genes responsible for α-globin production. α-Thalassemia can be divided into four categories:

- The silent carriers: Characterized by only one defective or deleted gene but three functional genes. These individuals have no health problems. In unusual cases of silent carrier, individuals carry one defective Constant Spring mutation but three functional genes. These individuals also have no health problems.
- α-Thalassemia trait: Characterized by two deleted or defective genes and two functional genes. These individuals may have mild anemia.
- α-Thalassemia major (Hb H disease): Characterized by three deleted or defective genes and only one functional gene. These patients have persistent anemia and significant health problems. When Hb H disease is combined with Hb Constant Spring, the severity of disease is more than that of Hb H disease alone. However, if a child inherits one Hb Constant Spring from the mother and one from the father, the child has homozygous Hb Constant Spring and the severity of the disease is similar to that of Hb H disease.
- Hydrops fetalis: Characterized by no functional α gene. These individuals have Hb Bart. This condition is not compatible with life unless intrauterine transfusion is initiated.

When an α gene is functional, it is denoted as "α"; when it is not functional or is deleted, it is denoted as " $-$." There is not much difference in impaired α-globin synthesis between a deleted gene and a nonfunctioning defective gene. With deletion or defect of one gene ($-/\alpha$, α/α), little clinical effect is observed because three α genes are sufficient to allow normal hemoglobin production. These patients are sometimes referred to as "silent carriers" because there are no clinical symptoms, but mean corpuscular volume (MCV) and mean corpuscular hemoglobin (MCH) may be slightly decreased. These individuals are diagnosed by deduction only when they have children with the thalassemia trait or Hb H disorder. An unusual case of silent carrier state is an individual carrying one hemoglobin Constant Spring mutation but three functional genes. Hemoglobin Constant Spring (a hemoglobin variant isolated from a family of ethnic Chinese background from the Constant Spring district of Jamaica) is a hemoglobin variant in which mutation of the α-globin gene produces an abnormally long α chain (172 amino acids instead of the normal 141 amino acids). Hemoglobin Constant Spring is due to non-deletion mutation of the α gene, which results in production of unstable α-globin. Moreover, this α-globin is produced in very low quantity (approximately 1% of normal expression level) and is found in people who live in or originate from Southeast Asia.

When two genes are defective or deleted, the α-thalassemia trait is present. There are two forms of the α-thalassemia trait. α-Thalassemia 1 ($-/-$, α/α) results from the *cis* deletion of both α genes on the same chromosome. This mutation is found in Southeast Asian populations. α-Thalassemia 2 ($-/\alpha$, $-/\alpha$) results from the transdeletion of α genes on two different chromosomes. This mutation is found in African and African American populations (the prevalence of disease is 28% in African Americans). Only in the case of *cis* deletion, ζ-globin is expressed in carriers. In the α-thalassemia trait, two functioning α genes are present, and as a result, erythropoiesis is almost normal in these individuals, but a mild microcytic hypochromic anemia (low MCV and MCH) may be observed. This form of the disease can mimic iron deficiency anemia. Therefore, it is essential to distinguish α-thalassemia from iron deficiency anemia.

If three genes are affected ($-/-$, $-/\alpha$), the disease is called Hb H disease, which is a severe form of α-thalassemia. Patients with severe anemia require blood transfusion. Because only one α gene is responsible for production of α-globin in Hb H disease, a high β-globin to α-globin ratio (two- to fivefold increase in β-globin production) may result in the formation of a tetramer containing only β, and this form of hemoglobin is called Hb H (four β chains). This form of hemoglobin cannot deliver oxygen in peripheral tissues because Hb H has a very high affinity for oxygen. A microcytic hypochromic anemia with target cells and Heinz bodies (which represents precipitated Hb H) is present in the peripheral blood smear of these patients. Moreover, red cells that contain Hb H are sensitive to oxidative stress and may be more susceptible to hemolysis, especially when oxidants such as sulfonamides are administered. More mature erythrocytes also contain increasing amounts of precipitated Hb H (Heinz bodies). These are removed from the circulation prematurely, which may also cause hemolysis. Therefore, clinically, these patients experience varying severity of chronic hemolytic anemia. Due to the subsequent increase in erythropoiesis, erythroid hyperplasia may result, causing bone structure abnormalities, with marrow hyperplasia, bone thinning, maxillary hyperplasia, and pathologic fractures. When Hb Constant Spring is associated with Hb H disease, a more severe form of anemia is observed requiring frequent transfusion [6]. However, when a child inherits one Hb Constant Spring gene from the father and one from the mother, Hb Constant Spring disease is present that is less severe than that of Hb H−Hb Constant Spring disease but comparable to that of Hb H disease. Patients with Hb H and related diseases require transfusion and chelation therapy to remove excess iron.

When four genes are defective or deleted $(-/-, -/-)$, the result is Hb Bart's disease, in which α-globin is absent because no gene is present to promote α-globin synthesis and as a result four γ chains form a tetramer. As in Hb H, the hemoglobin in Hb Bart's in unstable, which impairs the ability of the red cells to release oxygen to the surrounding tissues. The fetus usually cannot survive gestation, causing stillbirth with hydrops fetalis. However, currently, with the aid of intrauterine transfusion and the neonatal intensive care unit, survival may be possible, but survivors will have severe transfusion-dependent anemia like patients suffering from β-thalassemia major. Bone marrow transplant or cord blood transplant may be helpful.

4.3.2   β-Thalassemia

β-Thalassemia is due to a deficit or absent production of β-globin resulting in excess production of α-globin. Synthesis of β-globin may vary from near complete to absent, causing β-thalassemia of various degrees of severity due to mutation of genes (one gene each on chromosome 11); more than 200 point mutations have been reported. However, deletion of both genes is rare. β-Thalassemia can be broadly classified into three categories:

- β-Thalassemia trait: Characterized by one defective gene and one normal gene. Individuals may experience mild anemia but do not require transfusion.
- β-Thalassemia intermedia: Characterized by two defective genes, but some β-globin production is still observed in these individuals. However, some individuals may have significant health problems requiring intermittent transfusion.
- β-Thalassemia major (Cooley's anemia): Characterized by two defective genes but almost no function of either gene, leading to no synthesis of β-globin. These individuals have a severe form of disease requiring lifelong transfusion and may have shortened life span.

If a defective gene is incapable of producing any β-globin, it is characterized as "β^0," which causes the more severe form of β-thalassemia. However, if the mutated gene can retain some function, it is characterized as "β^+." In the case of one gene defect, β-thalassemia minor (trait; patients are β^0/β or β^+/β) is observed, and individuals are either normal or mildly anemic. These patients have increased Hb A_2. In addition, Hb F may also be elevated. MCV and MCH are low, but these patients are not transfusion dependent. If both genes are affected, resulting in severely impaired production of β-globin (β^0/β^0 or β^+/β^0), the disease is severe and is called β-thalassemia major (also known as Cooley's anemia). However, due to the presence of fetal hemoglobin, symptoms of β-thalassemia major are not observed prior to 6 months of age. Patients with β-thalassemia major have elevated Hb A_2 and Hb F (although Hb F may be normal in some individual). If production of β-globins is moderately hampered, then the disease is called β-thalassemia intermedia (β^0/β or β^+/β^+). These individuals have less severe disease than those with β-thalassemia major. In patients with β-thalassemia major, excess α-globulin chain precipitates leading to hemolytic anemia. These patients require lifelong transfusion and chelation therapy. Interestingly, having β^0 or β^+ does not predict the severity of disease because patients with both types have been diagnosed with β-thalassemia major or intermedia. Major features of α- and β-thalassemia are summarized in Table 4.2.

Table 4.2 Major Features of α- and β-Thalassemia

Disease	No. of Deleted Gene	Comments
α-Thalassemia silent carrier	One of four gene deletion	Asymptomatic
		May have low MCV, MCH
α-Thalassemia trait	Two of four gene deletion	Asymptomatic/mild symptoms
		Mild microcytic hypochromic anemia
Hemoglobin H disease	Three of four gene deletion	Microcytic hypochromic anemia and Hb H found in adults and Hb Bart's found in neonates
		Hb H may coexist with Hb Constant Spring, a more severe disease than Hb H
Hydrops fetalis	Four of four gene deletion	Hemoglobin Bart's disease
		Most severe form may cause stillbirth/hydrops fetalis
β-Thalassemia trait	One gene defect	Asymptomatic
β-Thalassemia intermedia	Both genes defective but retain some function	Variable degree of severity because some β-globin is still produced
β-Thalassemia major	Both genes defective	Severe impairment or no β-globin synthesis
		Severe disease with anemia, splenomegaly, requiring lifelong transfusion

4.3.3   δ-Thalassemia

δ-Thalassemia is due to mutation of genes responsible for synthesis of δ chain. A mutation that prevents formation of δ chain is called δ^0, and if a δ chain is formed, the mutation is termed as δ^+. If an individual inherits two δ^0 mutations, no δ chain is produced and no Hb A_2 can be detected in blood (normal level, <3.5%). However, if an individual inherits two δ^+ mutations, decreased Hb A_2 is observed. All patients with δ-thalassemia have normal hematological consequences, although the presence of δ mutation may obscure the diagnosis of β-thalassemia trait because in β-thalassemia, Hb A_2 is increased but the presence of δ mutation may reduce Hb A_2 concentration, masking the diagnosis of β-thalassemia trait.

δβ-Thalassemia is a rare hemoglobinopathy characterized by decreased or the total absence of production of δ- and β-globin. As a compensatory mechanism, γ-chain synthesis is increased, resulting in a significant amount of Hb F in blood, which is homogeneously distributed in red blood cells. This condition is found in many ethnic groups, but it is especially observed in individuals with Greek or Italian ancestry. Heterozygous individuals are asymptomatic with normal Hb A_2, but rarely reported homozygous individuals experience mild symptoms.

4.3.4   Sickle Cell Disease

The term "sickle cell disease"' includes all manifestations of abnormal Hb S, including sickle cell trait (Hb AS), homozygous sickle cell disease (Hb SS), and a range of mixed heterozygous hemoglobinopathies such as Hb SC disease, Hb SD disease, Hb SO Arab disease, and Hb S combined with β-thalassemia. Sickle cell disease affects millions of people throughout the world and is particularly common in people living in or migrating from Sub-Saharan Africa, South America, the Caribbean, Central America, Saudi Arabia, India, and Mediterranean countries such as Turkey, Greece, and Italy. Sickle cell disease is the most commonly observed hemoglobinopathy in the United States, affecting 1 in every 500 African American births and 1 in every 36,000 Hispanic American births. Sickle cell disease is a dangerous hemoglobinopathy, and the symptoms start before the age of 1 year with chronic hemolytic anemia, developmental disorder, crisis including extreme pain (sickle cell crisis), high susceptibility to various infections, spleen crisis, acute thoracic syndrome, and increased risk of stroke. Optimally treated individuals may have a life span of 50−60 years [7].

In sickle cell disease, the normal round shape of red blood cells (RBCs) is changed to a crescent shape; hence the name "sickle cell." In the heterozygous form (Hb AS), sickle cell disease protects against infection of *P. falciparum* malaria, but it does not do so in the more severe form of homozygous sickle cell disease (Hb SS). The genetic defect producing sickle hemoglobin is a single nucleotide substitution at codon 6 of the β-globin gene on chromosome 11 that results in a point mutation in the β-globin chain of hemoglobin (substitution of valine for glutamic acid at the sixth position). Hemoglobin S is formed when two normal α-globins combine with two mutant β-globins. Because of this hydrophobic amino acid substitution, Hb S polymerizes upon deoxygenation and multiple polymers bundle into a rodlike structure resulting in deformed RBCs. Possible diagnoses of patients with Hb S hemoglobinopathy include sickle cell trait (Hb AS), sickle cell disease (Hb SS), and sickle cell disease status post RBC transfusion/exchange. Patients with sickle cell trait may also have concomitant α-thalassemia, and the diagnosis of HbS/β-thalassemia $(0/+/++)$ is also occasionally made. Double heterozygous states of Hb SC, Hb SD, and Hb SO Arab are important sickling states that should not be missed. Hemoglobin C is formed due to substitution of glutamic acid residue with a lysine residue at the sixth position of β-globin. Individuals who are heterozygous with Hb C disease are asymptomatic with no apparent disease, but homozygous individuals have almost all hemoglobin (>95%) as Hb C and experience chronic hemolytic anemia and pain crisis. However, individuals who are heterozygous with both Hb C and Hb S (Hb SC disease) have weaker symptoms than those with sickle cell disease because Hb C does not polymerize as readily as Hb S.

Patients with Hb SS disease may have increased Hb F. The distribution of Hb F among the haplotypes of Hb SS are Hb F 5−7% in Bantu, Benin, or Cameroon; Hb F 7−10% in Senegal; and Hb F 10−25% in Arab−Indian. Hydroxyurea also causes an increase in Hb F. This is usually accompanied by macrocytosis. Hb F can also be increased in Hb S/HPFH (hereditary persistence of fetal hemoglobin). Hb A_2 values are typically increased in sickle cell disease and more so on high-performance liquid chromatography (HPLC) analysis. This is because the post-translational modification form of Hb S, Hb S1d, produces a peak in the A_2 window. This elevated value of Hb A_2 may produce diagnostic confusion with Hb SS disease and Hb S/β-thalassemia. It is important to remember that microcytosis is not a feature of Hb SS disease, and patients with Hb S/β-thalassemia typically exhibit microcytosis.

Hb SS patients and Hb $S/β^0$-thalassemia patients do not have any Hb A, unless they have been transfused or have undergone red cell exchange. Glycated Hb S has the same retention time (~ 2.5 min) as Hb A in HPLC. This will produce a small peak in the A window and raise the possibility of Hb $S/β^+$-thalassemia. Hb S/α-thalassemia is considered when the percentage of Hb S is lower than expected. Classical cases are 60% of Hb A and approximately 35−40% of Hb S. Cases of Hb S/α-thalassemia should have lower values of Hb S, typically below 30% with microcytosis. A similar picture will also be present in patients with sickle cell trait and iron deficiency. Various features of sickle cell disease are summarized in Table 4.3.

Table 4.3 Major Features of HbS Hemoglobinopathies

Disease	Hemoglobin Variants	Clinical Features
Sickle cell trait (heterozygous)	Hb AS	Hb S, 35–40%; Hb A_2, $\geq 3.5\%$ Normal hemoglobin No apparent illness
Sickle cell disease	Hb SS	Hb S, >90%; Hb A_2, <3.5%; Hb F, <10%; no Hb A Hemoglobin, 6–8 g/dL Severe disease with chronic hemolytic anemia
Sickle cell β^0-thalassemia	Hb Sβ^0	Hb S, >80%; Hb A_2, >3.5%; Hb F, <20%; no Hb A Hemoglobin, 7–9 g/dL Severe sickle cell disease
Sickle cell β^+-thalassemia	Hb Sβ^+	Hb S, >60%; Hb A_2, >3.5%; Hb F, >20%; Hb A, 5–30% Hemoglobin, 9–12 g/dL Variable mild to moderate sickle cell disease
Hemoglobin SC disease	Hb SC	Hb S, 50%; Hb C, 50%; Hb F, >5% Hemoglobin, 10–12 g/dL Moderate sickling disease but chronic hemolytic anemia may be present
Hemoglobin S/HPFH		Hb S, 60%; Hb A_2, <3.5%; Hb F, 30–40% Hemoglobin, 11–14 g/dL; no Hb A Mild sickling disease

4.3.5   Hereditary Persistence of Fetal Hemoglobin

In individuals with hereditary persistence of fetal hemoglobin (HPFH), significant amounts of Hb F can be detected well into adulthood. In normal adults, Hb F represents less than 1% of total hemoglobin, whereas in HPFH the percentage of HB F can be significantly elevated but Hb A_2 is also normal. HPFH is categorized into two major groups: deletional and nondeletional. Deletional HPFH is caused by variable-length deletion in the β-globin gene cluster, leading to decreased or absent β-globin synthesis and a compensatory increase in γ-globin synthesis with a pancellular or homogeneous distribution of Hb F in RBCs. Nondeletional HPFH is a broad category of related disorders with increased Hb F typically distributed heterocellularly. Heterocellular distribution is also seen in β-thalassemia and δβ-thalassemia.

Both homozygous and heterozygous HPFH individuals are asymptomatic with no clinical or significant hematological changes; individuals with homozygous HPFH may show up to 100% Hb F, whereas those with heterozygous HFPH typically show 20–28% Hb F. When HPFH is associated with sickle cells, the severity of disease may be reduced. Compound heterozygotes for sickle hemoglobin (Hb S) and HPFH have a high level of Hb F, but these individuals experience few, if any, sickle cell disease-related complications [8]. When HPFH is associated with thalassemia, individuals also experience less severe disease.

4.4 OTHER HEMOGLOBIN VARIANTS

Hemoglobin D (Hb D Punjab, also known as Hb D Los Angeles) is formed due to substitution of glutamine for glutamic acid, and Hb D Punjab is one of the most commonly observed abnormalities worldwide—found not only in the Punjab region of India but also in Italy, Belgium, Austria, and Turkey. Hemoglobin D disease can occur in four different forms: heterozygous Hb D trait, Hb D thalassemia, Hb SD disease, and, very rarely, homozygous Hb D disease. Heterozygous Hb D disease is a benign condition with no apparent illness, but when Hb D is associated with Hb S or β-thalassemia, clinical conditions such as sickling disease and moderate hemolytic anemia may be observed. Heterozygous Hb D is rare and usually presents with mild hemolytic anemia and mild to moderate splenomegaly [9].

Hemoglobin E is caused by point mutation of β-globin, which results in substitution of lysine for glutamic acid in position 26. As a result, production of β-globin is diminished, and Hb E also has structural defects and is a thalassemia-like phenotype. Hb E is unstable and can form Heinz bodies under oxidative stress. Hb E trait is associated with moderately severe microcytosis, but usually no anemia is present. However, individuals with Hb E homozygous, present with modest anemia similar to thalassemia trait. When β-thalassemia is combined with Hb E, such as in Hb E/β^0-thalassemia, patients may have significant anemia requiring transfusion, similarly to patients with β-thalassemia intermedia.

Hemoglobin O-Arab (Hb O-Arab; also known as Hb Egypt) is a rare abnormal hemoglobin variant in which, at position 121 of β-globin, normal glutamic acid is replaced by lysine. Hb O-Arab is found in people from the Balkans, Middle East, and Africa. Patients who are heterozygous for Hb O-Arab may experience mild anemia and microcytosis similarly to patients with β-thalassemia minor; those who are homozygous, which is extremely rare, may have anemia but despite the abnormal hemoglobin pattern may be mostly asymptomatic. However, patients with Hb S/Hb O-Arab may experience severe clinical symptoms similarly to individuals with Hb S/S. Similarly, patients with Hb O-Arab/β-thalassemia may experience severe anemia with a hemoglobin level between 6 and 8 g/dL and splenomegaly [10].

Hemoglobin Lepore is an unusual hemoglobin molecule that is composed of two α chains and two $\delta\beta$ chains as a result of fusion of δ and β genes. The $\delta\beta$ chains comprise the first 87 amino acids of the δ chain and 32 amino acids of the β chain. There are three common variants of hemoglobin Lepore: Hb Lepore Washington (also known as Hb Lepore Boston), Hb Lepore Baltimore, and Hb Lepore Hollandia. Hemoglobin Lepore is seen in individuals of Mediterranean descent. Individuals with Hb A/Hb Lepore are asymptomatic, with Hb Lepore representing 5−15% of hemoglobin; there is slightly elevated Hb F (2−3%), with low MCV as well as MCH. However, homozygous Lepore individuals suffer from severe anemia similarly to patients with β-thalassemia intermedia, with Hb Lepore representing 8−30% of hemoglobin and the remainder Hb F. Patients with Hb Lepore/β-thalassemia experience severe disease similarly to patients with β-thalassemia major.

Hemoglobin G Philadelphia (Hb G) is the most common α-chain defect, affecting 1 in 5000 African Americans, and is associated with α-thalassemia 2 deletions. Therefore, these individuals have only three functioning α genes, and Hb G represents one-third of total hemoglobin. Hb S is the most common β-chain defect observed in the African American population, whereas Hb G is the most common α-chain defect, again occurring most often in the African American population. Therefore, it is possible that an African American individual may have Hb S/Hb G, in which the hemoglobin molecule contains one normal α chain, one α G chain, one normal β chain, and one β S chain. This can result in detection of various hemoglobin in the blood, including Hb A (α_2, β_2), Hb S (α_2, β S$_2$), Hb G (α G$_2$, β_2), and HbS/G (α G$_2$, β S$_2$). In addition, Hb G$_2$ (α_2, δ_2), which is the counterpart of Hb A$_2$, is also present.

An increase in fetal hemoglobin percentage is associated with multiple pathologic states, including β-thalassemia, $\delta\beta$-thalassemia, and HPFH. β-Thalassemia is associated with high Hb A$_2$, and the latter two states are associated with normal Hb A$_2$ values. Hematologic malignancies are associated with increased Hb F and include acute erythroid leukemia (AML, M6), and juvenile myelomonocytic leukemia (JMML). Aplastic anemia is also associated with an increase in the percentage of Hb F. In elucidating the actual cause of high Hb F, it is important to consider the actual percentage of Hb F, Hb A$_2$ values, as well as the correlation with complete blood count (CBC) and peripheral smear. It is also important to note that drugs (hydroxyurea, sodium valproate, and erythropoietin) and stress erythropoiesis may also result in high Hb F. Hydroxyurea is used in sickle cell disease patients to increase the amount of Hb F, the presence of which may help to reduce the clinical effects of the disease. Measuring the level of Hb F may be useful in determining the appropriate dose of hydroxyurea. In 15−20% of cases involving pregnancy, Hb F may be increased by as much as 5%.

Other rarely reported hemoglobinopathies involve Hb I, Hb J, Hb Hope, and unstable hemoglobin such as Hb Koln, Hb Hasharon, and Hb Zurich (for these unstable hemoglobins, the isopropanol test is positive). Ceratin rarely reported hemoglobin variants are Hb Malmo, Hb Andrew, Hb Minneapolis,

Table 4.4 Various Other Common Hemoglobinopathies

Diagnosis	Hemoglobin/Hematological	Comments
Hb C trait (Hb AC)	Hb A, 60%; Hb C, 40% Normal/microcytic	Hb C implies ancestry from West Africa, clinically insignificant
Hb CC disease	No Hb A; Hb C almost 100% Mild microcytic	Mild chronic hemolytic anemia
Hb C trait/α-thalassemia	Hb A; major hemoglobin Hb C, <30%	
Hb C/β-thalassemia	Microcytic, hypochromatic	Moderate to severe anemia with splenomegaly
Hb E trait (Hb AE)	Hb A major; Hb E, 30–35% Normal/microcytic	No clinical significance; found in Cambodia, Laos, and Thailand (Hb E triangle, where Hb E trait is 50–60% of population) and Southeast Asia
Hb E disease	No Hb A; mostly Hb E Microcytic hypochromic red cells ± anemia	Usually asymptomatic
Hb E trait with α-thalassemia	Majority is A; Hb E, <25%	
Hb O trait (Hb AO)	Majority is A; Hb O, 30–40% Normal CBC	Clinically insignificant but Hb S/O is a sickling disorder
Hb D trait (Hb AD)	Hb A > Hb D Normal CBC	Clinically insignificant; Hb S/D is a sickling disorder
Hb G trait (Hb AG)	Hb A > Hb G Normal CBC	Clinically insignificant

Hb British Columbia, and Hb Kempsey. Patients with these rare hemoglobin variants experience erythrocytosis. Hb I is due to a single α-globin substitution (substitution of lysine at position 16 for glutamic acid). Hb I is clinically insignificant unless, on rare occasions, it is associated with α-thalassemia, in which approximately 70% of hemoglobin is Hb I. Hb J is characterized as a fast-moving band in hemoglobin electrophoresis (the band close to the anode, the farthest point from application of the sample), and more than 50 variants have been reported, including Hb J Cape Town and Hb J Chicago. However, heterozygous hemoglobinopathy involving Hb J is clinically insignificant. In Hb Hope, aspartic acid is substituted for glycine at position 136 of the β chain. Important other hemoglobinopathies are summarized in Table 4.4.

4.5 LABORATORY INVESTIGATION OF HEMOGLOBINOPATHIES

Multiple methodologies exist to detect hemoglobinopathies and thalassemias. Three methods that are routinely employed are gel electrophoresis, HPLC, and capillary electrophoresis. If any one method detects an abnormality, a second method must be used to confirm the abnormality. In addition, relevant clinical history, review of the CBC, and peripheral smear provide important correlation in the pursuit of an accurate diagnosis.

4.5.1 Gel Electrophoresis

In hemoglobin electrophoresis, red cell lysates are subjected to electric fields under alkaline (alkaline gel) and acidic (acid gel) pH. This can be carried out on filter paper, a cellulose acetate membrane, a starch gel, a citrate agar gel, or an agarose gel. Separation of different hemoglobins is largely, but not solely, dependent on the charge of the hemoglobin molecule. A change in the amino acid composition of the globin chains results in alteration of the charge of the hemoglobin molecule, thus resulting in a change in the speed of migration. In gel electrophoresis, different hemoglobins migrate at different speeds; the top lane is called the H lane and is mainly composed of Hb H and Hb I, and the point of origin is before the carbonic acid band (Table 4.5). On the alkaline gel in hemoglobin electrophoresis, the H is fast-migrating, and the band on the gel should be the same distance from J as A is from J in the opposite direction.

Table 4.5 Migration of Various Hemoglobin Bands in Alkaline Gel and Acid Gel Electrophoresis

Region	Hemoglobin Present
Alkaline gel electrophoresis	
Top band (farthest from origin: H Lane)	Hb H, Hb I
J Lane	Hb J
	Hb Bart's and Hb N are between Hb J and Hb H lanes
A Lane	Hb A
F Lane	Hb F
S Lane	Hb S, Hb D, Hb G, Hb Lepore
C Lane	Hb C, Hb E, Hb O, Hb A_2, Hb S/G hybrid
Carbonic anhydrase band (faint)	Hb G_2, Hb A_2', Hb CS
Acid gel electrophoresis	
Top band (fastest from origin) C Lane	Hb C
S Lane	Hb S, Hb S/G hybrid
	Hb O and Hb H are between S and A lanes
A lane	Hb A, Hb E, Hb A_2, Hb D, Hb G, Hb Lepore, Hb J, Hb I, Hb N, Hb H
F Lane	Hb F, Hb Hope, Hb Bart's

Table 4.6 Approximate Retention Times of Various Hemoglobins in HPLC Analysis

Approximate Retention Time (Min)	Hemoglobin
0.7 (Peak 1)	Acetylated Hb F, Hb H, Hb Bart's, bilirubin
1.1	Hb F
1.3 (Peak 2)	Hb A1c, Hb Hope
1.7 (Peak 3)	Aged Hb A (Hb A1d), Hb J, Hb N, Hb I
2.5	Hb A, Hb S1c
3.7	Hb A_2, Hb E, Hb Lepore, Hb S1d
3.9–4.2	Hb D, Hb G
4.5	Hb S, Hb A_2', Hb C1c Hb O Arab has a broad range from 4.5 to 5 min
4.6–4.7	Hb G_2
4.9	Hb C (preceding the main peak is a small peak, Hb C1d), Hb S/G hybrid, Hb CS (three peaks: 2–3%)

On the acid gel, the H migrates between the S and hemoglobins. The patterns of various bands in acid gel electrophoresis are summarized in Table 4.5.

4.5.2 High-Performance Liquid Chromatography

HPLC systems utilize a weak cation exchange column system. A sample of an RBC lysate in buffer is injected into the system, followed by application of a mobile phase so that various hemoglobins can partition (interact) between the stationary phase and the mobile phase. The time required for different hemoglobin molecules to elute is referred to as retention time. The eluted hemoglobin molecules are detected by light absorbance. HPLC permits the provisional identification of many more variant types of hemoglobins that cannot be distinguished by conventional gel electrophoresis. When HPLC is used, a recognized problem is carryover of specimen from one specimen to the next. For example, if the first specimen belongs to a patient with sickle cell disease (Hb SS), then a small peak may be seen at the "S" window in the next specimen. This can lead to diagnostic confusion as well as the sample being re-run. Approximate retention times of common hemoglobins in a typical HPLC analysis are summarized in Table 4.6.

4.5.3 Capillary Electrophoresis

In capillary electrophoresis, a thin capillary tube made of fused silica is used. When an electric filed is applied, the buffer solution within the capillary generates an electroendosmotic flow that moves toward the cathode. Separation of individual hemoglobins takes place due to differences in overall charges. Different hemoglobins are represented in different zones. Capillary zone

Table 4.7 Various Zones in which Common Hemoglobins Appear in Capillary Electrophoresis

Zone	Hemoglobin
Zone 1	Hb A$_2$'
Zone 2	Hb C, Hb CS
Zone 3	Hb A$_2$, Hb O-Arab
Zone 4	Hb E, Hb Koln
Zone 5	Hb S
Zone 6	Hb D Punjab/Los Angeles/Iran, Hb G Philadelphia
Zone 7	Hb F
Zone 8	Acetylated Hb F
Zone 9	Hb A
Zone 10	Hb Hope
Zone 11	Denatured Hb A
Zone 12	Hb Bart's
Zone 13	
Zone 14	
Zone 15	Hb H

electrophoresis has an advantage over HPLC in that hemoglobin adducts (glycated hemoglobins and the aging adduct Hb X1d) do not separate from the main hemoglobin peak in capillary electrophoresis, making interpretation easier. Common hemoglobin zones in capillary electrophoresis are given in Table 4.7.

Other less commonly used methodologies include isoelectric focusing, DNA analysis, and mass spectrometry. It is important to note that hemoglobinopathies may interfere with the measurement of glycosylated hemoglobin (Hb A1c), providing an unreliable result. When an Hb A1c result is inconsistent with a patient's clinical picture, the possibility of hemoglobinopathy must be considered. Depending on the methodology used for measurement of Hb A1c, such as HPLC or immunoassay, results may be falsely elevated or lower. Patients with Hb C trait particularly show variable results. In such cases, a test that is not affected by hemoglobinopathy, such as fructosamine measurement (which represents average blood glucose: 2 or 3 weeks), may be used [11].

4.6 DIAGNOSTIC TIPS FOR THALASSEMIAS, SICKLE CELL DISEASE, AND OTHER HEMOGLOBINOPATHY

For diagnosis of α-thalassemia, routine blood analysis (CBC) is the first step. MCV, MCH, and red cell distribution width (RDW) provide important clues for the diagnosis of not only thalassemias but also other hemoglobin disorders.

Thalassemias are characterized by hypochromatic and microcytic anemia, and it is important to differentiate thalassemia from iron deficiency anemia because iron supplementation has no benefit in patients with thalassemia. Often, silent carriers of α-thalassemia are diagnosed incidentally when their CBC shows a mild microcytic anemia. However, serum iron and serum ferritin levels are normal in a silent carrier of α-thalassemia but reduced in a patient with iron deficiency anemia. In addition, microcytic anemia with normal RDW indicates thalassemia trait. In hemoglobin H disease, MCV is further reduced, although in iron deficiency anemia MCV is rarely less than 80 fL. In addition, MCH is also reduced. For children, an MCV of less than 80 fL may be common, and Mentzer index (MCV/RBC count) is useful in differentiating thalassemia from iron deficiency anemia. In iron deficiency anemia, this ratio is usually greater than 13, but in thalassemia this value is less than 13. However, for accurate diagnosis of α-thalassemia, genetic testing is essential. Hemoglobin electrophoresis is not usually helpful for diagnosis of α-thalassemia except in infants in whom the presence of Hb Bart's or Hb H indicates α-thalassemia, but hemoglobin electrophoresis is usually normal in individuals with α-thalassemia trait. However, in an individual with Hb H disease, the presence of hemoglobin H in electrophoresis along with Hb Bart's is a useful diagnostic clue. In hydrops fetalis, newborns often die or are born with gross abnormalities. Circulating erythrocytes are markedly hypochromic, and anisopoikilocytosis is present. In addition, many nucleated erythroblasts are present in peripheral blood smear. Most of the hemoglobin observed in electrophoresis is Hb Bart's. Genetic testing of parents is essential for counseling of parents who may give birth to an infant with hydrops fetalis.

A patient with β-thalassemia major disease can be identified during infancy, but after 6 months of age, these patients present with irritability, growth retardation, abnormal swelling, and jaundice. Individuals with microcytic anemia but milder symptoms that start later in life are suffering from β-thalassemia intermedia. Hemoglobin electrophoresis of individuals with β-thalassemia trait usually shows reduced or absent Hb A, elevated levels of Hb A_2, and elevated levels of Hb F. Therefore, for the diagnosis of β-thalassemia trait, the proportion of Hb A_2 relative to the other hemoglobins is an important indicator. In certain cases, Hb A_2 variants may also be present. In such cases, the total Hb A_2 (Hb A_2 and Hb A_2 variant) needs to be considered for the diagnosis of β-thalassemia. Hb A_2' is the most common of the known Hb A_2 variants; it is reported in 1 or 2% of African Americans and detected in heterozygous and homozygous states and in combination with other Hb variants and thalassemia. The major clinical significance of HbA_2' is that for the diagnosis or exclusion of β-thalassemia minor, the sum of Hb A_2 and Hb A_2'' must be considered. When present, Hb A_2' accounts for a small percentage (1 or 2%) in heterozygotes and is difficult to detect by gel electrophoresis. However, it is easily detected by capillary electrophoresis and HPLC. In HPLC, Hb A_2' elutes in the "S" window. In Hb AS trait and HB SS disease, Hb A_2' may be masked by the presence of Hb S. In Hb AC trait and Hb CC disease, glycosylated Hb C will also elute in the "S" window. In these conditions, Hb A_2' will remain undetected. Conversely, sickle cell patients on chronic transfusion protocol or who have had recent, efficient RBC exchange may have in a very small percentage of Hb S, which the pathologist may interpret as Hb A_2'. It has been documented that the Hb A_2 concentration may be raised in HIV patients during treatment. Severe iron deficiency anemia can reduce Hb A_2 levels, and this may obscure diagnosis of β-thalassemia trait. Hematological features of α- and β-thalassemia are given in Table 4.8 [12].

Hb F quantification is useful in the diagnosis of β-thalassemia and other hemoglobinopathies. However, quantification of Hb F may be an issue when HPLC is used. Fast variants (e.g., Hb H or Hb Bart's) may not be quantified because they may elute off the column before the instrument begins to integrate in many systems designed for adult samples. This will affect the quantity of Hb F. An α-globin variant often separates from Hb A. Therefore, Hb F will not be adequately quantified. Hb F variants may also be due to mutation of the γ-globin chain; again, this may result in a separate peak and incorrect quantification. Some β-chain variants and adducts will not separate from Hb F, and this will lead to incorrect quantification. If Hb F appears to be greater than 10% on HPLC, its nature should be confirmed by an alternative method

Table 4.8 Hematological Features of α- and β-Thalassemia

Disease	CBC	Hemoglobin Electrophoresis
α-Thalassemia[a]		
Silent carrier	Hb, normal; MCH, <27 pg	Normal
Trait	Hb, normal; MCH, <26 pg; MCV, <75 fL	Normal
Hb H disease	Hb, 8–10 g/dL; MCH, <22 pg; MCV, low	Hb H, 10–20%
Hydrops fetalis	Hb, <6 g/dL; MCH, <20 pg	Hb Bart's, 80–90%
		Hb H, <1%
β-Thalassemia		
Minor	Hb, normal or low; MCV, 55–75 fL[b]; MCH, 19–25 pg	Hb A_2, >3.5%
Intermedia	Hb, 6–10 g/dL; MCV, 55–70 fL	Hb A_2, variable
	MCH, 15–23 pg	Hb F, up to 100%
Major	Hb, <7 g/dL; MCV, 50–60 fL; MCH, 14–20 pg	Hb A_2, variable
		Hb F, high

[a]Mentzer index for children is <13 for both α- and β-thalassemia.
[b]MCV: Abnormal: adult, <80 fL; children (7–12 years), <76 fL; children (6 months–6 years), <70 fL.

to exclude misidentification of Hb N or Hb J as Hb F. Characterization of patients with high Hb F includes evaluation of the following:

- Consider if Hb F is physiologically appropriate for age
- β-Thalassemia: Trait, intermedia (20−40%) or major (60−98%). Here, Hb A_2 will also be raised. Patients should have microcytic hypochromic anemia with normal RDW and disproportionately high RBC count. Peripheral smear should exhibit target cells.
- δβ-Thalassemia: Here, Hb A_2 is normal, but Hb F is increased due to an increase in γ chains. However, the increase in γ chains does not entirely compensate for the decreased β chains. Moreover, α chains are present in excess. Trait shows microcytosis without anemia. Homozygous patients have severity of disease comparable to that of thalassemia intermedia.

Hemoglobin electrophoresis is useful in diagnosis of sickle cell disease by identifying Hb S. The diagnostic approach to sickle cell disease is summarized in Table 4.9. However, the solubility test can also aid in the diagnosis of sickle cell disease. When a blood sample containing Hb S is added to a test solution containing saponin (to lyse cells) and sodium hydrosulfite (to deoxygenate the solution), a cloudy turbid suspension is formed if Hb S is present. If no Hb S is present, the solution remains clear. A false-negative result may be observed if Hb S is less than 10%, as is often observed in infants less than 3 months of age [13].

For diagnosis of Hb S/G hybrid on alkaline gel electrophoresis, one band is expected in the A lane, one band in the S lane (due to Hb S and Hb G), one band in the C lane (due to S/G hybrid), and one band in the carbonic anhydrase area (due to Hb G_2). Therefore, a total of four bands should be observed. If the band in the carbonic anhydrase is not prominent, at least three bands

Table 4.9 Diagnostic Approach to Sickle Cell Hemoglobinopathy

Hemoglobin Pattern	Diagnosis/Comments
Patient has Hb A and Hb S	Hb AS trait or Hb SS disease (post-transfusion) or Hb S/β^+-thalassemia or a normal person transfused from a donor with Hb AS trait. Transfusion history is essential for diagnosis.
	For a patient with Hb AS trait, Hb A is majority and Hb S is 30−40%; if donor was Hb S trait, then S% is usually between 0.8 and 14% of the total hemoglobin.
	In Hb S/β^+-thalassemia, Hb A_2 is expected to be high, and there should be microcytosis and hypochromia of the red cells. Hb A% is typically 5−25% depending on the severity of the genetic defect.
Patient has Hb S but no Hb A	Hb SS disease; Hb S/β^0-thalassemia; Hb A_2 is elevated with low MCV and MCH.
Patient has Hb S and high Hb F	Hb S/HPFH and Hb SS disease while patient is on hydroxyurea.
	High MCV favors hydroxyurea; medication history will be required.

should be seen. On acid gel electrophoresis, one band is expected in the A lane (due to Hb A, Hb G, and Hb G_2) and one band in the S lane (due to Hb S and Hb S/G hybrid). In electrophoresis, a band should be seen in zone 5 (Hb S) and a band in zone 6 (Hb G). It is important to emphasize that for hemoglobinopathies, the results of gel electrophoresis must be confirmed by a second method—HPLC or capillary electrophoresis.

In the presence of Hb S, if a higher value of Hb F is observed, then HbS/HPFH can be suspected. In this case, CBC should be normal and Hb F should be between 25 and 35%. However, with Hb S/β-thalassemia, Hb F could also be high. In HPFH and Hb S/HPFH, distribution of Hb F in red cells is normocellular, whereas in δ-thalassemia and Hb SS with high Hb F, it is heterocellular. Kleihauer—Betke tests or flow cytometry with anti-F antibody will illustrate the difference. Interpretations of various other hemoglobinopathies are given in Table 4.10.

In the USA, universal newborn screening for hemoglobinopathies is now required in all 50 states and the District of Columbia. In addition, the American College of Obstetricians and Gynecologists provides guidelines for screening of couples who may be at risk of having children with hemoglobinopathy. The diagnostic approaches for various hemoglobinopathies are summarized in Table 4.10. Persons of northern European, Japanese, Native American, or Korean descent are at low risk for hemoglobinopathies, but people with ancestors from Southeast Asia, Africa, or Mediterranean countries are at higher risk. A CBC should be done to accurately measure hemoglobin. If all parameters are normal and the couple belongs to a low-risk group, no further testing may be necessary. For higher risk couples, hemoglobin analysis by electrophoresis or an other method is recommended. Solubility test for sickle cell may be helpful. Genetic screening can help physicians to identify couples at risk of having children with hemoglobinopathy. Molecular protocols for hemoglobinopathies began in the 1970s using Southern blotting and restriction fragment-length polymorphism analysis for prenatal sickle cell disease. With the development of polymerase chain reaction, molecular testing for hemoglobinopathies now requires much less DNA for analysis [14]. However, currently, molecular testing that establishes a firm diagnosis of hemoglobinopathies, especially for α-thalassemia trait (direct gene analysis), is available in large academic medical centers and reference laboratories only.

4.7 APPARENT HEMOGLOBINOPATHY AFTER BLOOD TRANSFUSION

Blood transfusion history is essential in interpreting an abnormal hemoglobin pattern because small peaks of abnormal hemoglobin may appear

Table 4.10 Diagnostic Approach to Common Hemoglobinopathies

Diagnosis	Features
Diagnosis of Hb C	Band in the C lane in the alkaline gel: Possibilities are C, E, or O
	Band in the C lane in the acid gel
	HPLC shows a peak at approximately 5 min with a small peak just before this main peak (Hb C1d). A small peak may also be observed at 4.5 min (Hb C1c)
	Or
	Capillary electrophoresis shows a peak in zone 2
Diagnosis of Hb E	Band in the C lane in the alkaline gel: Possibilities are C, E, or O
	Band in A lane in acid gel
	HPLC shows a peak at 3.5 min and is greater than 10%
	Or
	Capillary electrophoresis shows a peak in zone 4
Diagnosis of Hb O	Band in the C lane in the alkaline gel: possibilities are C, E, or O
	Band between A and S lane in acid gel
	HPLC shows a peak between 4.5 and 5 min
	Or
	Capillary electrophoresis shows a peak in zone 3 (O Arab)
Diagnosis of Hb S	Band in the S lane in the alkaline gel: Possibilities are S, D, G, or Lepore
	Band in the S lane in the acid gel
	HPLC show a peak at 4.5 minutes
	Or
	Capillary electrophoresis shows a peak in zone 5
Diagnosis: Hb D	Band in the S lane in the alkaline gel: Possibilities are S, D, G, or Lepore
	Band in the A lane in acid gel
	HPLC shows a peak at 3.9–4.2 min; no additional peak
	Or
	Capillary electrophoresis shows a peak in zone 6
Diagnosis: Hb G	Band in the S lane in the alkaline gel: Possibilities are S, D, G, or Lepore
	Band in the A lane in acid gel
	HPLC shows a peak at 3.9–4.2 min and a small additional peak (G_2)
	Or
	Capillary electrophoresis shows a peak in zone 6
Diagnosis of Hb Lepore[a]	Band (faint) in the S lane in the alkaline gel: Possibilities are S, D, G, or Lepore
	Band in the A lane in acid gel
	HPLC shows a peak at 3.7 min (A_2 peak); quantity is lower than D or G or E. There is a small increase in Hb F%
	Or
	Capillary electrophoresis shows a peak in zone 6

[a]Hb Lepore band in the alkaline gel is faint.

from blood transfusion. Apparent hemoglobinopathy after blood transfusion is rarely reported, but it may cause diagnostic dilemmas resulting in repeated, unnecessary testing. Kozarski *et al.* reported 52 incidences of apparent hemo-globinopathies, of which 46 were Hb C, 4 were Hb S, and 2 were Hb O-Arab. The percentage of abnormal hemoglobin ranged from 0.8 to 14% (median, 5.6%). The authors recommended identifying and notifying the donor in such cases [15].

KEY POINTS

- The normal hemoglobin (Hb A) in adults contains two α chains and two β chains. Each α chain contains 141 amino acids, and each β chain contains 146 amino acids. Hb A$_2$ contains two α chains and two δ chains. The gene for the α chain is located in chromosome 16 (two genes in each chromosome, for a total of four genes), whereas genes for β (one gene in each chromosome, for a total of two genes), γ, and δ chains are located on chromosome 11.
- When hemoglobin is circulating with erythrocytes, glycosylation of the globin chains may occur. These are referred to as X1c (with X being any hemoglobin; e.g., Hb A1c). When the hemoglobin molecule ages, glutathione is bound to cysteine at the 93rd position of the β chain. This is Hb AIII or Hb A1d. Just like Hb A1c and Hb A1d, there can exist Hb C1c, Hb C1d, Hb S1c, and Hb S1d.
- Heme is synthesized in a complex manner involving enzymes in both mitochondrion and cytosol.
- Hemoglobinopathies can be divided into three major categories:
  - Quantitative disorders of hemoglobin synthesis: Production of structurally normal but decreased amounts of globin chains (thalassemia syndrome).
  - Disorder (qualitative) in hemoglobin structure: Production of structurally abnormal globulin chains such as Hb S, C, O, or E. Sickle cell syndrome is the most common example of such disease.
  - Failure to switch globin chain synthesis after birth: Hereditary persistence of Hb F, a relatively benign condition, may coexist with thalassemia or sickle cell disease, but there is decreased severity of such diseases (protective effect).
- Hemoglobinopathies are transmitted in an autosomal recessive manner.
- Disorders due to a β-chain defect, such as sickle cell disease, tend to manifest clinically after 6 months of age, whereas diseases due to α-chain defect are manifested *in utero* or following birth.

- The hemoglobin variants of most clinical significance are Hb S, C, and E. In West Africa, approximately 25% of individuals are heterozygous for the Hb S gene, which is related to sickle cell diseases. In addition, high frequencies of Hb S gene alleles are also found in people living in the Caribbean, southern and Central Africa, Mediterranean countries, the Arabian Peninsula, and East India. Hb C is found mostly in people living in or originating from West Africa. Hb E is widely distributed between East India and Southeast Asia, with highest prevalence in Thailand, Laos, and Cambodia, but it may be sporadically observed in areas of China and Indonesia. Thalassemia syndrome is not due to structural defects in the globin chain but, rather, to lack of sufficient synthesis of the globin chain and is also a genetically inherited disease. Thalassemic syndrome can be categorized as α-thalassemia and β-thalassemia. In general, β-thalassemia is observed in Mediterranean countries, the Arabian Peninsula, Turkey, Iran, West and Central Africa, India, and other Southeast Asian countries, whereas α-thalassemia is commonly observed in areas of Africa, the Mediterranean, the Middle East, and throughout Southeast Asia.
- α-Thalassemia occurs when there is a defect or deletion in one or more of four genes responsible for α-globin production. α-Thalassemia can be divided into four categories:
  - The silent carriers: Characterized by only one defective or deleted gene but three functional genes. These individuals have no health problems. In unusual cases of silent carrier, individuals carry one defective Constant Spring mutation but three functional genes. These individuals also have no health problems.
  - α-Thalassemia trait: Characterized by two deleted or defective genes and two functional genes. These individuals may have mild anemia.
  - α-Thalassemia major (Hb H disease): Characterized by three deleted or defective genes and only one functional gene. These patients have persistent anemia and significant health problems. When Hb H disease is combined with Hb Constant Spring, the severity of disease is more than that of Hb H disease alone. However, if a child inherits one Hb Constant Spring from the mother and one from the father, the child has homozygous Hb Constant Spring and the severity of the disease is similar to that of Hb H disease.
  - Hydrops fetalis: Characterized by no functional α gene. These individuals have Hb Bart. This condition is not compatible with life unless intrauterine transfusion is initiated.
- Hemoglobin Constant Spring (hemoglobin variant isolated from a family of ethnic Chinese background from the Constant Spring district of Jamaica) is a hemoglobin variant in which mutation of the α-globin gene produces an abnormally long α chain (172 amino acids instead of the normal 141 amino acids). Hemoglobin Constant Spring is due to

non-deletion mutation of the α gene, which results in the production of unstable α-globin. Moreover, this α-globin is produced in a very low quantity (approximately 1% of the normal expression level) and is found in people living in or originating from Southeast Asia.

- β-Thalassemia can be broadly classified into three categories:
  - β-Thalassemia trait: Characterized by one defective gene and one normal gene. Individuals may experience mild anemia but not be transfusion dependent.
  - β-Thalassemia intermedia: Characterized by two defective genes, but some β-globin production is still observed in these individuals. However, some individuals may have significant health problems requiring intermittent transfusion.
  - β-Thalassemia major (Cooley's anemia): Characterized by two defective genes but almost no function of either gene, leading to no synthesis of β-globin. These individuals have a severe form of disease requiring lifelong transfusion and may have shortened life span.
- Patients with β-thalassemia major have elevated Hb A_2 and Hb F (although in some individuals, Hb F may be normal).
- In the heterozygous form (Hb AS), sickle cell trait protects from infection of *P. falciparum* malaria but not in the more severe form of homozygous sickle cell disease (Hb SS). The genetic defect producing sickle hemoglobin is a single nucleotide substitution at codon 6 of the β-globin gene on chromosome 11 that results in a point mutation in the β-globin chain of hemoglobin (substitution of valine for glutamic acid at the sixth position).
- Double heterozygous states of Hb SC, Hb SD, and Hb SO Arab are important sickling states that should not be missed.
- Hemoglobin C is formed due to substitution of glutamic acid residue with a lysine residue at the sixth position of β-globin. Hemoglobin E is caused by point mutation of β-globin, which results in substitution of lysine for glutamic acid in position 26.
- Hemoglobin Lepore is an unusual hemoglobin molecule that is composed of two α chains and two $\delta\beta$ chains as a result of fusion of δ and β genes. The $\delta\beta$ chains have the first 87 amino acids of the δ chain and 32 amino acids of the β chain.
- Individuals with Hb A/Hb Lepore are asymptomatic, with Hb Lepore representing 5−15% of hemoglobin; there is slightly elevated Hb F (2−3%), with low MCV as well as MCH. However, homozygous Lepore individuals suffer from severe anemia similarly to patients with β-thalassemia intermedia, with Hb Lepore representing 8−30% of hemoglobin, the remainder being Hb F.
- Hemoglobin G Philadelphia (Hb G) is the most common α-chain defect, affecting 1 in 5000 African Americans, and is associated with α-thalassemia 2 deletions.

- It is possible that an African American individual may have Hb S/Hb G, in which the hemoglobin molecule contains one normal α chain, one α G chain, one normal β chain, and one β S chain. This can result in detection of various hemoglobin in the blood, including Hb A (α_2, β_2), Hb S (α_2, β S$_2$), Hb G (α G$_2$, β_2), and HbS/G (α G$_2$, β S$_2$). In addition, Hb G$_2$ (α_2, δ_2), which is the counterpart of Hb A$_2$, is also present.
- An increase in fetal hemoglobin percentage is associated with multiple pathologic states, including β-thalassemia, $\delta\beta$-thalassemia, and HPFH. β-Thalassemia is associated with high Hb A$_2$ and the latter two states are associated with normal Hb A$_2$ values. Hematologic malignancies are associated with increased Hb F and include acute erythroid leukemia (AML, M6) and juvenile myelomonocytic leukemia (JMML). Aplastic anemia is also associated with an increase in the percentage of Hb F. In elucidating the actual cause of high Hb F, it is important to consider the actual percentage of Hb F, Hb A$_2$ values, as well as the correlation with complete blood count (CBC) and peripheral smear. It is also important to note that drugs (hydroxyurea, sodium valproate, and erythropoietin) and stress erythropoiesis may also result in high Hb F. Hydroxyurea is used in sickle cell disease patients to increase the amount of Hb F, the presence of which may help to reduce the clinical effects of the disease. Measuring the level of Hb F may be useful in determining the appropriate dose of hydroxyurea. In 15−20% of cases involving pregnancy, Hb F may be increased by as much as 5%.

References

[1] Rahimi Z. Genetic, epidemiology, hematological and clinical features of hemoglobinopathies in Iran. Biomed Res Int 2013;2013:803487.

[2] Faulkner LB, Uderzo C, Masera G. International cooperation for the cure and prevention of severe hemoglobinopathies. J Pediatr Hematol Oncol 2013;35:419−23.

[3] Manning LR, Russell JR, Padovan JC, Chait BT, et al. Human embryonic, fetal and adult hemoglobins have different subunit interface strength: correlation with lifespan in the red cell. Protein Sci 2007;16:1641−58.

[4] Giordano PC. Strategies for basic laboratory diagnostics of the hemoglobinopathies in multi-ethnic societies: interpretation of results and pitfalls. Int J Lab Hematol 2013;35:465−79.

[5] Rappaport VJ, Velazquez M, Williams K. Hemoglobinopathies in pregnancy. Obstet Gynecol Clin North Am 2004;31:287−317.

[6] Sriiam S, Leecharoenkiat A, Lithanatudom P, Wannatung T, et al. Proteomic analysis of hemoglobin H Constant Spring (Hb H-CS) erythroblasts. Blood Cells Mol Dis 2012;48:77−85.

[7] Kohne E. Hemoglobinopathies: clinical manifestations, diagnosis and treatment. Dtsch Arztebl Int 2011;108:532−40.

[8] Ngo DA, Aygun B, Akinsheye I, Hankins JS. Fetal hemoglobin levels and hematological characteristics of compound heterozygous for hemoglobin S and deletional hereditary persistence of fetal hemoglobin. Br J Haematol 2012;156:259−64.

[9] Pandey S, Mishra RM, Pandey S, Shah V, et al. Molecular characterization of hemoglobin D Punjab traits and clinical−hematological profile of patients. Sao Paulo Med J 2012;130: 248−51.

[10] Dror S. Clinical and hematological features of homozygous hemoglobin O-Arab [beta 121 Glu→Lys]. Pediatr Blood Cancer 2013;60:506−7.

[11] Smaldone A. Glycemic control and hemoglobinopathy: when A1C may not be reliable. Diabetes Spectrum 2008;21:46−9.

[12] Muncie H, Campbell JS. Alpha and beta thalassemia. Am Fam Physician 2009;339:344−71.

[13] Lubin B, Witkowska E, Kleman K. Laboratory diagnosis of hemoglobinopathies. Clin Biochem 1991;24:363−74.

[14] Benson JA, Therell BL. History and current status of newborn screening for hemoglobinopathies. Semin Perinatol 2010;34:134−44.

[15] Kozarski TB, Howanitz PJ, Howanitz JH, Lilic N, et al. Blood transfusions leading to apparent hemoglobin C, S, and O-Arab hemoglobinopathies. Arch Pathol Lab Med 2006;130: 1830−3.

CHAPTER 5

Benign White Blood Cell and Platelet Disorders

5.1 INTRODUCTION

Automated hematology analyzers can rapidly analyze whole blood specimens for the complete blood count (CBC). Results include red blood cell (RBC) count, white blood cell (WBC) count, platelet count, hemoglobin concentration, hematocrit, RBC indices, and a leukocyte differential. Less sophisticated automated hematology analyzers in a physician's office setting may sometimes provide a limited CBC, using older technology for whole blood analysis (e.g., impedance technology) that will generate only a three-part leukocyte differential. A three-part leukocyte differential provides values for neutrophils, lymphocytes, and all other white cells together. More modern hematology analyzers are capable of analyzing all leukocytes using flow cytometry-based methods, some in combination with cytochemistry or fluorescence or conductivity, to count all the different types of WBCs, including neutrophils, lymphocytes, monocytes, basophils, and eosinophils (five-part differential). Nucleated RBCs are also detected. Leukocytosis or elevated WBC count is a common laboratory finding. For example, a WBC count of 30×10^9/L (30,000/μL) is abnormal in an adult but normal in a newborn within the first few days of life. Normal WBC differential also changes with age, and proper normal ranges must be established for each laboratory performing such tests. Leukocytosis is also a feature of leukemias; thus, distinguishing leukemias from other causes of leukocytosis is crucial. Examination of peripheral blood smear along with review of CBC analysis is essential for such differentiation, and if necessary, further analysis such as flow cytometry, molecular studies, and possible bone marrow examination must be undertaken [1]. Similarly, platelet disorders such as thrombocytopenia or thrombocytosis may be a benign condition or may indicate a serious condition such as severe thrombocytopenia observed in patients with acute leukemias. In this chapter, benign WBC and platelet disorders are reviewed.

CONTENTS

A. Wahed and A. Dasgupta: Hematology and Coagulation. DOI: http://dx.doi.org/10.1016/B978-0-12-800241-4.00005-X

5.2 HEREDITARY VARIATION IN WHITE BLOOD CELL MORPHOLOGY

A number of heredity-mediated variations in WBC morphology have been described that are mostly clinically benign except for Chediak−Higashi syndrome. Pelger−Huët anomaly is a rare disorder transmitted in an autosomal dominant pattern in which more than 75% of neutrophils are bilobed (neutrophils with a hyposegmented nucleus). This is a benign condition due to an inherited defect of terminal neutrophil differentiation as a result of mutations in the lamin B receptor (*LBR*) gene. The presence of occasional bilobed neutrophils is also seen in myelodysplastic syndrome, and these are referred to as pseudo-Pelger−Huët cells. Distinguishing the benign form of Pelger−Huët anomaly from the acquired or pseudo-Pelger−Huët anomaly is important.

May−Hegglin anomaly, another rare disorder, is also transmitted in an autosomal dominant pattern and is characterized by thrombocytopenia, giant platelets, and leukocyte defect causing the appearance of very small rods (2−5 μm) in the cytoplasm (Dohle-like bodies). Actual Dohle bodies are seen with reactive or toxic polymorphonuclear neutrophils (PMNs). Reactive or toxic PMNs also have prominent azurophilic granules and vacuoles. These features are not seen in May−Hegglin anomaly. May−Hegglin anomaly is a benign condition because most patients do not appear to have any significant bleeding problems for which treatment may be required.

Chediak−Higashi syndrome is a rare multisystemic disorder transmitted as autosomal recessive, characterized by hypopigmentation of the skin, eyes, and hair (silver hair); prolonged bleeding time; easy bruisability; and immunodeficiency. The pathological feature of this rare disease is the presence of massive lysosomal inclusion bodies, which are formed through a combination of fusion, cytoplasmic injury, and phagocytosis, in WBCs. These abnormal inclusion bodies may be responsible for most of the impaired leukocyte and other blood cell functions in these patients. In addition, dysfunction of natural killer cells is also observed. Approximately 85% of affected individuals develop the accelerated phase of this disease—a lymphoproliferative infiltration of bone marrow and the reticuloendothelial system—mostly during childhood [2]. This is a potentially fatal condition if not treated. Hematopoietic stem cell transplantation may be curative [3].

Alder−Reilly anomaly, a clinically benign rare condition, is transmitted as autosomal recessive and is characterized by large azurophilic granules (partially degraded protein−carbohydrate complexes known as mucopolysaccharides) in neutrophils and others granulocytes, monocytes, and lymphocytes. However, similar abnormalities are seen in mucopolysaccharidoses.

In Maroteaux–Lamy syndrome, abnormal granulation of granulocytes and monocytes with lymphocytes occurs with vacuolation, as seen in mucopolysaccharidosis VI.

5.3 CHANGES IN WHITE CELL COUNTS

Changes in white cell count can be observed in various conditions in which values may be increased or decreased. Leukocytosis is a common clinical observation in which an increase in WBC count above two standard deviations of the mean is observed ($>11,000/\mu L$). Leukocytosis is mostly a benign condition in which elevated WBC count reflects the normal response of bone marrow to infection, an inflammatory process, or drugs. However, leukocytosis may also be due to bone marrow abnormality related to leukemia or myeloproliferative disease. In this section, various changes in WBC count due to benign conditions are discussed.

5.3.1 Neutrophilia

Neutrophilia is the most common cause of leukocytosis. Various causes of neutrophilia are summarized in Box 5.1. The most common cause is infection or inflammation; there are congenital forms of neutrophilia, but such conditions are rarely encountered. For example, leukocyte adhesion deficiency is a rare autosomal recessive immunodeficiency disease characterized by severe recurrent bacterial infection due to an inability of neutrophils to adhere to endothelial cell walls and migrate to the site of infection. Depending on the genetic effect, hematopoietic stem cell transplantation is often the only cure [4].

Various morphological changes are observed in neutrophilia. Such morphologic changes seen in reactive neutrophils are summarized as follows:

- Dohle bodies (represent endoplasmic reticulum)
- Prominent 1° (azurophilic) granules in cytoplasm
- Vacuoles in cytoplasm.

BOX 5.1 VARIOUS CAUSES OF NEUTROPHILIA

- Infection
- Inflammation
- Any form of stress
- Splenectomy, hyposplenism
- Drugs: Corticosteroids

- Acute hemorrhage
- Hemolytic anemia
- Congenital forms (rare): Leukocyte adhesion deficiency, chronic idiopathic neutrophilia, hereditary neutrophilia

Significant neutrophilic leukocytosis may result in a picture resembling chronic myeloid leukemia (CML) that is referred to as leukemoid reaction. Basophilia and eosinophilia are not significant. The leukocyte alkaline phosphatase (LAP) score is high in such conditions. In contrast, the LAP score in CML is low. Another cause of low LAP, although unrelated, is paroxysmal nocturnal hemoglobinuria.

In order to determine LAP score, PMNs are stained for alkaline phosphatase, and each neutrophil is given a score from 0 to 4. One hundred cells are counted. The total score is the actual score.

5.3.2 Eosinophilia and Monocytosis

Eosinophils are WBCs that participate in immunological and allergic events. Eosinophilia is usually defined as eosinophil count greater than 500 cells/μL (0.5×10^9/L). Parasitic infections are often responsible for eosinophilia in pediatric patients. Infections such as scarlet fever, chorea, and genitourinary infection may also cause eosinophilia. Skin rash, chronic inflammation such as rheumatoid arthritis and lupus erythematosus, and adrenal insufficiency such as Addison's disease, may also cause eosinophilia. Pleural and pulmonary conditions such as Loffler's syndrome may also cause eosinophilia. Eosinophilia—myalgia is a disorder associated with dietary supplementation of tryptophan. Other causes of eosinophilia include malignancies affecting the immune system such as Hodgkin's lymphoma and non-Hodgkin's lymphoma [5].

Hypereosinophilic syndrome is a rare and heterogeneous group of hematological and systemic disorders characterized by eosinophil count greater than 1.5×10^9/L (1500 cells/μL) lasting more than 6 months in the absence of other known causes of eosinophilia. This syndrome may cause end organ damage, primarily the heart, causing eosinophilic endomyocardial fibrosis. Patients with hypereosinophilic syndrome do not typically have asthma [6].

Monocytes are WBCs that give rise to macrophages and dendritic cells in the immune system. Causes of monocytosis include various infections (tuberculosis, brucellosis, typhoid, typhus, Rocky Mountain spotted fever, malaria, etc.), chronic myelomonocytic leukemia (CMML), juvenile myelomonocytic leukemia (JMML), Hodgkin lymphoma, acute myeloid leukemia (AML) M4/M5, and autoimmune diseases.

5.3.3 Basophilia

Basophils are inflammatory mediators of substance such as histamine, and along with mast cells, they have receptors for IgE. Basophilia is an uncommon situation. Causes include viral infections (varicella or chicken pox), inflammatory conditions (ulcerative colitis), CML or myeloproliferative disorder, and myxedema, as well as endocrinological causes (hypothyroidism and ovulation) [5].

5.3.4 Neutropenia

Neutropenia is defined as an absolute neutrophil count that is more than two standard deviations below the normal neutrophil count. Mild to moderate neutropenia may not predispose an individual to an increased susceptibility to life-threatening infection, but patients with severe neutropenia (neutrophil count $<0.5 \times 10^9/L$) may be prone to severe, even life-threatening, infection. Severe neutropenia accompanied by fever of recent onset is a medical emergency requiring immediate investigation and treatment. Various causes of neutropenia (inherited and acquired) are summarized in Box 5.2. Often, neutropenia in adults is due to acquired causes such as drug induced or arises post infection. Felty's syndrome is a well-characterized clinical abnormality consisting of rheumatoid arthritis, splenomegaly, and severe neutropenia. The cause could be attributed to increased neutrophil margination and inhibition of granulopoiesis mediated by antibodies to neutrophils or by T cells.

Neutropenia may also be inherited. Yemenite Jews and other populations, including Ethiopian Jews and Bedouins, have low neutrophil counts, and this condition is called ethnic benign neutropenia because it is not associated with an increased risk of infection. This condition may also be found in populations throughout the world, including African, African American, and African Caribbean [7]. Cyclical neutropenia is a rare disorder caused by a stem cell regulatory defect characterized by a transient severe neutropenia occurring approximately every 21 days. The familial form seems to be inherited in an autosomal dominant pattern. Kostmann's syndrome is inherited in an autosomal recessive

---

BOX 5.2 VARIOUS CAUSES OF SELECTIVE NEUTROPENIA

Acquired Causes
- Drug induced (e.g., methimazole, sulfasalazine, trimethoprim—sulfamethoxazole, Bactrim)
- Postinfection
- Autoimmune disease (e.g., systemic lupus erythematosus)
- Paroxysmal nocturnal hemoglobinuria
- Felty's syndrome (triad of rheumatoid arthritis, splenomegaly, and neutropenia)
- Splenomegaly

Inherited Forms
- Ethnic familial neutropenia
- Cyclical neutropenia
- Kostmann's syndrome
- Shwachman—Diamond syndrome

pattern, and affected children develop frequent life-threatening infection due to severe neutropenia. This disorder is due to point mutations in the gene coding for granulocyte colony-stimulating factor (G-CSF).

Shwachman–Diamond syndrome is also transmitted as autosomal recessive and is characterized by exocrine pancreatic deficiency and neutropenia, thrombocytopenia, short stature, and mental retardation. This disorder may progress to myelodysplastic syndrome (MDS) and AML.

5.3.5   Lymphocytosis and Infectious Mononucleosis

Lymphocytosis occurs most commonly after viral infections (e.g., cytomegalovirus, mumps, varicella, influenza, and rubella), with lymphoid leukemias and lymphomas, and with smoking. Lymphocytosis is rarely observed in bacterial infection, with an exception being *Bordetella pertussis* infection. With lymphocytosis, reactive lymphocytes may be seen in the peripheral smear. Reactive lymphocytes are also referred to as Downey cells. There are three types of Downey cells:

- Type I: Small cells with minimum cytoplasm, indented nucleus/ irregular nuclear membrane, and condensed chromatin.
- Type II: Larger cells with abundant cytoplasm; the lymphocyte cytoplasm seems to hug the red cells. Type II is the most common type of Downey cell.
- Type III: Cells with large moderate basophilic cytoplasm and nucleus with coarse chromatin. Nucleoli are apparent.

In infectious mononucleosis caused by Epstein–Barr virus, there is lymphocytosis with characteristic large atypical lymphocytes in the blood. The virus infects B lymphocytes, and T lymphocytes attack the virally infected B lymphocytes. T lymphocytes are the reactive lymphocytes. Features of infectious mononucleosis, from the peripheral blood, include the following:

- Fifty percent or more of the white cells are mononuclear cells.
- At least 10% of the lymphocytes exhibit reactive changes.
- There is lymphocytic morphologic heterogeneity (i.e., different types of reactive lymphocytes are seen).

5.3.6   Lymphocytopenia

Lymphocytes consist of T lymphocytes, B lymphocytes, and natural killer cells. The term lymphocytopenia (lymphopenia) refers to less than 1000 lymphocytes/μL of blood in adults or less than 3000 lymphocytes/μL of blood in children. Severe combined immunodeficiency is a heterogeneous disorder characterized by severe deficiency of T and B lymphocytes as well as natural killer cells. The X-linked inherited form is most commonly characterized by the absence of T lymphocytes and natural killer cells but poorly functioning B cells. A deficiency of adenosine deaminase is present in 30−40% of cases of the autosomal recessive form of this inherited disorder. The different forms of severe combined immunodeficiency are clinically indistinguishable and observed in early infancy manifested by severe infection. Treatment includes correction of the defect by stem cell transplant or enzyme replacement with adenosine deaminase. Lymphocytopenia may also be acquired, for example in patients with HIV infection [8].

5.4 PLATELET DISORDERS

Common platelet disorders include thrombocytopenia, thrombocytopathia, and thrombocytosis. Thrombocytopenia is often discovered incidentally during a patient office visit when CBC is ordered along with other tests. However, severe thrombocytopenia may be a reflection of a severe disease. Similarly, thrombocytosis is a common finding during a routine blood test and may represent a benign condition. However, like severe thrombocytopenia, severe thrombocytosis may indicate a serious clinical condition requiring further investigation.

5.4.1 Thrombocytopenias

Thrombocytopenia is defined as a platelet count less than 150,000/μL of blood (150×10^9/L), but even patients with a platelet count of 50,000/μL or more may be asymptomatic. However, counts from 10,000 to 30,000/μL may be associated with bleeding, and patients with a platelet count less than 10,000/μL are very sensitive to spontaneous bleeding. Thrombocytopenia may be due to decreased platelet production (congenital or acquired), increased destruction of platelets, increased platelet consumption, or sequestration. Various causes of thrombocytopenias, except inherited forms, are given in Box 5.3. Inherited forms of thrombocytopenia are summarized in Table 5.1.

Congenital thrombocytopenia can be broadly divided into three groups: cytopenia with small platelets, cytopenia with normal platelets, and cytopenia with large platelets.

Kasabach−Merritt syndrome is a rare, locally aggressive vascular tumor characterized by a rapidly enlarging vascular anomaly, consumption coagulopathy, thrombocytopenia, prolonged bleeding time, hypofibrinogenemia, and the presence of D dimer and fibrin split products, with or without microangiopathic hemolytic anemia. Prognosis is poor because few treatment options are available [9].

Congenital thrombocytopenia with small platelets may be related to Wiskott−Aldrich syndrome (WAS), which is characterized by eczema, immunodeficiency, and thrombocytopenia. This disease is transmitted as X-linked

BOX 5.3 VARIOUS CAUSES OF THROMBOCYTOPENIA

Decreased Production (Any Cause of Bone Marrow Suppression/Failure)

- Bone marrow failure (aplastic anemia, paroxysmal nocturnal hemoglobinuria, etc.)
- Bone marrow suppression due to medication, chemotherapy, or radiation therapy
- Infection (cytomegalovirus, HIV, parvovirus B19, hepatitis C, etc.)
- Myelodysplastic syndrome
- Neoplastic marrow infiltration
- Inherited forms summarized in Table 5.1

Increased Platelet Consumption

- Disseminated intravascular coagulation (DIC)
- Thrombotic thrombocytopenic purpura (TTP)
- Hemolytic uremic syndrome (HUS)

Increased Platelet Destruction

- Immune thrombocytopenic purpura (ITP)
- Mechanical destruction (e.g., mechanical valves, extracorporeal bypass)

Thrombocytopenias Due to Sequestration

- Sequestration in hemangiomas (Kasabach−Merritt syndrome)

Table 5.1 Various Types of Congenital Thrombocytopenia

Congenital Thrombocytopenia with Small Platelets	Congenital Thrombocytopenia with Normal-Sized Platelets	Congenital Thrombocytopenia with Large Platelets
■ Wiskott−Aldrich syndrome (WAS): X-linked recessive related to mutation of WASP gene ■ X-linked thrombocytopenia: Isolated thrombocytopenia related also to mutation of WASP gene, but disease is milder than WAS ■ Inherited microthrombocytes: Transmitted as autosomal dominant with normal platelet function	■ Thrombocytopenia with absent radii (TAR syndrome) ■ Amegakaryocytic thrombocytopenia due to mutation of *MPL* gene ■ Fanconi's anemia: Autosomal recessive	■ Bernard−Soulier syndrome: Autosomal recessive ■ May−Hegglin anomaly: Autosomal dominant ■ Sebastian syndrome: Autosomal dominant ■ Epstein syndrome: Autosomal dominant ■ Fechtner syndrome: Autosomal dominant ■ Gray platelet syndrome: Autosomal recessive ■ DiGeorge and velocardiofacial syndrome: Autosomal dominant

recessive due to inheritance of the *WASP* gene, located at Xp11 encoding WAS protein (WASP). Platelets are dysfunctional. X-linked thrombocytopenia, a congenital disorder characterized by isolated thrombocytopenia and small platelets but in general without other complications as seen in WAS, is a mild allelic variant also caused by mutation of the *WASP* gene [10]. Inherited microthrombocytes is a disorder transmitted as autosomal dominant with normal platelet function.

Congenital thrombocytopenia with normal-sized platelets can be related to thrombocytopenia with absent radii (TAR syndrome); amegakaryocytic thrombocytopenia, which is due to mutation of the *MPL* gene that encodes thrombopoietin receptor (thrombopoietin is required for maturation of megakaryoblasts to megakaryocytes); or Fanconi's anemia, which is transmitted in an autosomal recessive pattern.

Congenital thrombocytopenia with large platelets can be related to Bernard–Soulier syndrome, which is transmitted in an autosomal recessive pattern. In the early phase of primary hemostasis, platelets adhere to damaged vessel walls by binding via the platelet glycoprotein (GP)Ib-V-IX complex to von Willebrand factor exposed on the subendothelium. In this disorder, the (GP)Ib-V-IX complex is abnormal. Platelet aggregation studies show impaired aggregation to ristocetin. Some cases of Bernard–Soulier syndrome are due to defects of the *GpIbβ* gene, located on chromosome 22. This gene may be affected in velocardiofacial syndrome or DiGeorge syndrome associated with deletion of 22q11.2

Congenital thrombocytopenia with large platelets can also be due to May–Hegglin anomaly, another inherited disorder that is transmitted in an autosomal dominant pattern. This disorder is due to defective myosin heavy chain 9 gene at 22q11. Neutrophils have Dohle-like bodies. Sebastian syndrome is transmitted in an autosomal dominant pattern and is due to defective myosin heavy chain 9 gene at 22q11. Epstein syndrome is transmitted in an autosomal dominant pattern and is related to defective myosin heavy chain 9 gene at 22q11. Patients have Alport-like syndrome with features of nephritis, sensorineural deafness, and cataract. Fechtner syndrome is transmitted as autosomal dominant and also due to defective myosin heavy chain 9 gene at 22q11. Patients have Alport-like syndrome with features of nephritis, sensorineural deafness, cataract, as well as Dohle-like bodies in neutrophils. Gray platelet syndrome, a rare congenital autosomal recessive bleeding disorder, is due to hypogranular platelets that are dysfunctional. DiGeorge and velocardiofacial syndromes are transmitted as autosomal dominant with loss of function of the *GP1BB* gene at 22q11. In this disorder, cardiac, parathyroid, and thymus abnormalities may be observed.

5.4.2 Thrombocytosis

Thrombocytosis is defined as platelet count exceeding $450,000/\mu L$ ($450 \times 10^9/L$). This abnormality is termed primary thrombocytosis if platelet increase is related to alterations targeting the hematopoietic cells in the bone marrow. Examples of such states are essential thrombocythemia or thrombocytosis seen in other myeloproliferative disorders. The disorder is considered as secondary (also called reactive thrombocytosis) if platelet increase is due to an external cause, such as infection, inflammation, neoplasms, or iron deficiency. Secondary thrombocytosis can also be due to redistribution such as observed post splenectomy.

There are several examples of primary thrombocytosis that are inherited disorders. Familial thrombocytosis is transmitted in an autosomal dominant pattern and is due to mutation in the thrombopoietin receptor gene (myeloproliferative leukemia virus oncogene MPL gene; first identified from the murine myeloproliferative leukemia virus that was capable of immortalizing bone marrow hematopoietic cells from different lineages). Several mutations of this gene causing thrombocytosis have been reported. Inherited primary thrombocytopenia may also be due to mutation of the promoter of the thrombopoietin (*TPO*) gene, which encodes thrombopoietin. Thrombopoietin is the primary humoral regulator of platelet production.

5.4.3   Thrombocytopathia

Thrombocytopathia is any of several hematological disorders characterized by dysfunctional platelets (thrombocytes) that lead to prolonged bleeding time, defective clot formation, and a tendency for hemorrhage. Thrombocytopathia may be congenital or acquired.

Congenital causes of thrombocytopathia include the following:

- Disorders of platelet adhesion: von Willebrand's disease, Bernard–Soulier syndrome
- Disorders of platelet activation: Storage pool disorders, Chediak–Higashi syndrome, Hermansky–Pudlak syndrome
- Disorders of platelet aggregation: Glanzmann's syndrome.

Acquired causes of thrombocytopathia include the following:

- Drugs: Aspirin, nonsteroidal anti-inflammatory drugs (NSAIDs)
- Uremia
- Acquired Von Willebrand's disease
- Myeloproliferative diseases
- Anti-platelet antibodies.

KEY POINTS

- Pelger–Huët anomaly is transmitted in an autosomal dominant pattern in which more than 75% of neutrophils are bilobed. The presence of occasional bilobed neutrophils is also seen in myelodysplastic syndrome, and these are referred to as pseudo-Pelger–Huët cells.

- May–Hegglin anomaly is transmitted in an autosomal dominant pattern and is characterized by thrombocytopenia, giant platelets, and prominent Dohle-like bodies.
- Chediak–Higashi syndrome is transmitted in an autosomal recessive pattern and is characterized by giant neutrophilic granules (due to fusion of lysosomes), defective chemotaxis and phagocytosis, immunodeficiency, and oculocutaneous albinism.
- Alder–Reilly anomaly is transmitted in an autosomal recessive pattern and is characterized by large azurophilic granules in neutrophils and other granulocytes, monocytes, and lymphocytes; the condition is clinically benign.
- Kostmann's syndrome is an example of congenital neutropenia that is inherited in an autosomal recessive pattern due to point mutations in the gene coding for G-CSF.
- Shwachman–Diamond syndrome is another example of congenital neutropenia that is also transmitted in an autosomal recessive pattern. This disorder is characterized by exocrine pancreatic deficiency and neutropenia, thrombocytopenia, short stature, and mental retardation; it may progress to MDS and AML.
- Reactive lymphocytes are also referred to as Downey cells. There are three types of Downey cells:
  - Type I: Small cells with minimum cytoplasm, indented nucleus/ irregular nuclear membrane, and condensed chromatin.
  - Type II: larger cells with abundant cytoplasm; the lymphocyte cytoplasm seems to hug the red cells. This is the most common type of Downey cell.
  - Type III: Cells with large moderate basophilic cytoplasm and nucleus with coarse chromatin. Nucleoli are apparent.
- Features of infectious mononucleosis, from the peripheral blood: 50% or more of the white cells are mononuclear cells; at least 10% of the lymphocytes exhibit reactive changes; and there is lymphocytic morphologic heterogeneity (i.e., different types of reactive lymphocytes are seen).
- Wiskott–Aldrich syndrome is characterized by eczema, immunodeficiency, and thrombocytopenia. This disorder is transmitted as X-linked recessive due to inheritance of the *WASP* gene, located at Xp11. Platelets are dysfunctional.
- Bernard–Soulier syndrome is transmitted as autosomal recessive. The (Gp)Ib-IX-V complex is abnormal. Platelet aggregation studies show impaired aggregation to ristocetin. Some cases of Bernard–Soulier syndrome are due to defects of the *GpIbβ* gene, located on chromosome 22. This gene may be affected in velocardiofacial syndrome or DiGeorge syndrome associated with deletion of 22q11.2.

■ Causes of thrombocytopenia with giant platelets and associated with defective myosin heavy chain 9 gene at 22q11 are Bernard−Soulier syndrome, May−Hegglin anomaly, Sebastian syndrome, Epstein syndrome, and Fechtner syndrome.

References

[1] George TI. Malignant or benign leukocytosis. Hematol Am Soc Hematol Educ Program 2012;2012:475−84.

[2] Roy A, Kar R, Basu D, Srivani S, et al. Clinico-hematological profile of Chediak−Higashi syndrome: experience from a tertiary care center in South India. Indian J Pathol Microbiol 2011;54:547−51.

[3] Nagai K, Ochi F, Maeda M, Ohga S, et al. Clinical characteristics and outcomes of Chediak−Higashi syndrome: a nationwide survey in Japan. Pediatr Blood Cancer 2013; 60:1582−6.

[4] Van de Vijver E, van den Berg TK, Kuijpers TW. Leukocyte adhesion deficiencies. Hematol Oncol Clin North Am 2013;27:101−16.

[5] Abramson N, Melton B. Leukocytosis: basics of clinical assessment. Am Fam Physician 2000;62:2053−60.

[6] Kahn JE, Bletry O, Guillevin L. Hypereosinophilic syndrome. Best Pract Res Clin Rheumatol 2008;22:863−82.

[7] Paz Z, Nails M, Ziv E. The genetics of benign neutropenia. Isr Med Assoc J 2011;13:625−9.

[8] Stock W, Hoffman R. White blood cell 1: non-malignant disorders. Lancet 2000;355: 1351−7.

[9] Acharya S, Pillai K, Francis A, Criton S, et al. Kasabach−Merritt syndrome: management with interferon. Indian J Dermatol 2010;55:281−3.

[10] Imai K, Morio T, Zhu Y, Jin Y, et al. Clinical course of patients with WASP gene mutations. Blood 2004;103:456−64.

CHAPTER 6

Myeloid Neoplasms

6.1 INTRODUCTION

Mature blood cells (white blood cells, red blood cells, and platelets) are created from one of two types of hematopoietic stem cells present in the bone marrow—lymphoid stem cells and myeloid stem cells. Lymphoid stem cells mature to lymphocytes, and myeloid stem cells mature into granulocytes, monocytes, red blood cells (RBCs), and platelets. In patients with leukemia, bone marrow produces either abnormal lymphoid cells or abnormal myeloid cells. This chapter focuses on myeloid neoplasms only.

The term "myeloid" includes all cells belonging to the granulocyte (neutrophils, eosinophil, and basophil), monocyte/macrophage, erythroid, megakaryocyte, and mast cell lineages. Myeloid neoplasms (myeloid malignancies) are clonal diseases of hematopoietic stem or progenitor cells due to genetic or epigenetic factors that perturb the key process of normal development of such cells. The disease state could be acute, such as acute myeloid leukemia (AML), or chronic, such as myelodysplastic syndrome. Mutations responsible for these disorders occur in several genes whose encoded proteins principally belong to five classes: signaling pathway proteins, transcription factors, epigenetic regulators, tumor suppressors, and components of spliceosome [1].

6.2 CLASSIFICATION OF MYELOID NEOPLASM

The World Health Organization (WHO) categorizes myeloid malignancies into five primary types:

- Myeloproliferative neoplasms (MPNs): These are typically chronic processes in which, in peripheral blood, counts for one or more lineages are high, but the presence of blasts in the peripheral blood is not a typical feature. However, bone marrow is usually hypercellular; again, the

CONTENTS

A. Wahed and A. Dasgupta: Hematology and Coagulation. DOI: http://dx.doi.org/10.1016/B978-0-12-800241-4.00006-1

percentage of blasts is normal or slightly high. Dysplasia is typically absent.

- Myelodysplastic syndrome (MDS): Another chronic process characterized by the observation of cytopenia in peripheral blood, and bone marrow is typically hypercellular. Dysplasia is a dominant feature. In some subtypes, blasts may be significantly increased.
- MDS/MPN: This type has mixed features of the two previously mentioned processes and is typically a chronic process. In the peripheral blood, white cell counts are usually increased. Bone marrow is hypercellular. Dysplasia is present.
- Myeloid/lymphoid neoplasms associated with eosinophilia and abnormalities of growth factor receptors derived from platelets or fibroblasts (platelet-derived growth factor receptor, α polypeptide (PDGFRA); platelet-derived growth factor receptor, β polypeptide (PDGFRB); or fibroblast growth factor receptor 1 (FGFR1)). This is a chronic disease with eosinophilia.
- Acute myeloid leukemia (AML): Acute illness in which blasts are 20% or higher, but less than 20% blasts may be encountered in AML if it is associated with chromosomal translocations including t(15;17), t(8;21), t(16;16), or inv16.

6.3   MYELOPROLIFERATIVE NEOPLASM

The myeloproliferative neoplasms can be further classified as chronic myelogenous leukemia (CML; which may also be $BCR-ABL1^+$); chronic neutrophilic leukemia; polycythemia vera; primary myelofibrosis; essential thrombocythemia; chronic eosinophilic leukemia, not otherwise specified; mastocytosis; and myeloproliferative neoplasm that is unclassifiable.

6.3.1   Chronic Myelogenous Leukemia, $BCR-ABL1^+$

CML is the most common form of the myeloid neoplasms, with median onset observed in the fifth or sixth decade of life. This type of leukemia is defined by the presence of the Philadelphia chromosome and/or the BCR–ABL1 fusion gene. The Philadelphia chromosome was first described in 1960 by Peter C. Nowell, a faculty member of the University of Pennsylvania, Philadelphia, as an unusual small chromosome present in leukocytes of patients with CML [2]. As methodologies for cytogenetics improved, this chromosomal abnormality was determined to be due to reciprocal translocation involving the long arms of chromosome 9 and 22 t(9; 22)(q34; q11), and this translocation results in the formation of the BCR–ABL1 fusion gene (the ABL1 gene from chromosome 9 fuses with the BCR gene on chromosome 22; BCR stands for "break point cluster region," and ABL1 is an oncogene known as Abelson murine leukemia viral oncogene, which encodes and enhances tyrosine kinase activity). The abnormal fusion gene is present in all myeloid lineages as well as in some lymphoid cells. This disease has three clinical phases: the chronic phase, the accelerated phase, and blast crisis. Most often, this disease is diagnosed in the chronic phase. Patients may present with weakness and organomegaly. The chronic phase can last for years before transforming to the accelerated phase or directly to blast crisis. The Philadelphia chromosome is present in 90−95% of cases of CML [3]. In cases negative for the Philadelphia chromosome, the *BCR−ABL1* fusion gene can be detected by florescence *in situ* hybridization (FISH), polymerase chain reaction (PCR), or Southern blot techniques. A diagnosis of CML requires the presence of the Philadelphia chromosome or documentation of the presence of the *BCR−ABL1* fusion gene. The discovery of imatinib mesylate as a tyrosine kinase inhibitor has resulted in a dramatic improvement in the lives of patients with CML.

The site of the breakpoint in the *BCR* gene may vary, and this may also have an influence on the phenotype of the disease. In the vast majority of cases, the breakpoint is referred to as the major breakpoint cluster region (M-BCR). The resultant abnormal fusion protein, p210, has increased tyrosinase kinase activity. If the breakpoint is in the region referred to as the minor region, then the fusion protein is p190. p190 is most frequently associated with Philadelphia chromosome-positive acute lymphoblastic leukemia (ALL). If present in CML, these patients have an increased number of monocytes, and this can resemble chronic myelomonocytic leukemia (CMML). If the breakpoint is in the region referred to as the mu region, then the fusion protein is p230. These patients have prominent neutrophilic maturation and/or thrombocytosis.

Characteristics of peripheral blood in the chronic phase of CML include the following:

- Leukocytosis: Peak in myelocytes and polymorphonuclear neutrophils (PMNs), absolute basophilia (invariably present), absolute eosinophilia (commonly present), and absolute monocytosis (with normal percentage). Orderly differentiation is preserved.
- Platelets: Normal or increased (low platelets are uncommon).
- RBCs: Nucleated red blood cells (NRBCs) may be seen.

Characteristic features of bone marrow examination (chronic phase) include the following:

- Increased myeloid cells to erythroid cells (M:E) ratio, usually greater than 10:1; blast population up to 9%
- Paratrabecular cuff of immature myeloid cells (5−10 cells thick; normal, 2−3)
- Megakaryocytes: Small, hypolobated (micromegakaryocyte)

- Increased reticulin fibrosis
- Pseudo-Gaucher cells and sea blue histiocytes (seen due to increased cell turnover)
- No significant dysplasia.

The accelerated phase of CML has the following features:

- Persistent or increasing white blood cell (WBC) count ($>$10,000/mm^3 or $>$10,000/μL) and/or persistent splenomegaly unresponsive to therapy (specific for CML)
- Persistent thrombocytosis ($>$1 million/μL) despite therapy
- Persistent thrombocytopenia ($<$100,000/μL) unrelated to therapy
- Clonal cytogenetic evolution (additional chromosomal abnormalities)
- Basophilia 20% or greater
- Myeloblasts 10−19% in peripheral blood or bone marrow.

Blast crisis of CML has the following features:

- Blasts in the peripheral blood or bone marrow 20% or greater
- Extramedullary blast proliferation or cluster of blasts in the bone marrow.

In two-third of cases, the blast lineage is myeloid, and in 20−30% of cases the blasts are lymphoblasts.

6.3.2 Chronic Neutrophilic Leukemia

Chronic neutrophilic leukemia is a rare myeloproliferative neoplasm with approximately 150 reported cases. This disease is difficult to diagnose because diagnosis is contingent on exclusion of underlying causes of reactive neutrophilia, particularly if evidence of clonality is lacking. The Philadelphia chromosome or BCR−ABL1 fusion gene must also be absent, thus excluding the diagnosis of CML. The recent discovery of specific oncogenic mutation in the colony-stimulating factor 3 receptor gene (CSF3R) in these patients has opened a new era for diagnosis and treatment [4]. Characteristics of chronic neutrophilic leukemia include the following:

- Persistent peripheral blood leukocytosis (WBC $>$25,000 with blasts $<$1%, bands and PMNs $>$80%, and immature cells [promyelocytes, myelocytes, and metamyelocytes] $<$10%) without any other causes of neutrophilia is a diagnostic criterion of chronic neutrophilic leukemia
- Neutrophils having toxic granulation
- Hypercellular marrow with increased neutrophils that have normal maturation and blasts $<$5%
- Hepatosplenomegaly
- Association with multiple myeloma.

6.3.3 Polycythemia Vera, Primary Myelofibrosis, and Essential Thrombocythemia

Polycythemia vera (PV), primary myelofibrosis, and essential thrombocythemia are Philadelphia chromosome-negative clonal stem cell disorders. In PV, the bone marrow is hypercellular, and proliferation of all cell lines may be seen. It is characterized by an increase in red cells and thus hemoglobin. *JAK2* (Janus kinase 2) gene gain-of-function somatic mutation occurs in PV. In the vast majority of cases of PV, *JAK2 V617F* mutation is documented, and in approximately 3% of cases, *JAK2* exon 12 mutations are seen. In primary myelofibrosis, there occurs megakaryocytic and also granulocytic proliferation with ultimate fibrosis formation. Extramedullary hematopoiesis may thus occur in the liver and spleen, with resultant enlargement of these organs. In essential thrombocythemia, the proliferation is typically restricted to megakaryocytes. Platelets counts are thus increased. All three conditions (PV, primary myelofibrosis, and essential thrombocythemia) may ultimately result in bone marrow fibrosis, as well as convert to acute leukemia, especially AML. The *JAK2* gene and the *MPL* gene (myeloproliferative leukemia virus oncogene) play an important role in cell signaling and proliferation. Mutations in these genes confer activation of the JAK−STAT pathway (STAT: signal transducer and activator of transcription—a gene regulatory protein) and other pathways promoting differentiation and proliferation of various lineages. The *JAK2−V617* mutation is present in approximately 95% of patients with PV, in 58% patients with primary myelofibrosis, and in 50% of patients with essential thrombocythemia [5]. Various characteristics and diagnostic features of these diseases are summarized in Table 6.1.

6.3.4 Chronic Eosinophilic Leukemia

Chronic eosinophilic leukemia (CEL) is a chronic myeloproliferative disease with unknown etiology in which a clonal proliferation of eosinophilic precursor leads to increased eosinophils in blood, bone marrow, or peripheral tissues. Eventually, this leads to eosinophilic filtration and functional damage of peripheral organs. Although a very rare disease, a tyrosine kinase inhibitor such as imatinib can be used for patient management. Diagnostic criteria of CEL include the following:

- There is marked eosinophilia ($>1500/mm^3$ or 1.5×10^9/L of blood and persisting for more than 6 months) with evidence of clonality of eosinophils or an increase in myeloblasts in peripheral blood or bone marrow.
- Other causes of eosinophilia are excluded, including neoplasms such as Hodgkin lymphoma, ALL, T cell lymphomas, and systemic mastocytosis.

Table 6.1 Key Features of Polycythemia Vera, Primary Myelofibrosis, and Essential Thrombocythemia

Polycythemia Vera	Primary Myelofibrosis	Essential Thrombocythemia
Genetics		
Almost all cases have *JAK2* mutation; most common mutation is *JAK2 V617F*.	Approximately 58% of cases with *JAK2* mutation.	Approximately 50% of cases with *JAK2* mutation.
Characteristics		
Characterized by increased erythropoiesis. There may be proliferation of granulocytes and megakaryocytes. Three phases of disease: prodromal, clinically overt, and post-polycythemic fibrosis. Clinical features include plethora, hypertension, and thrombosis.	Extramedullary hematopoiesis with resultant hepatosplenomegaly is present. Thrombocytosis may be a feature.	Patients may be asymptomatic. Unexplained thrombocytosis with or without thrombosis or hemorrhage.
Diagnosis		
Diagnostic criteria include hemoglobin >18.5 g/dL (male) and >16.5 g/dL (female) (or increased red cell mass); low erythropoietin; splenomegaly; and hypercellular marrow (all cell lines).	Peripheral blood: Leukoerythroblastic anemia with teardrop red cells. Bone marrow: Increased megakaryocytes with atypia; increased fibrosis; osteosclerosis; and intrasinusoidal hematopoiesis.	In peripheral blood, platelets >450,000/mm^3 (>450,000/μL). Bone marrow: Increased megakaryocytes with hyperlobulated nuclei and increased mature cytoplasm.

- Signs and symptoms of organ involvement are present. Organs that may be involved include the heart (endomyocardial fibrosis, restrictive cardiomyopathy, and valvular dysfunction). There may also be neurological involvement (peripheral neuropathy and central nervous system dysfunction), lung infiltrates, and diarrhea.
- Patients with Philadelphia chromosome and/or the *BCR−ABL1* fusion gene should be excluded. In addition, patients with rearrangement of *PDGFRA*, *PDGFRB*, or *FGFR1* genes should also be excluded, although there are isolated reports that in a minority of patients with chronic eosinophilic leukemia, *PDGFRA*, *PDGFRB*, or *FGFR1* genes may be present [6].
- If clonality of eosinophils cannot be proven or there is no increase in myeloblasts, then the diagnosis is idiopathic hypereosinophilic syndrome, in which eosinophilia is present for at least 6 months with features of organ involvement and dysfunction.

6.3.5   Mastocytosis

This is a clonal neoplastic disorder of mast cells involving one or more organs. In these involved organs, multifocal clusters or aggregates of abnormal mast cells are present. Classifications of mastocytosis include cutaneous mastocytosis (most common, usually observed in children and with good prognosis), extracutaneous mastocytoma, systemic mastocytosis (various types: indolent, aggressive, and systemic mastocytosis associated with clonal hematologic non-mast cell lineage disease), mast cell leukemia, and mast cell sarcoma.

Histological and immunohistochemical evaluation of a bone marrow trephine biopsy is used for diagnosis, and staining of mast cells with Giemsa and chloroacetate esterase (CAE) is helpful. Moreover, almost all mast cells, regardless of their origin, stage of maturation, and activation of sublineage status, express tryptase. Other mast cell-related antigens—such as chymase, CD117, and, for neoplastic mast cells, CD2 and CD25—can be used for diagnosis. The diagnosis of systemic mastocytosis requires the presence of the major criterion and one minor criterion or, in the absence of the major criterion, the presence of three minor criteria:

- Major criterion: Multifocal, dense infiltrates of mast cells (>15 mast cells in aggregates) in bone marrow or extracutaneous organs should be present.
- Minor criteria: (1) In bone marrow or extracutaneous organs, greater than 25% of the mast cells are spindle shaped or have atypical morphology, (2) mast cells are positive for CD2 and/or CD25, (3) serum total tryptase is high, and (4) point mutation at codon 816 of KIT is detected.

In aggressive systemic mastocytosis, features that may be present include cytopenias, hepatomegaly with liver dysfunctions, splenomegaly with hypersplenism, skeletal osteolytic lesions, and gastrointestinal infiltration with malabsorption.

6.4 MYELOID AND LYMPHOID NEOPLASM ASSOCIATED WITH EOSINOPHILIA

Myeloid and lymphoid neoplasms with eosinophilia and abnormalities of *PDGFRA*, *PDGFRB*, and *FGFR1* are a group of hematological neoplasms resulting from abnormal fusion genes that encode constitutively activated tyrosine kinase. In 2008, WHO classified these disorders as myeloid and lymphoid neoplasm with eosinophilia, and all patients with these three diseases may present with MPN. Conventional cytogenetics (karyotyping) will detect the majority of abnormalities involving *PDGFRB* and *FGFR1*, but FISH molecular studies may be needed to detect abnormalities involving *PDGFRA* [7]. Various features of these disorders include the following:

- Patients with *PDGFRA* can present with acute myeloid leukemia or as T lymphoblastic leukemia with eosinophil, and patients respond to imatinib mesylate (first-line therapy).
- Patients with *PDGFRB* present more often with CMML with eosinophilia. Again, imatinib mesylate is the first-line therapy.
- For patients with *FGFR1*, lymphomatous presentation is common, especially with T lymphoblastic lymphoma with eosinophilia. Unfortunately, these patients do not respond to therapy with imatinib mesylate, and they have a poor prognosis [7].

6.5   MYELODYSPLASTIC/MYELOPROLIFERATIVE NEOPLASMS

Myelodysplastic/myeloproliferative neoplasms can be classified into four categories:

- Chronic myelomonocytic leukemia (CMML)
- Atypical chronic myeloid leukemia, *BCR−ABL1*⁻
- Juvenile myelomonocytic leukemia (JMML)
- Myelodysplastic/myeloproliferative neoplasm, unclassifiable.

CMML is characterized by monocytosis and dysplasia. As expected, blasts have to be less than 20%; otherwise, this would be classified as acute leukemia. Also, the Philadelphia chromosome should be absent, and the *BCR−ABL* gene rearrangement test should be negative. CMML can be further classified into various subtypes: CMML-1 (blasts <5% in blood and <10% in bone marrow), CMML-2 (blasts 5−19% in blood or 10−19% in bone marrow or with Auer rods), and CMML with eosinophilia. However, in CMML with eosinophilia, rearrangements of *PDGFRA* and *PDGFRB* are absent.

Atypical chronic myeloid leukemia is characterized by leukocytosis resembling CML. However, dysplasia is present, but basophilia is absent. The Philadelphia chromosome is also absent, or tests for *BCR−ABL1* gene rearrangement are negative. Unlike CMML, monocytosis is not a feature.

JMML is characterized by leukocytosis and monocytosis in individuals younger than age 14 years with features of dysplasia. The hemoglobin F (Hb F) percentage may be increased in JMML. JMML is associated with neurofibromatosis-1.

6.6   MYELODYSPLASTIC SYNDROME

In MDS, the bone marrow produces cells that are of inadequate quality. The poor quality cells are not good enough to be released into the circulation

and thus are destroyed within the bone marrow; this process is called ineffective hematopoiesis. In normal marrow, there can be up to 10% ineffective
hematopoiesis, but in MDS this percentage is significantly higher. As a result,
the bone marrow becomes hypercellular, with cytopenia observed in the
peripheral blood. As expected, in MDS, evidence of abnormality with regard
to morphology is present in cells in the peripheral blood as well as in the
bone marrow. A major feature of MDS is that it is a clonal stem cell disease
that occurs principally in older adults with a median age of 70 years. MDS is
characterized by cytopenia in the peripheral blood with a hypercellular marrow, and morphological evidence of dysplasia is present. In addition, there
are certain cytogenetic abnormalities associated with MDS. Patients with
MDS are also at an increased risk of developing AML.

WHO classifies MDS as follows:

- Refractory cytopenia with unilineage dysplasia (RCUD)
- Refractory anemia with ringed sideroblasts (RARS)
- Refractory cytopenia with multilineage dysplasia (RCMD)
- Refractory anemia with excess of blasts-1 (RAEB-1)
- Refractory anemia with excess of blasts-2 (RAEB-2)
- Myelodysplastic syndrome—unclassified (MDS-U)
- MDS associated with isolated del(5q).

6.6.1 Features of Dysplasia in Red Cells, Erythroid Precursors, Granulocytes, and Megakaryocytes

Features of dysplasia in red cells and erythroid precursors include dimorphic
red cells, macrocytic red cells, nuclear budding, internuclear bridging, bi/
multinucleated cells, karyorrhexis, megaloblastoid changes, cytoplasmic fray-
ing, nuclear cytoplasmic dyssynchrony, cytoplasmic vacuolization, ringed
sideroblasts, and periodic acid−Schiff (PAS) positivity.

It is important to note that there are many causes of dimorphic and macrocytic red cells, and these are considered soft signs. Cytoplasmic vacuolization
can also be seen with prolonged alcohol intake. Vitamin B_{12} and folate deficiency may result in the features of dyserythropoiesis listed previously.
Therefore, a diagnosis of MDS requires exclusion of folate and vitamin B_{12}
deficiency.

Features of dysplasia in granulocytes include nuclear hypolobulation (pseudo-
Pelger−Huët cells), hypersegmented PMNs, abnormally small or large granulocytes, hypogranulation, and the presence of Auer rods.

When the majority of neutrophils (usually 75% or more) have two nuclear
segments, this is referred to as Pelger−Huët abnormality. This is inherited as

autosomal dominant and is clinically insignificant. In MDS, pseudo-Pelger–Huët cells have two nuclear segments and are hypogranular; a small number of cells have this morphology.

Features of dysplasia in megakaryocytes include micromegakaryocyte, nuclear hypolobulation, and multinucleation.

However, it is important to note that micromegakaryocytes are also seen in CML. When dealing with a case of MDS, it is important to review the peripheral smear and bone marrow slides. Flow cytometry may or may not contribute to the diagnosis, but cytogenetic study is essential.

6.6.2   Arriving at a Diagnosis of MDS and Subclassifying MDS

The approach to the assessment of peripheral smear and bone marrow in patients with suspected MDS is as follows:

- Is unicytopenia, bicytopenia, or pancytopenia present in the peripheral blood?
- What is the blast count in the peripheral blood and/or in the bone marrow?
- Are Auer rods present in the peripheral blood and/or the bone marrow?
- In the bone marrow, dysplasia should be greater than 10% in each cell line.
- In the bone marrow, how many cell lines are affected by dysplasia?
- Is more than 15% of ringed sideroblasts (as a percentage of all erythroid cells) present? This will be evident in the bone marrow aspirate smear stained for iron.

It is important to note that if the blast count is 20% or more, this is not MDS but, rather, AML.

If there are Auer rods in the peripheral blood or bone marrow, then this is RAEB-2. The blast count must be less than 20%. If the blast count in the peripheral blood is 5–19% or the blast count in the bone marrow is 10–19%, then this is also RAEB-2. The degree of cytopenias or the number of cell lines affected by dysplasia is not important. The presence of ringed sideroblasts is also not important. RAEB-1 is similar to RAEB-2, only the percentage of blasts observed is lower. In the peripheral blood, the blast count is less than 5%, and in the bone marrow it is 5–9%.

If there is unicytopenia or bicytopenia in the peripheral blood with less than 1% blasts and unilineage dysplasia (with >10% cells dysplastic), less than 5% blasts, and less than 15% ringed sideroblasts in the bone marrow, this is refractory cytopenia with unilineage dysplasia (RCUD). If there is anemia, with erythroid dysplasia and greater than 15% of ringed sideroblasts, this is refractory anemia with ringed sideroblasts (RARS).

If dysplasia is present in at least two cell lines with less than 1% blasts in peripheral blood and it is not RAEB-1 or RAEB-2, then this is refractory cytopenia with multilineage dysplasia (RCMD). Greater than 15% of ringed sideroblasts may or may not be present.

If there is pancytopenia with other features of RCUD *or* features of RCUD and RCMD but the blast count in the peripheral blood is 1−5% *or* there is less than 10% dysplasia in the cell line(s) but there is a cytogenetic abnormality, which is considered as presumptive evidence of MDS, then this is MDS-U (MDS-unclassifiable).

Finally, there is MDS associated with isolated del(5q), which is characterized by the following:

- Increased incidence in females
- Anemia, with normal or increased platelet counts
- Splenomegaly
- Normal or increased megakaryocytes with hypolobated nuclei
- Blasts less than 1% in the peripheral blood and less than 5% in the bone marrow
- Isolated del(5q) cytogenetic abnormality.

It is important to note that for all of the previously mentioned criteria, the blast count in peripheral blood should be less than 1% and in the bone marrow less than 5%, with the exception of RAEB-1 and RAEB-2. Thus, it is the percentage of blasts and Auer rods that dictates RAEB-1 or -2. Once RAEB-1 and -2 are excluded, then a key point is whether there is dysplasia in the bone marrow affecting one or two cell lines. If two cell lines are affected, then it is RCMD. If only one cell line is affected, then this is RCUD. RARS is unique in the sense that the erythroid cell line is affected with ringed sideroblasts. Ringed sideroblasts may also be seen in RCMD. MDS associated with del(5q) is also unique, with features mentioned previously.

RCUD may be refractory anemia (RA), refractory neutropenia (RN), or refractory thrombocytopenia (RT). Siderocytes are mature red cells with iron. Sideroblasts are red cell precursors with iron. Both of these types of cells are physiological. Ringed sideroblasts are red cell precursors with iron distributed in a ring-like manner around the nucleus. It does not have to be a complete ring. One-third of the circumference of the nucleus is enough to fulfill the criteria. The iron is present within the mitochondria.

The risk of AML is highest with RAEB-2 and low with RA (an example of RCUD) and MDS associated with isolated del(5q).

Previously, MDS was categorized according to the French–American–British (FAB) classification into five classes: RA, RARS, RAEB, refractory anemia with excess of blasts in transformation (RAEB-t), and CMML. However, CMML is now part of the group MPN/MDS as per the WHO classification of myeloproliferative neoplasms.

RAEB-t (refractory anemia with excess blasts in transformation) was used with blasts count between 20% and 30%; previously AML required a blast count of 30% or higher. Now, with AML being diagnosed with blast count of 20% or higher, RAEB-t is an obsolete term.

6.6.3   Abnormal Localization of Immature Precursors

This term is used when immature myeloid cell clusters (five to eight cells) are present in the central portion of marrow, away from usual locations (paratrabecular or perivascular). Three or more such foci fulfill the criteria for ALIP. ALIP is frequently present in cases of RAEB and is associated with a rapid evolution to AML. If ALIP is found in other subtypes of MDS, the case should be re-evaluated.

6.6.4   Cytogenetic Abnormalities Associated with MDS

Approximately 30–50% of patients with primary MDS have clonal cytogenetic abnormalities, whereas 80% of therapy-related MDS cases have cytogenetic abnormalities. Therefore, cytogenetic study is useful in the diagnosis of MDS and in the prediction of prognosis of individual patients, using the International Prognostic Scoring System [8]. Many cytogenetic abnormalities have been described in patients with MDS, but common abnormalities include the following:

- Unbalanced: -7 or del(7q), -5 or del(5q), i(17q), or t(17p)
- Balanced: t(11;16), t(3;21).

However, there are many other abnormalities. Interestingly, MDS patients with +8 (trisomy 8) as sole aberration are predominately male. Certain aberrations are associated with good prognosis, whereas others are associated with intermediate or poor prognosis. The following are examples:

- Karyotypes associated with good prognosis: Normal, $-Y$, del(5q), and del(20q)
- Karyotypes associated with intermediate prognosis: Trisomy 8 ($+8$) and more than one abnormality
- Karyotypes associated with poor prognosis: Complex (three or more abnormalities) or chromosome 7 abnormalities.

6.6.5 Unusual Situations in MDS

In approximately 10% of cases, the bone marrow in MDS can be hypocellular. Differential diagnosis in such situations includes aplastic anemia. In approximately 10% of cases of MDS, evidence of significant fibrosis is present. These cases are referred to as MDS with fibrosis.

6.7 ACUTE LEUKEMIA

In acute leukemia, immature cells proliferate and fail to mature within the bone marrow. The bone marrow cannot produce normal cells, and when approximately 1 trillion (1 million × 1 million) leukemic cells accumulate in the bone marrow, these cells start to spill over into the peripheral blood. The immature leukemic cells are thus now present in the peripheral blood. The term used for the leukemic process before the leukemic cells are found in the peripheral blood is subleukemic leukemia. In acute leukemia, typically patients present with blasts in the peripheral blood. Blasts are not a normal finding in the peripheral blood.

Causes of blasts in the peripheral blood include the following:

- Acute leukemia
- Myelodysplastic syndrome
- Neupogen effect
- Neonates with a reactive condition
- Leukoerythroblastic blood picture
- Leukemoid reaction
- CML.

In acute leukemia, typically anemia, thrombocytopenia, and leukocytosis with increased blasts are observed. For diagnosis of acute leukemia, it is important to determine if the patient is suffering from AML or ALL. If the patient is suspected to have AML, then it is important to decide whether it is acute promyelocytic leukemia (APL; a subtype of AML in which immune granulocytes called promyelocytes accumulate) or not because APL patients are especially prone to disseminated intravascular coagulation (DIC) and therapy with all-*trans* retinoic acid (ATRA; also known as tretinoin) should be started as soon as possible. To make a diagnosis, the age of the patient should be considered because ALL is seen most often in children, whereas AML is seen in all age groups. Moreover, the morphology of blasts should be carefully observed, and the presence of Auer rods should be investigated.

6.7.1   Blasts

All blasts have a high nuclear to cytoplasmic ratio and fine chromatin, and are with or without nucleoli. Myeloblasts tend to be larger than lymphoblasts; the chromatin is finer; and nucleoli, if present, tend to be increased in number (more than two). The cytoplasm may have fine azurophilic granules. Sometimes these granules condense to form Auer rods. There are three types of myeloblasts: Type I (rare granules), type II (few granules, 5−20), and type III (abundant granules). Auer rods are mostly seen with type III myeloblasts.

Lymphoblasts are smaller blasts; the chromatin is not as fine as that of myeloblasts. Nucleoli are absent or, if present, there are less than two. Typically, lymphoblasts have no granules.

When a conclusive diagnosis of AML or ALL cannot be made, ancillary tests such as cytochemistry and flow cytometry are helpful. Typically, in all cases of acute leukemia, flow cytometry and cytogenetic and molecular studies should be done. These studies can be performed on the peripheral blood as well as the bone marrow specimen.

6.7.2   Cytochemistry

Various staining methods can be used for cytochemistry analysis for diagnosis. These include the following:

- Myeloperoxidase (MPO): Stains myeloblasts in a Golgi distribution; monoblasts are negative; promonocytes or monocytes may have cytoplasmic stain.
- Nonspecific esterase (NSE; e.g., butyrate) stain: Stains cytoplasm of cells of monocytic lineage
- Chloroacetate esterase (CAE): Stains cells of neutrophilic lineage and mast cells; also stains abnormal eosinophils (seen in M4 Eo)
- Sudan black: Parallels MPO but less specific and thus not used
- Periodic acid−Schiff (PAS): Stains FAB M6 and M7 subtypes of AML; may also stain ALL. Thus, it is not useful for reliably differentiating AML and ALL.

6.7.3   Classification of AML and Diagnosis

AML is currently classified according to the WHO classification. The previous classification was the FAB classification (Table 6.2).

For diagnosis of AML, the presence of 20% or more blasts in the bone marrow is required. If the blast count is lower, the diagnosis of MDS should be considered. An exception to the 20% blast count rule is AML with recurrent

---

Table 6.2 World Health Organization (WHO) and French—American—British (FAB) Classification of AML

WHO Classification of AML	FAB Classification of AML
■ AML with recurrent genetic abnormalities ■ AML with MDS-related changes: History of MDS; AML with MDS-related cytogenetics; AML with multilineage dysplasia (50%) ■ Therapy-related myeloid neoplasms (alkylating agents, topoisomerase II inhibitors, radiation) ■ AML (not otherwise specified): AML with minimal differentiation, AML without maturation, AML with maturation, acute myelomonocytic leukemia, acute monoblastic and monocytic leukemia, acute erythroid leukemia, acute megakaryoblastic leukemia, acute basophilic leukemia, acute panmyelosis with myelofibrosis ■ Myeloid sarcoma ■ Myeloid proliferations related to Down syndrome ■ Blastic plasmacytoid dendritic cell neoplasm	■ M0: AML with minimum differentiation ■ M1: AML without maturation (here promyelocytes, myelocytes, metamyelocytes, and bands/PMNs are <10%) ■ M2: AML with maturation (here promyelocytes, myelocytes, metamyelocytes, and bands/PMNs are >10%) ■ M3: Acute promyelocytic leukemia (classical and microgranular) ■ M4: Acute myelomonocytic leukemia (where cells of monocytic lineage are >20%) (M4 has a variant called M4 Eo) ■ M5: Acute monocytic leukemia (where cells of monocytic lineage are >80%) (subdivided into M5a and M5b) ■ M6: Acute erythroleukemia (subdivided into M6a and M6b) ■ M7: Acute megakaryoblastic leukemia

---

genetic abnormalities. Here, if an individual has any of these cytogenetic abnormalities, then he or she has AML, irrespective of the blast count.

The recurrent genetic abnormalities are t(15;17), t(8;21), t(16;16), and inv16. In fact, genetic abnormality of t(15;17) is actually the case of acute promyelocytic leukemia. In the WHO classification, it is included in the category of recurrent genetic abnormalities. In the FAB classification, it is M3. APL is characterized by proliferation and maturation arrest of abnormal promyelocytes. Patients are at increased risk of DIC. In the setting of AML, the abnormal promyelocytes may be counted as blasts.

There are two types of APL: classical and microgranular. In the classical case, abnormal promyelocytes with abundant granules are seen. Cells with Auer rods are readily apparent. Cells with multiple Auer rods are referred to as Faggot cells. In the microgranular variant, the granules of the abnormal promyelocytes are present but not visible by light microscopy. The morphologic clue is the appearance of the nuclei of the abnormal promyelocytes. The nuclei are highly irregular, having a bilobed appearance that is sometimes referred to as an apple core appearance. The WBC count in the microgranular variant of APL is also usually very high.

Individuals with t(8;21) may have typical M2 morphology (FAB classification), which includes blasts with salmon-colored granules, slender Auer rods, and cytoplasmic vacuoles. Individuals with t(16;16) or inv16 may have the

M4 Eo subtype of AML (FAB classification). These patients have eosinophilia with the presence of abnormal eosinophils. These abnormal eosinophils have basophilic granules and stain with CAE.

AML with MDS-related changes and therapy-related myeloid neoplasms is fairly straightforward. If a patient with AML has a prior history of MDS or has MDS-related cytogenetics or has AML with multilineage dysplasia, then the patient should be diagnosed as suffering from AML with MDS-related changes. Multilineage dysplasia in this context means at least two bone marrow cell lines are dysplastic and at least 50% of each cell line is dysplastic. Therapy-related myeloid neoplasms are seen in patients who have received chemotherapy (e.g., alkylating agents and topoisomerase II inhibitors) or radiation.

AML (not otherwise specified; NOS), according to the WHO classification, corresponds to the FAB classification with the exception of acute basophilic leukemia and acute panmyelosis with myelofibrosis, which do not exist in the FAB classification. Also, APL does not belong in the AML, NOS category because it is part of AML with recurrent genetic abnormalities. Features of various subtypes of AML according to the FAB classification are summarized here:

- AML with minimum differentiation, M0: Blasts with no granules or few granules are predominantly present. Auer rods are not seen. Special staining for MPO is negative. Thus, it is not possible to determine whether the blasts are myeloblasts or lymphoblasts. Flow cytometry is crucial in this case. The blasts express myeloid antigens, and thus the case is labeled as AML.
- AML without maturation, M1: Myeloblasts dominate the picture (promyelocytes, myelocytes, metamyelocytes, and bands/PMNs are <10%). However, it is possible to designate the blasts as myeloblasts.
- AML with maturation, M2: Promyelocytes, myelocytes, metamyelocytes, and bands/PMNs are >10%.
- Acute promyelocytic leukemia, M3: This was discussed previously.
- Acute myelomonocytic leukemia, M4: Cells of monocytic lineage are greater than 20% but less than 80%. A variant of M4 is M4 Eo (discussed previously).
- Acute monocytic leukemia, FAB M5: Cells of monocytic lineage are greater than 80%. If the majority of the cells are monoblasts (typically 80% or more), then this may be categorized as M5a (acute monoblastic leukemia). In M5b, the majority of the monocytic cells are promonocytes. It is important to note that in the setting of AML, promonocytes are counted as blasts.
- Acute erythroid leukemia, M6: Fifty percent or more of the bone marrow cells are erythroid. Of the nonerythroid population, at least

20% are myeloblasts. The erythroid cells are typically dysplastic and also may have vacuoles. These vacuoles may be positive for PAS stain. Pure erythroid leukemia is characterized by numerous proerythroblasts.

- Acute megakaryoblastic leukemia, M7: At least half of the blast population is of megakaryocytic lineage. Megakaryoblasts typically have cytoplasmic blebs. Features of dysplasia may also be present in the myeloid cell lines.

6.7.4   AML and Flow Cytometry

Individuals with AML should have flow blasts positive for myeloid markers such as CD13, CD33, and CD117. Cells that denote immaturity are CD117, human leukocyte antigen (HLA DR), and CD34. Thus, in AML with minimum differentiation (FAB classification M0), positivity for CD13, CD33, CD117, HLA DR, and CD34 should be observed, but MPO should be negative. In AML without maturation (FAB classification M1), positivity for the previously mentioned markers as well as positivity for MPO should be observed.

In AML with maturation (FAB classification M2), positivity for CD13, CD33, CD117, HLA DR, CD34, and MPO should be observed, as well as positivity for mature markers such as CD11b, CD15, and CD65.

In APL, the abnormal promyelocytes are CD34$^-$; completely or partially negative for HLA DR, CD11b, and CD15; and positive for CD13 (heterogeneous), CD33 (homogeneous and bright), and CD117. The abnormal promyelocytes lack CD10 and CD16, which are mature myeloid markers. It has been reported that the combination of the absence of CD11b, CD11c, and HLA DR identifies 100% of APLs [5]. CD2 positivity is a feature of APL.

In acute myelomonocytic leukemia (FAB classification M4), positivity for mature myeloid markers such as CD15 and CD 65 should be seen. Positivity for monocytic markers such as CD4, CD11b, CD11c, CD14, CD36, and CD64 should also be observed, and at least 20% myeloblasts positive for myeloid markers should also be present. In acute monocytic leukemia (FAB classification M5), positivity for monocytic markers will be seen.

In acute erythroid leukemia (FAB classification M6), positivity for glycophorin and hemoglobin A (Hb A) should be observed, but MPO may be negative. HLA DR and CD34 may also be negative. The leukemic cells are positive for CD235a and CD36, but negative for CD64 (CD36 positivity is seen in monocytes and erythroid cells; if the same cells are negative for CD64, another monocytic marker, it can be concluded that these cells are actually erythroid).

In acute megakaryoblastic leukemia (FAB classification M7), positivity for CD41 and CD61 is expected. HLA DR and CD34 may again be negative. CD41 is the glycoprotein IIb/IIIa antigen, and CD61 is the glycoprotein IIIa antigen.

6.7.5 Cytogenetics and AML

In addition to the genetic abnormalities mentioned for the category of recurrent genetic abnormalities, other genetic abnormalities in AML may also be seen. Therefore, cytogenetics analysis is useful for diagnosis. Common abnormalities include the following:

- t(1;22): Usually associated with megakaryocytic lineage
- T(3;3) or inv3: Usually associated with abnormal megs, multilineage dysplasia, and increased platelets
- t(6;9): Usually associated with monocytic lineage, basophilia, and multilineage dysplasia
- t(9;11): Usually associated with monocytic lineage
- AML with gene mutations include mutations of fms-related tyrosine kinase 3 (FLT3), nucleophosmin (NPM1), and, less commonly, mutations of the CEBPA gene, KIT, MLL, WT1, NRAS, and KRAS.

FLT3 is located at 13q12 and encodes a tyrosine kinase receptor involved in hematopoietic stem cell differentiation and proliferation. FLT3 mutations are seen in AML and MDS. When FLT3 mutation is seen in AML, it is most commonly seen with t(6;9), APL, and with normal karyotype. FLT3 mutations impart an adverse outcome. Mutations of NPM1 and CEBPA in the absence of FLT3 mutation in AML with normal karyotype indicate a favorable prognosis.

KEY POINTS

- CML is defined by the presence of the Philadelphia chromosome and/or the BCR−ABL1 fusion gene. The abnormal fusion gene is present in all myeloid lineages as well as in some lymphoid cells.
- The Philadelphia (Ph) chromosome is due to reciprocal translocation resulting in t(9;22)(q34;q11). The ABL1 gene from chromosome 9 fuses with the BCR gene on chromosome 22. BCR stands for "break point cluster region." ABL1 is an oncogene. The Philadelphia chromosome is present in 90−95% of cases of CML. In the Philadelphia chromosome negative cases, the BCR−ABL1 fusion gene can be detected by FISH, PCR, or Southern blot techniques.
- CML has three clinical phases: The chronic phase, the accelerated phase, and blast crisis.

- The site of the breakpoint in the *BCR* gene may vary, and this may also have an influence on the phenotype of the disease. In the vast majority of cases, the breakpoint is referred to as the major breakpoint cluster region (M-BCR). The resultant abnormal fusion protein, p210, has increased tyrosinase kinase activity. If the breakpoint is in the region referred to as the minor region, then the fusion protein is p190. p190 is most frequently associated with Ph-positive ALL. If present in CML, these patients have an increased number of monocytes, and this can resemble CMML. If the breakpoint is in the region referred to as the mu region, then the fusion protein is p230. These patients have prominent neutrophilic maturation and/or thrombocytosis.
- The chronic phase of CML is characterized by the following in the peripheral blood:
  - Leukocytosis: Peak in myelocytes and PMNs, absolute basophilia (invariably present), absolute eosinophilia (commonly present), and absolute monocytosis (with normal percentage). Orderly differentiation is preserved.
  - Platelets: Normal or increased (low platelets uncommon).
  - RBCs: NRBCs may be seen.
- The chronic phase of CML is characterized by the following in the bone marrow:
  - Increased M:E ratio, usually greater than 10:1
  - Paratrabecular cuff of immature myeloid cells (5–10 cells thick; normal, 2–3)
  - Megakaryocytes: Small, hypolobated (micromegakaryocyte)
  - Increased reticulin fibrosis
  - Pseudo-Gaucher cells and sea blue histiocytes (seen due to increased cell turnover)
  - No significant dysplasia.
- The accelerated phase of CML is characterized by persistent or increasing WBC ($>10,000/mm^3$) and/or persistent splenomegaly unresponsive to therapy (specific for CML), persistent thrombocytosis (>1 million/mm^3) despite therapy, persistent thrombocytopenia ($<100,000/mm^3$) unrelated to therapy, clonal cytogenetic abnormalities (additional chromosomal defects), and basophilia 20% or greater. Myeloblasts should represent 10–19% in peripheral blood or bone marrow.
- Blast crisis of CML is characterized by blasts in the peripheral blood or bone marrow 20% or greater; extramedullary blast proliferation; and in two-thirds of cases, the blast lineage is myeloid, and in 20–30% of cases the blasts are lymphoblasts.
- Polycythemia vera: Almost all cases have *JAK2* mutation; the most common mutation is *JAK2 V617F*, and diagnostic criteria include

hemoglobin greater than 18.5 g/dL (male) and greater than 16.5 g/dL (female) (or increased red cell mass), low erythropoietin, splenomegaly, and hypercellular marrow (all cell lines).

- Primary myelofibrosis: A total of 58% with *JAK2* mutation and key features include leukoerythroblastic anemia with teardrop red cells (peripheral blood); increased megakaryocytes with atypia, increased fibrosis, osteosclerosis, and intrasinusoidal hematopoiesis (bone marrow).

- Essential thrombocythemia: A total of 50% with *JAK2* mutation, and key features include platelets greater than 450,000/mm^3 with increased megakaryocytes with hyperlobulated nuclei and increased mature cytoplasm.

- Mastocytosis is a clonal neoplastic disorder of mast cells involving one or more organs. In involved organs, multifocal clusters or aggregates of abnormal mast cells are present. Mast cells can be stained with Giemsa, CAE, tryptase/chymase, CD117, and, for neoplastic mast cells, CD2 and CD25. The diagnosis of systemic mastocytosis requires the presence of the major criterion and one minor criterion or, in the absence of the major criterion, the presence of three minor criteria.

- CMML is characterized by monocytosis and dysplasia. Blasts have to be less than 20%; otherwise, this would be acute leukemia. Also, the Philadelphia chromosome is absent. A test for *BCR–ABL* gene rearrangement is negative.

- Atypical CML is characterized by leukocytosis resembling CML. However, dysplasia is present. Basophilia is not present. The Philadelphia chromosome is absent, or a test for the *BCR–ABL1* gene rearrangement is negative.

- JMML is characterized by leukocytosis and monocytosis in individuals less than 14 years of age with features of dysplasia. The Hb F percentage may be increased in JMML.

- Key features of MDS include the following: MDS is a clonal stem cell disease; MDS occurs principally in older adults with a median age of 70 years; MDS is characterized by cytopenia in the peripheral blood with a hypercellular marrow and morphological evidence of dysplasia is present; and there are certain cytogenetic abnormalities associated with MDS. There is also an increased risk of AML.

- Subclassification of MDS has several features. If there are Auer rods in the peripheral blood or bone marrow, then this is RAEB-2. The blast count must be less than 20%. If the blast count in the peripheral blood is 5–19% or the blast count in the bone marrow is 10–19%, then this is also RAEB-2. The degree of cytopenias or how many cell lines are affected by dysplasia is not important. The presence of ringed sideroblasts is not important. RAEB-1 is similar to RAEB-2, except that the percentage of blasts observed is lower. In the peripheral blood, the blast count is less than 5%, and in the bone marrow it is 5−9%. If there is unicytopenia or bicytopenia in the peripheral blood with less than 1% blasts and unilineage dysplasia (with >10% cells dysplastic), less than 5% blasts, and less than 15% ringed sideroblasts in the bone marrow, this is RCUD. If there is anemia, with erythroid dysplasia and greater than 15% ringed sideroblasts, this is RARS. If dysplasia is present in at least two cell lines with less than 1% blasts in peripheral blood and it is not RAEB-1 or RAEB-2, then this is RCMD. Greater than 15% of ringed sideroblasts may or may not be present. If there is pancytopenia with other features of RCUD *or* features of RCUD and RCMD but the blast count in the peripheral blood is 1−5% *or* there is less than 10% dysplasia in the cell line(s) but there is a cytogenetic abnormality that is considered presumptive evidence of MDS, then this is MDS-U.

■ MDS is associated with isolated del(5q), which is characterized by increased incidence in females; anemia, with normal or increased platelet counts; splenomegaly; normal or increased megakaryocytes with hypolobated nuclei; blasts less than 1% in the peripheral blood and less than 5% in the bone marrow; and isolated del(5q) cytogenetic abnormality.

■ It is important to note that the following karyotypes are associated with good prognosis of MDS: Normal, −Y, del(5q), and del(20q). The following are associated with a poor prognosis of MDS: Complex (three or more abnormalities) or chromosome 7 abnormalities. All other abnormalities impart intermediate prognosis.

■ Diagnosis of AML requires the presence of 20% or more blasts in the bone marrow. If the blast count is lower, the patient probably is suffering from MDS. An exception to the 20% blast count rule is AML with recurrent genetic abnormalities. Here, if an individual has any of these cytogenetic abnormalities, then he or she has AML irrespective of the blast count. The recurrent genetic abnormalities are t(15;17), t(8;21), t(16;16), and inv16.

■ Cytogenetic abnormality t(15;17): Cases are actually acute promyelocytic leukemia. In the WHO classification, it is included in the category of recurrent genetic abnormalities. In the FAB classification, it is M3. Acute promyelocytic leukemia is characterized by proliferation and maturation arrest of abnormal promyelocytes. Patients are at increased risk of DIC. In the setting of AML, the abnormal promyelocytes may be counted as blasts. There are two types of APL: classical and microgranular. In the classical case, abnormal promyelocytes with abundant granules are seen. Cells with Auer rods are easily apparent. Cells with multiple Auer rods are referred to as Faggot cells. In the microgranular variant, the granules of the abnormal promyelocytes are present but not visible by light microscopy. The morphologic clue is the appearance of the nuclei of the abnormal promyelocytes. The nuclei are highly irregular, having a bilobed appearance sometimes referred to as an apple core appearance. The WBC count in the microgranular variant of APL is also usually very high.

- Individuals with t(8;21) may have typical FAB M2 morphology, which includes blasts with salmon-colored granules, slender Auer rods, and cytoplasmic vacuoles.
- Individuals with t(16;16) or inv16 may have the FAB M4 Eo subtype of AML. Here, there is eosinophilia with the presence of abnormal eosinophils. These abnormal eosinophils have basophilic granules and stain with CAE.
- AML with MDS-related changes and therapy-related myeloid neoplasms is fairly straightforward. If a patient with AML has a prior history of MDS or has MDS-related cytogenetics or has AML with multilineage dysplasia, then he or she should be diagnosed as having AML with MDS-related changes. Multilineage dysplasia in this context means at least two bone marrow cell lines are dysplastic, and at least 50% of each cell line is dysplastic. Therapy-related myeloid neoplasms are seen in patients who have received chemotherapy (e.g., alkylating agents and topoisomerase II inhibitors) or radiation.
- AML, NOS, according to the WHO classification, corresponds to the FAB classification with the exception of acute basophilic leukemia and acute panmyelosis with myelofibrosis, which do not exist in the FAB classification. Also, APL does not belong in the AML, NOS category because it is part of AML with recurrent genetic abnormalities.
- AML with minimum differentiation, FAB M0: Blasts are predominantly present. The blasts typically have no or few granules. Auer rods are not seen. Special staining for MPO is negative. Thus, it is not possible to determine whether the blasts are myeloblasts or lymphoblasts. Flow cytometry is crucial in this case. The blasts express myeloid antigens, and thus the case is labeled as AML.
- AML without maturation, FAB M1: Myeloblasts dominate the picture (here, promyelocytes, myelocytes, metamyelocytes, and bands/PMNs are <10%). However, it is possible to designate the blasts as myeloblasts.
- AML with maturation, FAB M2: Promyelocytes, myelocytes, metamyelocytes, and bands/PMNs are greater than 10%.
- Acute myelomonocytic leukemia, FAB M4: Cells of monocytic lineage are greater than 20% but less than 80%. A variant of M4 is M4 Eo.
- Acute monocytic leukemia, FAB M5: Cells of monocytic lineage are greater than 80%. If the majority of the cells are monoblasts (typically

80% or greater), then this may be categorized as M5a (acute monoblastic leukemia). In M5b, the majority of the monocytic cells are promonocytes. Note that in the setting of AML, promonocytes are counted as blasts.

- Acute erythroid leukemia, M6: Half or more of the bone marrow cells are erythroid. Of the nonerythroid population, at least 20% are myeloblasts. The erythroid cells are typically dysplastic and also may have vacuoles. These vacuoles may be positive for PAS. Pure erythroid leukemia is characterized by numerous proerythroblasts.
- Acute megakaryoblastic leukemia, M7: At least half of the blast population is of megakaryocytic lineage. Megakaryoblasts typically have cytoplasmic blebs. Features of dysplasia may also be present in the myeloid cell lines.

References

[1] Murati A, Brecqueville B, Devillier R, Mozziconacci MJ, et al. Myeloid malignancies: mutations, models and management. BMC Cancer 2012;12:304.

[2] Koretzky G. The legacy of the Philadelphia chromosome. J Clin Invest 2007;117:2030−2.

[3] Keung YK, Beaty M, Powell B, Molnar I, et al. Philadelphia chromosome positive myelodysplastic syndrome and acute myeloid leukemia: retrospective study and review of the literature. Leukemia Res 2004;28:579−86.

[4] Elliott MA, Tefferi A. The molecular genetics of chronic neutrophilic leukemia: defining a new era in diagnosis and therapy. Curr Opin Hematol 2014;21:148−54.

[5] Dos Santos LC, Ribeiro JC, Silva NP, Cerutti NP, et al. Cytogenetics, JAK2 and MPL mutations in polycythemia vera, primary myelofibrosis and essential thrombocythemia. Rev Bras Hematol Hemoter 2011;33:417−24.

[6] Gotlib J. World Health Organization defined eosinophilic disorders: 2014 updates on diagnosis, risk stratification and management. Am J Hematol 2014;89:325−37.

[7] Savage N, George TI, Gotlib J. Myeloid neoplasms associated with eosinophilia and rearrangement of PDGFRA, PDGFRB and FGFR1: a review. Int J Lab Hematol 2013;35:491−500.

[8] Kawankar N, Vundinti BR. Cytogenetic abnormalities in myelodysplastic syndrome: an overview. Hematology 2011;16:131−8.

CHAPTER 7

Monoclonal Gammopathy and Its Detection

7.1 INTRODUCTION

Monoclonal gammopathy is present in a patient when a monoclonal protein (also referred as M protein or paraprotein) is identified in the patient's serum, urine, or both. Monoclonal gammopathy may be of undetermined significance or due to myeloma or a B cell lymphoproliferative disorder producing this paraprotein. Multiple myeloma, a malignant disorder of bone marrow, is the most common form of myeloma. This disease is called multiple myeloma because it affects multiple organs in the body. In multiple myeloma, plasma cells that proliferate at a low rate become malignant with a massive clonal expression resulting in a high rate of production of monoclonal immunoglobulin in the circulation. Monoclonal gammopathy of undetermined significance was first described in 1978 and is a precancerous condition affecting approximately 3% of people above age 50 years [1]. Annually, this condition may progress to multiple myeloma in 1% of these individuals. A variant of monoclonal gammopathy of undetermined significance is asymptomatic or smoldering plasma cell myeloma, in which the diagnostic criteria for multiple myeloma are present but no related organ damage is observed. Circulating micro RNAs may be a potential biomarker for distinguishing healthy people from those with multiple myeloma or related conditions [2]. The risk of malignant transformation of monoclonal gammopathy of undetermined significance into multiple myeloma is higher in females than in males. It is also higher in individuals with IgA paraprotein compared to individuals with IgG paraprotein [3].

The paraprotein can be an intact immunoglobulin, only light chains (light-chain myeloma, light-chain deposition disease, or amyloid light-chain amyloidosis), or rarely found only as heavy chains (heavy-chain disease). Paraproteins can be detected in the serum and can also be excreted into the urine. Sometimes, if the paraprotein is only light chain (light-chain disease), then this is detected in the urine alone and not in the serum. The serum may

A. Wahed and A. Dasgupta: Hematology and Coagulation. DOI: http://dx.doi.org/10.1016/B978-0-12-800241-4.00007-3

paradoxically exhibit only hypogammaglobulinemia. It is important to note that the presence of paraprotein in serum, urine, or both indicates monoclonal gammopathy and not necessarily the presence of multiple myeloma in a patient. Multiple myeloma is one of the causes of monoclonal gammopathy. Transient monoclonal gammopathy may be observed in an immunocompromised patient with infection due to an opportunistic pathogen such as cytomegalovirus [4]. Monoclonal gammopathy is usually observed in patients above age 50 years and is rare in children. Gerritsen *et al.* studied 4000 pediatric patients during a 10-year period and observed monoclonal gammopathy in only 155 children, but such gammopathies were found most frequently in patients with primary and secondary immunodeficiency, hematological malignancies, autoimmune disease, and severe aplastic anemia. Follow-up analysis revealed that most of these monoclonal gammopathies were transient [5].

7.2 DIAGNOSTIC APPROACH TO MONOCLONAL GAMMOPATHY USING ELECTROPHORESIS

Serum protein electrophoresis, urine electrophoresis, serum immunofixation, and urine immunofixation are all performed primarily to investigate suspicion of monoclonal gammopathy. Agarose gel electrophoresis and capillary electrophoresis are the principal methods employed in screening for paraproteins. Both methods are applicable for both serum and urine specimens. After a paraprotein is detected, confirmation and the isotyping of paraprotein, which is usually achieved by immunofixation, are essential. For urine immunofixation, the best practice is to utilize a 24-hr urine specimen that has been concentrated; this technique allows for detection of even a faint band.

7.2.1 Serum Protein Electrophoresis

Serum protein electrophoresis is an inexpensive, easy-to-perform screening procedure for initial identification of monoclonal bands. Monoclonal bands are usually seen in the γ zone but may be seen in proximity of the β band or, rarely, in the α_2 region. Blood is collected in a tube with clot activator. After separation from blood components, serum is placed on paper treated with agarose gel, followed by exposure to an electric current in the presence of a buffer solution (electrophoretic cell). Various serum proteins are then separated based on charge. After a predetermined exposure time to an electric field, the paper is removed, dried, and placed on a fixative to prevent further diffusion of specimen components, followed by staining to visualize various protein bands. Coomassie brilliant blue is a common staining agent used to visualize bands in serum protein electrophoresis. Then using a densitometer, each fraction is quantitated. The serum protein components are separated into five major fractions:

- Albumin
- α_1 Globulins (α_1 zone)
- α_2 Globulins (α_2 zone)
- β Globulins (β zone often splits into β_1 and β_2 bands)
- γ Globulins (γ zone).

Albumin and globulins are two major fractions of the electrophoresis pattern. Albumin, the largest band, lies closest to the positive electrode (anode) and has a molecular weight of approximately 67 kDa. Reduced intensity of this band is observed in inflammation, liver dysfunction, uremia, nephrotic syndrome, and other conditions that lead to hypoalbuminemia, such as critical illness and pregnancy. A smear observed in front of the albumin band may be due to hyperbilirubinemia or the presence of certain drugs. A band in front of the albumin band may be due to prealbumin (a carrier for thyroxine and vitamin A), which is commonly seen in cerebrospinal fluid specimens or serum specimens in patients with malnutrition. Two, rather than one, albumin bands may represent bisalbuminemia. This is a familial abnormality with no clinical significance. Analbuminemia is a genetically inherited metabolic disorder first described in 1954, but this disorder is rare, affecting less than 1 in 1 million births. The condition is benign because low albumin levels are compensated for by high levels of non-albumin proteins and circulatory adaptation. Hyperlipidemia is usually observed in these patients. Pseudo-analbuminemia due to the presence of a slow-moving albumin variant appearing in the α_1 region of serum protein electrophoresis has also been reported [6].

Moving toward the negative portion of the gel (cathode), the α zone is the next band after albumin. The α zone can be subdivided into two zones: the α_1 band and the α_2 band. The α_1 band consists mostly of α_1 antitrypsin (AT) (90%), α_1 chymotrypsin, and thyroid binding globulin. α_1 AT is an acute phase reactant, and its concentration is increased in inflammation and other conditions. The α_1 AT band is decreased in patients with α_1 AT deficiency or, in patients with severe liver disease, with decreased production of globulin. At the leading edge of this band, a haze due to high-density lipoprotein (HDL) may be observed, although different stains are used (Sudan Red 7B or Oil Red O) for lipoprotein analysis using electrophoresis. The α_2 band consists of α_2-macroglobulin, haptoglobin, and ceruloplasmin. Because both haptoglobin and ceruloplasmin are acute phase reactants, this band is increased in inflammatory states. α_2-Macroglobulin is increased in nephrotic syndrome and cirrhosis of liver.

The β zone may consist of two bands, β_1 and β_2. β_1 is mostly composed of transferrin and low density lipoprotein (LDL). An increased β_1 band is observed in iron deficiency anemia due to an increased level of free transferrin. This band may also be elevated in pregnant women. Very-low-density lipoprotein (VLDL) usually appears in the pre-β zone. The β_2 band is mostly composed of complement proteins. If two bands are observed in the β_2 region, this implies electrophoresis of the plasma specimen (fibrinogen band) instead of the serum specimen or IgA paraprotein.

Much of the clinical interest in serum protein electrophoresis is focused on the γ zone because immunoglobulins mostly migrate to this region. Usually, the C-reactive protein band is found between the β region and the γ region. Serum protein electrophoresis is most commonly ordered when multiple myeloma is suspected and observation of a monoclonal band (paraprotein) indicates that monoclonal gammopathy may be present in the patient. If a monoclonal band or paraprotein is observed in serum protein electrophoresis, the following steps are performed:

- The monoclonal band is measured quantitatively using densitometric scan of the gel.
- Serum and/or urine immunofixation is conducted to confirm the presence of the paraprotein as well as determine the isotype of the paraprotein.
- A serum light-chain assay is conducted or recommended to the ordering clinician.

Monoclonal gammopathy can be due to various underlying diseases, including multiple myeloma. In approximately 5% of cases, two paraproteins may be detected. This is referred to as biclonal gammopathy. A patient may also have nonsecretory myeloma, as in the case of a plasma cell neoplasm in which the clonal cells are either not producing or not secreting M proteins. The most commonly observed paraprotein is IgG, followed by IgA light chain and, rarely, IgD. When a monoclonal band is identified using serum protein electrophoresis, serum immunofixation and 24-hr urine immunofixation is typically recommended. There are situations in which a band may be apparent but it is not a monoclonal band. Examples include the following:

- Fibrinogen is seen as a discrete band when electrophoresis is performed on plasma instead of serum specimen. This fibrinogen band is seen between the β and γ regions. If the electrophoresis is repeated after the addition of thrombin, this band should disappear. In addition, immunofixation study should be negative.
- Intravascular hemolysis results in the release of free hemoglobin in to the circulation, which binds to haptoglobin. The hemoglobin–haptoglobin

complex may appear as a large band in the α_2 area. Serum immunofixation studies should be negative in such cases.

- In patients with iron deficiency anemia, concentrations of transferrin may be high, which may result in a band in the β region. Again, immunofixation should be negative.
- Patients with nephrotic syndrome usually show low albumin and total protein, but this condition may also produce increased α_2 and β fractions. Bands in either of these regions may mimic a monoclonal band.
- When performing gel electrophoresis, a band may be visible at the point of application. Typically, this band is present in all samples performed at the same time using the same agarose gel support material.

Common problems associated with the interpretation of serum protein electrophoresis are summarized in Box 7.1. A low concentration of a paraprotein may not be detected by serum electrophoresis. There are also situations in which a false-negative interpretation could be made on serum electrophoresis, including the following:

- A clear band is not seen in cases of α heavy-chain disease (HCD). This is presumably due to the tendency of these chains to polymerize or due to their high carbohydrate content. HCDs are rare B cell lymphoproliferative neoplasms characterized by the production of a monoclonal component consisting of monoclonal immunoglobulin heavy chain without associated light chain.
- In μ-HCD, a localized band is found in only 40% of cases. Panhypogammaglobulinemia is a prominent feature in such patients.

BOX 7.1 COMMON PROBLEMS ASSOCIATED WITH SERUM PROTEIN ELECTROPHORESIS

- Serum protein electrophoresis performed using plasma instead of serum produces an additional distinct band between the β and γ zones due to fibrinogen, but such a band is absent in subsequent immunofixation study.
- A band may be seen at the point of application. Typically, this band is present in all samples performed at the same time.
- If the concentration of transferrin is high (e.g., iron deficiency), a strong band in the β region is observed.

- In nephrotic syndrome, prominent bands may be seen in α_2 and β regions that are not due to monoclonal proteins.
- Hemoglobin—haptoglobin complexes (seen in intravascular hemolysis) may produce a band in the α_2 region.
- Paraproteins may form dimers, pentamers, polymers, or aggregates with each other, resulting in a broad smear rather than a distinct band.
- In light-chain myeloma, light chains are rapidly excreted in the urine and no corresponding band may be present in serum protein electrophoresis.

- In occasional cases of γ-HCD, a localized band may not be seen.
- When paraproteins form dimers, pentamers, polymers, or aggregates with each other, or form complexes with other plasma components, a broad smear may be visible instead of a distinct band.
- Some patients may produce only light chains, which are rapidly excreted in the urine, and no distinct band may be present in the serum protein electrophoresis. Urine protein electrophoresis is more appropriate for diagnosis of light-chain disease. When light chains cause nephropathy, resulting in renal insufficiency, excretion of the light chains is hampered and a band may be seen in serum electrophoresis.
- In some patients with IgD myeloma, the paraprotein band may be very faint.

Hypogammaglobulinemia may be congenital or acquired. The acquired causes include multiple myeloma and primary amyloidosis. Panhypogammaglobulinemia can occur in approximately 10% of cases of multiple myeloma. Most of the patients affected have Bence−Jones protein in the urine but lack intact immunoglobulins in the serum. Bence−Jones proteins are monoclonal free κ or λ light chains in the urine. Detection of Bence−Jones protein may be suggestive of multiple myeloma or Waldenström macroglobulinemia. Panhypogammaglobulinemia can also be seen in 20% of cases of primary amyloidosis. It is important to recommend urine immunofixation studies when panhypogammaglobulinemia is present in serum protein electrophoresis.

Although monoclonal gammopathy is the major reason for serum protein electrophoresis, polyclonal gammopathy may also be observed in some patients. Monoclonal gammopathies are associated with a clonal process that is malignant or potentially malignant. However, polyclonal gammopathy, in which there is a nonspecific increase in γ-globulins, may not be associated with malignancies. Many conditions may lead to polyclonal gammopathies. Serum protein electrophoresis may also exhibit changes that imply specific underlying clinical conditions other than monoclonal gammopathy. Common features of serum protein electrophoresis in various disease states other than monoclonal gammopathy include the following:

- Inflammation: Increased intensity of α_1 and α_2 with a sharp leading edge of α_1 may be observed; however, with chronic inflammation, the albumin band may be decreased with increased γ zone due to polyclonal gammopathy.
- Nephrotic syndrome: In nephrotic syndrome, the albumin band is decreased due to hypoalbuminemia. In addition, the α_2 band may be more distinct.

- Cirrhosis or chronic liver disease: A low albumin band due to significant hypoalbuminemia with a prominent β_2 band and beta-gamma bridging are characteristic features of liver cirrhosis or chronic liver disease. In addition, polyclonal hypergammaglobulinemia is observed.

7.2.2 Urine Electrophoresis

Urine protein electrophoresis is analogous to serum protein electrophoresis and is used to detect monoclonal proteins in the urine. Ideally, it should be performed on a 24-hr urine sample (concentrated 50−100 times). Molecules less than 15 kDa are filtered through a glomerular filtration process and are excreted freely into urine. In contrast, only selected molecules with a molecular weight between 16 and 69 kDa can be filtered by the kidney and may appear in the urine. Albumin is approximately 67 kDa. Therefore, trace albumin in urine is physiological.

The molecular weight of the protein, the concentration of the protein in the blood, charge, and hydrostatic pressure all regulate passage of a protein through the glomerular filtration process.

Proteins that pass through glomerular filtration include albumin, α_1 acid glycoprotein (orosomucoid), α_1 microglobulin, β_2 microglobulin, retinol binding protein, and trace amounts of γ globulins. However, 90% of these are reabsorbed, and only a small amount may be excreted in the urine. Normally, total urinary protein is less than 150 mg/24 hr and consists of mostly albumin and Tamm−Horsfall protein (secreted from the ascending limb of the loop of Henle). The extent of proteinuria can be assessed by quantifying the amount of proteinuria as well as expressing it as protein-to-creatinine ratio. The normal protein-to-creatinine ratio is less than 0.5 in children 6 months to 2 years of age, less than 0.25 in children more than 2 years of age, and less than 0.2 in adults.

Proteinuria with minor injury (typically only albumin is lost in urine) can be related to vigorous physical exercise, congestive heart failure, pregnancy, alcohol abuse, or hyperthermia. Overflow proteinuria can be seen in patients with myeloma or massive hemolysis of crush injury (myoglobin in urine). In addition, β_2 microglobulin, eosinophil-derived neurotoxin, and lysozyme can produce bands in urine electrophoresis. Therefore, immunofixation studies are required to document true M proteins and rule out the presence of other proteins in urine electrophoresis.

Proteinuria can be classified as glomerular, tubular, or combined proteinuria. Glomerular proteinuria can be subclassified as selective glomerular proteinuria (urine will have albumin and transferrin bands) or nonselective glomerular proteinuria (all different types of proteins will be present in urine).

In glomerular proteinuria, albumin is always the dominant protein present. In tubular proteinuria, albumin is a minor component. The presence of α_1 microglobulin and β_2 microglobulin is an indicator of tubular damage.

7.2.3 Immunofixation Studies

In immunofixation, electrophoresis of one specimen from a patient suspected of monoclonal gammopathy is performed using five separate lanes. Then, each sample is overlaid with different monoclonal antibodies: anti-γ (to detect γ heavy chain), anti-μ (to detect μ heavy chain), anti-α (to detect α heavy chain), anti-κ (to detect κ light chain), and anti-λ (to detect λ light chain). Antigen antibody reaction should take place. After washing to remove unbound antibodies, the gel paper is stained, which allows identification of specific isotype of the monoclonal protein. A normal serum protein electrophoresis does not exclude diagnosis of myeloma because approximately 11% of myeloma patients may have normal serum protein electrophoresis. Therefore, serum and urine immunofixation studies should be performed regardless of serum electrophoresis results if clinical suspicion is high. It is also important to note that an M band or paraprotein in serum protein electrophoresis may not be a true band unless it is identified using serum or urine immunofixation because these tests are more sensitive than serum protein electrophoresis. In addition, the immunofixation technique can also determine the particular isotype of the monoclonal protein. However, the immunofixation technique cannot be used to estimate the quantity of the M protein. In contrast, serum protein electrophoresis is capable of estimating the concentration of an M protein.

In multiple myeloma, sometimes only free light chains are produced. The concentration of the light chains in serum may be so low that these light chains remain undetected using serum protein electrophoresis and even serum immunofixation. In such cases, immunofixation on a 24-hr urine sample is useful. Another available test is detection of serum free light chains by immunoassay, which allows for calculating the ratio of κ to λ free light chains. This test is more sensitive than urine immunofixation.

One source of possible error in urine immunofixation studies is the "stepladder" pattern. Here, multiple bands are seen in the κ (more often) or λ lanes and are indicative of polyclonal spillage rather than monoclonal spillage into the urine. During urine immunofixation, five or six faint, regular diffuse bands with hazy background staining between bands may be seen. This is more often seen in the κ lane than in the λ lane. This is referred to as the stepladder pattern and is a feature of polyclonal hypergammaglobulinemia with spillage into the urine.

7.2.4 Capillary Zone Electrophoresis

Capillary zone electrophoresis is an alternative method of performing serum protein electrophoresis. Protein stains are not required, and a point of application is not observed. It is considered to be faster and more sensitive compared to agarose gel electrophoresis, in which a classic case of monoclonal gammopathy produces a peak, typically in the γ zone. However, subtle changes in the γ zones may also represent underlying monoclonal gammopathy. Interpretation can be subjective, and a relatively high percentage of cases may be referred for ancillary studies such as immunofixation depending on the preference of the pathologist interpreting the result. However, disregarding a subtle change in capillary zone electrophoresis may potentially result in missing a case.

7.2.5 Free Light-chain Assay

Patients with monoclonal gammopathy may have negative serum protein electrophoresis and serum immunofixation studies. This may be due to very low levels of paraproteins and light-chain gammopathy in which the light chains are very rapidly cleared from the serum by the kidneys. Because of this, urine electrophoresis and urine immunofixation is part of the workup for cases in which monoclonal gammopathy is a clinical consideration. Urine electrophoresis and urine immunofixation studies are also performed to document the amount (if any) of potentially nephrotoxic light chains being excreted in the urine in a case of monoclonal gammopathy.

Quantitative serum assays for κ and λ free light-chain disease have increased the sensitivity of serum testing strategies for identifying monoclonal gammopathies, especially the light-chain diseases. Cases that may appear as nonsecretory myeloma can actually be cases of light-chain myeloma. Free light-chain assays allow disease monitoring as well as provide prognostic information for monoclonal gammopathy of undetermined significance (MGUS), smoldering myeloma.

The rapid clearance of light chains by the kidney is reduced in renal failure. Levels may be 20−30 times higher than normal in end stage renal disease. In addition, the κ:λ ratio may be as high as 3:1 in renal failure (normal, 0.26−1.65). Therefore, patients with renal failure may be misdiagnosed as having κ light-chain monoclonal gammopathy. If a patient has λ light-chain monoclonal gammopathy, with the relative increase in κ light chain in renal failure, the ratio may become normal. Thus, a case of λ light-chain monoclonal gammopathy may be missed.

7.2.6   Paraprotein Interference in Clinical Laboratory Tests

Interference of paraprotein can produce both false-positive and false-negative test results, depending on the analyte and the analyzer. However, the magnitude of interference may not correlate with the amount of paraprotein present in serum. The most common causes of interference include falsely low HDL cholesterol, falsely high bilirubin, and altered values of inorganic phosphate. Other tests in which altered results may occur include LDL cholesterol, C-reactive protein, creatinine, glucose, urea nitrogen, inorganic calcium, and blood count. There is a poor correlation between the concentration or type of paraprotein and the likelihood of interference [7].

7.3   PLASMA CELL NEOPLASM

Plasma cell neoplasm is characterized by the clonal expansion of terminally differentiated B lymphoid cells that have undergone somatic hypermutation that results in the production of monoclonal immunoglobulin protein. Plasma cell neoplasm may cause benign (nonmalignant) or malignant disease and encompass a spectrum of diseases, including monoclonal gammopathy, plasma cell myeloma, solitary plasmacytoma of bone marrow or soft tissue, primary amyloidosis, and monoclonal immunoglobulin disposition disease. The following is a classification of various plasma cell neoplastic disorders (see also Box 7.2):

- Monoclonal gammopathy of undetermined significance (MGUS), a disorder of plasma cells, may be benign but in some patients may progress to a malignant disorder. The diagnostic criteria are M protein concentration less than 30 g/L, plasma cells in bone marrow less than 10%, and the absence of organ or tissue damage (CRAB: hypercalcemia, renal dysfunction, anemia, and bone lesions). Risk of progression is approximately 1% per year. There are two subtypes: IgM MGUS may progress to lymphoplasmacytic lymphoma/Waldenström macroglobulinemia (LPL/WM), and non-IgM MGUS may progress to myeloma.
- POEMS (polyneuropathy, organomegaly, endocrinopathy, myeloma, and skin changes) syndrome is also known as osteosclerotic myeloma.

BOX 7.2 DISEASES RELATED TO PLASMA CELL DYSCRASIA

- Myeloma (solitary with bone or soft tissue involvement, or multiple myeloma)
- Monoclonal gammopathy of undetermined significance
- Amyloidosis
- Light-chain deposition disease (rare disorder)
- Heavy-chain deposition disease (rare disorder)

- Plasmacytoma: Solitary plasmacytoma of bone and extraosseous plasmacytoma (most often in upper respiratory tract).
- Monoclonal immunoglobulin deposition disease: In this disease, tissue deposition of monoclonal Ig, Ig chain, or fragment takes place. In Ig-related primary amyloidosis, Congo red staining of amyloid protein in tissue can be used for diagnostic purposes. Under electron microscopy, amyloid deposits appear to be composed of linear, nonbranching, aggregated fibrils that are 7.5–10 nm thick and of indefinite length arranged in a loose meshwork. Monoclonal immunoglobulin deposition disease can be related to systemic light-chain or heavy-chain deposition disease.
- Plasma cell myeloma: In classical or symptomatic plasma cell myeloma, M protein is found in serum or urine (typically >30 g/L of IgG or >25 g/L of IgA or >1 g/24 hr of urine light chain), bone marrow shows clonal plasma cells (usually >10% of nucleated cells in bone marrow), or there is evidence of plasmacytoma. In addition, there is evidence of organ or tissue damage (CRAB). Other variants include (1) smoldering—criteria met but no CRAB; (2) nonsecretory (85% nonsecretors have Igs in the cytoplasm but are not secreting them, and 15% nonproducers); and (3) plasma cell leukemia (>2000 plasma cells/mm^3 or 20% by differential count).

7.3.1 Morphology of Plasma Cells in Myeloma

In myeloma, plasma cells are found as interstitial clusters or focal nodules or in diffuse sheets in the bone marrow. The cells may appear mature or immature, plasmablastic or pleomorphic. The following descriptive terms have been used, but none is diagnostic for myeloma:

- Mott cells: Plasma cells with grape-like clusters of cytoplasmic inclusions
- Russell bodies: Cherry red cytoplasmic inclusions found in plasma cells
- Dutcher bodies: Intranuclear inclusions found in plasma cells
- Flame-shaped cells: Plasma cells with vermilion staining glycogen-rich IgA
- Thesaurocytes: Plasma cells with ground glass cytoplasm.

7.3.2 Immunophenotype of Neoplastic Plasma Cells

Plasma cells (normal and neoplastic) express CD38(bright) and CD138. CD138 is more specific but less sensitive. Normal peripheral blood plasma cells are CD45$^+$. In bone marrow, there are two subsets of plasma cells: one major subset positive for CD45 and a smaller negative one. Aberrant CD56 expression is identified in most patients with myeloma. Bright CD38 expression with coexpression of CD56 is used to detect abnormal populations of plasma cells by flow cytometry. Abnormal plasma cells are also CD19$^-$, in contrast to normal plasma cells, which are CD19$^+$. Normal and abnormal plasma cells are negative for CD20.

The abnormal population of plasma cells once identified should also exhibit cytoplasmic light-chain restriction.

7.4 CYTOGENETICS IN MYELOMA DIAGNOSIS

Conventional cytogenetics documents abnormality in approximately 30% of cases; with fluorescence *in situ* hybridization, this increases to 90%. 14q32 translocations involving the heavy-chain locus are the most common abnormality.

Genetic abnormalities associated with an unfavorable prognosis are hypodiploidy, del13, del17p, t(4;14), and t(14;16). Genetic abnormalities associated with a favorable prognosis are hyperdiploidy and t(11;14). Therefore, cytogenetics and molecular profiling have value in determining the prognosis of multiple myeloma [8].

It is interesting to note that high serum β_2 microglobulin and low serum albumin are also associated with a less favorable prognosis.

KEY POINTS

- Serum protein electrophoresis, urine electrophoresis, serum immunofixation, and urine immunofixation are all performed primarily to investigate suspicion of monoclonal gammopathy in patients. Monoclonal gammopathy is present when a monoclonal protein (also called paraprotein or M protein) is identified in a patient's serum, urine, or both. The paraprotein can be an intact immunoglobulin, only light chains (light-chain myeloma, light-chain deposition disease, or amyloid light-chain amyloidosis), or, rarely, found only as heavy chains (heavy-chain disease).
- Agarose gel electrophoresis and capillary electrophoresis are two principal methods employed in screening for paraproteins. Both methods are applicable for both serum and urine specimens. Paraproteins are seen usually in the γ region of the electrophoresis but also may be present in the β region or, rarely, the α_2 region.
- Once a paraprotein is detected, confirmation and the isotyping of paraprotein is essential, which is usually achieved by immunofixation. In approximately 5% of cases, two paraproteins may be detected. This is referred to as biclonal gammopathy. A patient may also have

nonsecretory myeloma, as in the case of a plasma cell neoplasm in which the clonal cells are either not producing or not secreting M proteins. The most commonly observed paraprotein is IgG, followed by IgA light chain and, rarely, IgD. A normal serum protein electrophoresis does not exclude diagnosis of myeloma because approximately 11% of myeloma patients may have normal serum protein electrophoresis. Therefore, serum and urine immunofixation studies should be performed, regardless of serum electrophoresis results, if clinical suspicion is high.

- Other components of serum protein electrophoresis include albumin, α_1 globulins (α_1 zone), α_2 globulins (α_2 zone), β globulins (β zone is often divided into β_1 and β_2 bands), and γ globulins (γ zone).
- Reduced intensity of the albumin band is observed in inflammation, liver dysfunction, uremia, nephrotic syndrome, and other conditions that lead to hypoalbuminemia. A smear observed in front of the albumin band may be due to hyperbilirubinemia or the presence of certain drugs. A band in front of the albumin band may be due to prealbumin (a carrier for thyroxine and vitamin A), which is commonly seen in cerebrospinal fluid specimens or serum specimens in patients with malnutrition. Two, rather than one, albumin bands may represent bisalbuminemia. This is a familial abnormality with no clinical significance.
- The α_1 band mostly consists of α_1 AT (90%), α_1 chymotrypsin, and thyroid binding globulin. α_1 AT is an acute phase reactant, and its concentration is increased in inflammation and other conditions. The α_1 AT band is decreased in patients with α_1 AT deficiency or, in patients with severe liver disease, with decreased production of globulin. At the leading edge of this band, a haze due to HDL may be observed. The α_2 band consists of α_2 macroglobulin, haptoglobin, and ceruloplasmin. Because both haptoglobin and ceruloplasmin are acute phase reactants, this band is increased in inflammatory states. α_2 Macroglobulin is increased in nephrotic syndrome and cirrhosis of the liver.
- The β zone may consist of two bands, β_1 and β_2. β_1 is mostly composed of transferrin and LDL. An increased β_1 band is observed in iron deficiency anemia due to an increased level of free transferrin. This band may also be elevated in pregnant women. VLDL usually appears in the pre-β zone. The β_2 band is mostly composed of complement proteins. If two bands are observed in the β_2 region, it implies electrophoresis of plasma specimen (fibrinogen band) instead of serum specimen or IgA paraprotein.
- There are situations in which a band may be apparent but, in reality, it is not a monoclonal band. For example, fibrinogen is seen as a discrete band between β and γ regions when electrophoresis is performed on

plasma instead of serum specimen. If the electrophoresis is repeated after the addition of thrombin, this band should disappear. In addition, immunofixation study should be negative. Intravascular hemolysis results in the release of free hemoglobin in to the circulation, which binds to haptoglobin. The hemoglobin−haptoglobin complex may appear as a large band in the α_2 area. Serum immunofixation studies should be negative in such cases.

- In patients with iron deficiency anemia, concentrations of transferrin may be high, which may result in a band in the β region. Again, immunofixation should be negative.
- Patients with nephrotic syndrome usually show low albumin and total protein, but this condition may also produce increased α_2 and β fractions. Bands in either of these regions may mimic a monoclonal band.
- Hypogammaglobulinemia may be congenital or acquired. Acquired causes include multiple myeloma and primary amyloidosis. Panhypogammaglobulinemia can occur in approximately 10% of cases of multiple myeloma. Most of these patients have a Bence−Jones protein in the urine but lack intact immunoglobulins in the serum.
- Common features of serum protein electrophoresis in various disease states other than monoclonal gammopathy include the following:
  - Inflammation: Increased intensity of α_1 and α_2 with a sharp leading edge of α_1 may be observed; however, with chronic inflammation, the albumin band may be decreased with increased γ zone due to polyclonal gammopathy.
  - Nephrotic syndrome: In nephrotic syndrome, the albumin band is decreased due to hypoalbuminemia. In addition, the α_2 band may be more distinct.
  - Cirrhosis or chronic liver disease: A low albumin band due to significant hypoalbuminemia with a prominent β_2 band and $\beta\gamma$ bridging are characteristic features of liver cirrhosis or chronic liver disease. In addition, polyclonal hypergammaglobulinemia is observed.
- Proteinuria can be classified as glomerular, tubular, or combined proteinuria. Glomerular proteinuria can be subclassified as selective glomerular proteinuria (urine will have albumin and transferrin bands) or nonselective glomerular proteinuria (urine will have the presence of all different types of proteins). In glomerular proteinuria, the dominant protein present is always albumin. Albumin is a minor component in tubular proteinuria. The presence of α_1 microglobulin and β_2 microglobulin is an indicator of tubular damage.
- One source of possible error in urine immunofixation studies is the "stepladder" pattern. Here, multiple bands are seen in the κ (more often)

or λ lanes and are indicative of polyclonal spillage rather than monoclonal spillage into the urine. During urine immunofixation, five or six faint, regular diffuse bands with hazy background staining between bands may be seen. This is more often seen in the κ lane than in the λ lane. This is referred to as the stepladder pattern and is a feature of polyclonal hypergammaglobulinemia with spillage into the urine.

- The rapid clearance of light chains by the kidney is reduced in renal failure. Levels may be 20−30 times higher than normal in end stage renal disease. In addition, the $\kappa{:}\lambda$ ratio may be as high as 3:1 in renal failure (normal, 0.26−1.65). Therefore, patients with renal failure may be misdiagnosed as having κ light-chain monoclonal gammopathy. If a patient has λ light-chain monoclonal gammopathy, with the relative increase in κ light chain in renal failure, the ratio may become normal. Thus, a case of λ light-chain monoclonal gammopathy may be missed.
- Plasma cell dyscrasias are MGUS, POEMS, plasmacytoma, Ig deposition diseases, and plasma cell myeloma.
- Plasma cells (normal and neoplastic) express CD38(bright) and CD138. CD138 is more specific but less sensitive. Normal peripheral blood plasma cells are $CD45^{+}$. In bone marrow, there are two subsets of plasma cells: one major subset positive for CD45 and a smaller negative one. Aberrant CD56 expression is identified in most patients with myeloma. Bright CD38 expression with coexpression of CD56 is used to detect abnormal populations of plasma cells by flow cytometry. Abnormal plasma cells are also $CD19^{-}$, in contrast to normal plasma cells, which are $CD19^{+}$. Normal and abnormal plasma cells are negative for CD20.

References

[1] Agarwal A, Ghobrial IM. Monoclonal gammopathy of undetermined significance and smoldering multiple myeloma: a review of current understanding of epidemiology, biology, risk stratification, and management of myeloma precursor disease. Clin Cancer Res 2013;19:985−94.

[2] Jones CI, Zabolotskaya MV, King AJ, Stewart HJ, et al. Identification of circulating micro RNAs as diagnostic biomarkers for use in multiple myeloma. Br J Cancer 2012;107:1987−96.

[3] Gregersen H, Mellemkjaer L, Ibsen JS, Dahlerup JF, et al. The impact of M component type and immunoglobulin concentration on the risk of malignant transformation in patients with monoclonal gammopathy of undetermined significance. Haematologica 2001; 86:1172−9.

[4] Vodopick H, Chaskes SJ, Solomon A, Stewart JA. Transient monoclonal gammopathy associated with cytomegalovirus infection. Blood 1974;44:189−95.

[5] Gerritsen E, Vossen J, van Tol M, Jol-van der Zijde C, et al. Monoclonal gammopathies in children. J Clin Immunol 1999;9:296–305.

[6] Gras J, Padros R, Marti I, Gomez-Acha JA. Pseudo-analbuminemia due to the presence of a slow albumin variant moving into the α_1 zone. Clin Chim Acta 1980;104:125–8.

[7] Roy V. Artifactual laboratory abnormalities in patients with paraproteinemia. South Med J 2009;102:167–70.

[8] Sawyer JR. The prognostic significance of cytogenetics and molecular profiling in multiple myeloma. Cancer Genet 2011;204:3–12.

CHAPTER 8

Application of Flow Cytometry in the Diagnosis of Hematological Disorders

8.1 INTRODUCTION

Flow cytometry (cyto = single cell; metry = measurement) is a laser-based biophysical method capable of providing both qualitative and quantitative measurements of multiple characteristics of a single cell or any other particle by measuring a particle's optical and fluorescence characteristics. Flow cytometry can provide information not only on cell size but also on cytoplasmic complexity, DNA or RNA content, and a wide range of membrane-bound and intracellular proteins. Fluorescent dye can bind or intercalate with DNA or RNA, and antibodies conjugated with fluorescent dyes can bind specific proteins on the cell membrane or inside the cell. When labeled cells are passed through a light source, fluorescent molecules are excited to a higher energy state, and upon returning to their ground state, the fluorochromes emit light that is measured by the flow cytometer. The use of several fluorochromes, each with similar excitation wavelengths but different emission wavelengths (emitted light that is measured; emitted light is at a higher wavelength than excitation wavelength), allows measuring various cells and/or cell parameters simultaneously. Commonly used dyes include propidium iodide, phycoerythrin, and fluorescein, but many other dyes are commercially available, including tandem dyes with internal fluorescence energy transfer capability [1]. The first impedance-based flow cytometric device was discovered by Wallace H. Coulter, and the first fluorescence-based flow cytometer was developed by Wolfgang Gohde in Germany. In addition to hematology, flow cytometry is applied for diagnosis in immunology (histocompatibility cross-matching), oncology, blood banking, and diagnosis of certain genetic disorders. This technique is also used outside medicine for research, such as in marine biology.

The cluster of differentiation (also known as cluster of designation) is abbreviated as CD (CD nomenclature) and is a protocol for identification of cell surface molecules providing targets for immunophenotyping of cells.

CONTENTS

133

A. Wahed and A. Dasgupta: Hematology and Coagulation. DOI: http://dx.doi.org/10.1016/B978-0-12-800241-4.00008-5

These cell surface molecules are glycoproteins with complex functions, including acting as receptors and cell signaling. The precise functions of many cell surface glycoproteins are still unknown. The CD nomenclature was established by the First International Workshop and Conference on Human Leukocyte Differentiation Antigen (HLDA) in 1984 for classification of many monoclonal antibodies generated by different research laboratories world-wide against epitopes of these surface glycoprotein molecules on leukocytes. Since then, its application has been extended to many other cell types, and more than 300 unique clusters or subclusters have been identified. The proposed surface molecule is assigned a unique number once two monoclonal antibodies are shown to bind with the surface antigen. The 9th HLDA International Congress was the most recent congress, held in 2010 in Barcelona, Spain, where 64 new antibodies were tested [2].

The following questions should be considered before drawing conclusions based on flow cytometry:

- Are the cells that are being studied viable?
- Are the negative controls within acceptable limits?
- Are you sure that the cells that are considered as debris (mainly red cells) are actually so?
- For acute leukemia, are the cells gated blasts?

These important questions should be addressed keeping in mind the following points:

- Dead cells can be identified by using a specific dye. For flow cytometry analysis, one such dye that is used is 7-amino actinomycin D (7-AAD).
- An isotype control (i.e., where an antibody is used that has the same Ig isotype as the test antibody but a different specificity that is known to be irrelevant to the sample being analyzed) is used to determine whether fluorescence that is observed is due to nonspecific binding of the fluorescent antibody. This can be checked with CD71, a marker for erythroid cells. The cells being analyzed should be negative for CD71.
- Typically, a plot of side scatter versus CD45 is used. Blasts should have low side scatter (side scatter represents complexity/granularity and blasts are cells with a large nucleus and thus a low complex cell population) and weak CD45. Lymphocytes have brighter CD45 expression than blasts.

8.2 FLOW CYTOMETRY AND MATURE B CELL LYMPHOID NEOPLASMS

Mature B cell lymphoid neoplasms are distinguished from non-neoplastic cells by immunoglobulin light-chain class restriction and aberrant antigen expression. Mature B lymphoid cells exhibit surface immunoglobulins. If the cells are neoplastic, then they express either κ or λ classes of light chains. This is termed light-chain restriction. Light-chain restriction can be seen in the following:

- Mature B cell lymphoid neoplasms
- Rare reactive B cell populations, such as in the tonsils of children (which have been shown to exhibit λ light-chain restriction) and multicentric Castleman disease
- Some cases of florid follicular hyperplasia.

Therefore, light-chain restriction is not synonymous with neoplasm, and findings of flow cytometry should be interpreted in conjunction with other findings.

Aberrant antigen expression may also be observed—for example, expression of CD5 in B cells. However, CD5 expression in B cells can also be a normal phenomenon and is seen in some B cells in peripheral blood, mantle zone cells in lymph nodes, and a subset of hematogones in bone marrow. Nonetheless, CD5$^+$ B cell lymphomas can be observed in the following:

- Chronic lymphocytic leukemia (CLL)/small lymphocytic lymphoma (SLL): Both diseases are almost identical both morphologically and clinically
- Mantle cell lymphoma
- Miscellaneous: B cell prolymphocytic leukemia (B-PLL; sometimes), diffuse large B cell lymphoma (DLBCL; rarely), marginal zone lymphoma (MZL; rarely), and lymphoplasmacytic lymphoma (LPL; very rarely).

8.2.1　B Cell Markers

There are various markers of B cells, which are described here:

- CD19: This expression is restricted to B cells and is present throughout B cell maturation, from B lymphoblasts to plasma cells. However, neoplastic plasma cells are typically CD19$^-$.
- CD20: Expressed by mature B cells and can also be expressed weakly on a subset of mature T cells. Because CD20 is expressed later during the maturation process of B cells in acute B lymphoblastic leukemia (B-ALL), more CD19$^+$ cells should be observed compared to CD20 cells. Normal plasma cells are negative for CD20, and plasma cells of plasma cell neoplasms are also CD20$^-$; thus, CD20 should not be used to distinguish normal and abnormal plasma cells. Bright CD20 staining can be seen in normal follicular center cells, in cells of follicular

lymphoma, and in hairy cell leukemia. Weak intensity (dim) staining for CD20 can be seen in CLL/SLL. Patients taking rituximab will show lack of staining for CD20.

- CD22: Early B cells have cytoplasmic expression of CD22, and with maturation (similar to CD20), surface expression of CD22 is observed. In addition, similarly to CD20, weak (dim) staining of CD22 is observed in CLL/SLL and bright staining in cases with hairy cell leukemia.
- CD23: Weakly expressed in resting B cells and increased expression with activation. CD23 is positive in CLL/SLL cases.
- CD10: Normally expressed by immature B cells, immature T cells, follicular germinal center cells, and neutrophils. CD10$^+$ B cell neoplasms include B-ALL, follicular lymphoma, Burkitt lymphoma, diffuse large B cell lymphoma, and hairy cell leukemia (uncommon).

8.3 FLOW CYTOMETRY AND MATURE T AND NATURAL KILLER CELL LYMPHOID NEOPLASMS

Mature T and natural killer (NK) cell lymphoid neoplasms can be identified with flow cytometry by detection of aberrant antigen expression. This can be achieved by observing the loss of one or more pan T cell markers (e.g., the loss of CD5 or CD7) or the presence of antigens not normally expressed (e.g., expression of CD15, CD13, CD33, or even CD20; NK cells expressing CD5). For analysis of NK cells by flow cytometry, it is important to note that these cells are negative for surface antigen CD3 but always positive for CD16, CD56, and CD57. In addition, NK cells are also positive for CD2, cCD3, CD7, CD8 (some), and CD26.

Characteristics of T and NK cell markers include the following:

- CD1a: Expressed in immature T cells and therefore positive in T cell cases of acute lymphoblastic leukemia (T-ALL).
- CD2: Expressed in T and NK cells; may be aberrantly expressed in acute myeloid leukemia (AML).
- sCD3: Expressed in more mature T cells; may be lost and not observed in T cell lymphomas.
- cCD3: Expressed in all T cells and also NK cells.
- CD4: Expressed in a subset of T cells as well as monocytes (in monocytic leukemias, a disproportionate increase in CD4$^+$ cells is observed).
- CD5: Expressed in T cells and some B cells; may be lost in T cell lymphomas and may be aberrantly expressed in some B cell lymphomas.

- CD7: Expressed in T and NK cells; may be lost and not observed in T cell lymphomas.
- CD8: Expressed in T cell subset and some NK cells.
- CD10: Normally expressed by immature B cells, immature T cells, follicular germinal center cells, and neutrophils. Expressed in angioimmunoblastic T cell lymphoma (AITCL).
- CD16: Expressed in NK cells and maturing neutrophils.
- CD26: Expressed in immature and activated T cells, NK cells. Most CD4$^+$ cells are also CD26$^+$. In cutaneous T cell lymphoma/Sézary syndrome, the tumor cells are CD4$^+$ and CD26$^-$.
- CD30: Expressed in activated T and B cells and monocytes; expressed in anaplastic large cell lymphoma (ALCL).
- CD56 and CD57: Expressed in NK cells.

8.3.1 Detection of Clonal or Restricted Populations of T and NK Cells

The enormous diversity of T cell receptor specificities is created by germline variable (V), diversity (D), junctional (J), and constant (C) region genes. Of the many V segments available in the germline configuration, only one is incorporated into each chain of rearranged receptors [3]. Therefore, normal T cells usually show a mixture of cells with variable expression of Vβ family subtypes. In T cell lymphomas, the neoplastic cells will demonstrate restricted Vβ expression, which may be identified by flow cytometry. Clonal T cell receptor gene rearrangement studies by polymerase chain reaction can also be performed. However, false-positive and false-negative results may be observed using both methods. NK cells lack T cell receptor gene expression. Therefore, neither of the previously mentioned tests is feasible. However, flow cytometric analysis of NK receptor expression has been developed to provide evidence of NK cell clonality. Various CD$^+$ T cell lymphomas are listed in Table 8.1.

Table 8.1 T Cell Lymphomas and CD4 and CD8 Positivity

CD4$^+$ T Cell Lymphomas	CD8$^+$ T Cell Lymphomas	CD4$^-$ and CD8$^-$ T Cell Lymphomas
• Peripheral T cell lymphoma, NOS (can be CD4$^+$ and CD8$^+$) • Anaplastic large cell lymphoma (ALCL) • Angioimmunoblastic T cell lymphoma (AITL) • Adult T cell leukemia/lymphoma (can be CD4$^+$ and CD8$^+$) • T cell polymorphocytic leukemia (can be CD4$^+$ and CD8$^+$) • Cutaneous T cell lymphoma/Sézary syndrome	• T cell large granular lymphocyte leukemia • Subcutaneous panniculitis-like T cell lymphoma • Hepatosplenic T cell lymphoma (can be CD4$^-$ and CD8$^-$)	• Enteropathy-associated T cell lymphoma • Hepatosplenic T cell lymphoma

8.4 PLASMA CELL DYSCRASIAS

Plasma cell dyscrasias are a heterogeneous group of disorders caused by the monoclonal proliferation of lymphoplasmacytic cells in the bone marrow. Multiple myeloma is the most serious and prevalent plasma cell dyscrasia, with a median age of onset of 60 years. Symptoms result from lytic bone disease, anemia, renal failure, and immunodeficiency. Monoclonal gammopathy of undetermined significance affecting up to 3.2% of patients above age 50 years is related to plasma cell dyscrasia. Although it is a benign condition, it may progress to cancer in some patients. Most patients with multiple myeloma show evidence of bone marrow plasmacytosis and a monoclonal gammopathy in serum or urine, and lytic bone lesions may be present in up to 60% of patients [4].

Flow cytometry is also useful in the diagnosis of plasma cell dyscrasias. Plasma cells are $CD38^+$ (bright) and $CD138^+$ (although less sensitive). Plasma cells in the blood and tonsils are $CD45^+$, whereas in the bone marrow plasma cells may be either $CD45^+$ positive or $CD45^-$. Normal plasma cells are $CD19^+$ and $CD20^-$. Abnormal plasma cells are $CD19^-$ and $CD20^-$. These cells are typically also $CD56^+$. Up to 20% of cases of neoplastic plasma cell neoplasms are $CD117^+$. If neoplastic plasma cells are negative for CD56, CD27, or CD45, it is thought to impart a worse prognosis.

8.5 FLOW CYTOMETRY AND ACUTE LEUKEMIA

Blasts are identified as having low side scatter and weak CD45, which are present at left of the lymphocyte population in scattergrams, CD45 versus side scatter. Basophils, having lost granules during processing, may also be found in the same area. Myeloblasts should also demonstrate markers of immaturity—CD34 (immature myeloid and lymphoid marker) and CD117 (immature myeloid marker)—pan myeloid markers (myeloperoxidase (MPO), CD13, and CD33), but lack of mature myeloid markers such as CD11b, CD15, and CD16. In addition, lymphoblasts should demonstrate markers of immaturity—CD34 and TdT (terminal deoxynucleotidyl transferase). B lymphoblasts will also lack surface immunoglobulin and show lack of CD20 (which is a more mature B cell marker; more cells will be positive for CD19 than for CD20). These lymphoblasts are also positive for cCD22. T lymphoblasts will show CD1a (immature T cell marker) and lack surface CD3. However, these cells are positive for cCD3.

In AML there may be aberrant expression of lymphoid antigens, and those that are frequently expressed include CD2, CD7, CD19, and CD56. Similarly, in ALL, there may be aberrant expression of one or two myeloid antigens.

8.5.1 Flow Cytometry and Subtypes of Acute Myeloid Leukemia

Flow cytometry is useful in the classification of subtypes of AML. The following are features of subtypes of AML determined by flow cytometry:

- In acute undifferentiated myeloid leukemia (French−American−British classification (FAB), M0), the blasts are positive for CD34 and human leukocyte antigen-DR (HLA-DR). Myeloid-specific antigens such as CD13 and CD33 are positive.
- In AML without maturation (FAB, M1), the flow appearance is similar to that of AML, M0.
- In AML with maturation (FAB, M2), there is a reduced percentage of blasts with evidence of maturation. CD45 side scatter shows a continuum of cells from the myeloblast region to the maturing myeloid cell regions. Markers indicating maturing myeloid cells, such as CD15 and CD11b, will be expressed. Expression of CD19 and, less often, CD56 in cases of M2 is associated with the presence of t8;21.
- In acute promyelocytic leukemia (APL), the abnormal promyelocytes are CD34$^-$; completely or partially negative for HLA-DR, CD11b, and CD15; and positive for CD13 (heterogeneous), CD33 (homogeneous and bright), and CD117. The abnormal promyelocytes lack CD10 and CD16, which are mature myeloid markers. It has been reported that the combination of the absence of CD11b, CD11c, and HLA-DR identifies 100% of APLs [5]. CD2 positivity is a feature of APL.
- Acute myelomonocytic and monocytic leukemia is positive for monocytic markers CD11b, CD14, CD15, CD16, CD36, and CD64. Monocytic cells are positive for CD4. A significant population of cells positive for CD4, with the same cells negative for CD3, indicates that the cells are monocytic. The presence of CD2 correlates with M4 Eo.
- Acute erythroid leukemia is positive for CD235a (glycophorin) and for CD36 in the absence of CD64. CD36 is positive for monocytes, erythroid cells, and megakaryocytes. If the case is not megakaryocytic (i.e., CD41$^-$ and CD61$^-$) and CD64 is negative (i.e., not monocytic), then it favors the diagnosis of erythroid leukemia.
- In acute megakaryocytics leukemia, megakaryoblasts are positive for CD41 and CD61.

The following are markers for acute leukemia:

- CD1a: Expressed in immature T cells and, as expected, positive in T-ALL.
- CD2: Expressed in T and NK cells and may be aberrantly expressed in AML.
- sCD3 (surface antigen CD3): Expressed in more mature T cells; may be lost in T cell lymphomas.

- cCD3 (cytoplasmic antigen CD3): Expressed in all T cells and also NK cells.
- CD4: Expressed in a subset of T cells as well as monocytes (in monocytic leukemias, there is disproportionate increase in $CD4^+$ cells).
- CD5: Expressed in T cells and some B cells; may be lost in T cell lymphomas; may be aberrantly expressed in some B cell lymphomas.
- CD7: Expressed in T and NK cells; may be lost in T cell lymphomas.
- CD8: Expressed in T cell subset and some NK cells.
- CD10: Also known as CALLA (common acute lymphoblastic leukemia antigen); if present, indicates good prognosis.
- CD11b: Expressed only in maturing neutrophils and monocytes.
- CD13: Expressed in neutrophils and monocytes.
- CD14: Expressed only in monocytes.
- CD15: Expressed only in maturing neutrophils and monocytes.
- CD16: Expressed in maturing neutrophils and monocytes but also in NK cells.
- CD19: Expression of this antigen is restricted to B cells, and it is present throughout B cell maturation, from B lymphoblasts to plasma cells. Plasma cell neoplasms are typically $CD19^-$
- CD20: Expressed by mature B cells and can also be expressed weakly on a subset of mature T cells. Normal plasma cells are typically negative for CD20. Because CD20 is expressed later in the maturation process of B cells, B-ALL cases should have more CD19 cells than CD20 cells.
- CD22: B cells during the early maturation state have cytoplasmic expression of CD22, and with maturation, CD22 is expressed as a surface antigen. This is similar to CD20, in which there is dim expression observed in CLL/SLL and bright expression in cases of hairy cell leukemia.
- CD23: Weakly expressed in resting B cells and increased expression with activation.
- CD33: Expressed in neutrophils and monocytes.
- CD34: Expressed in immature B and T cells and myeloblasts.
- CD36: Expressed in monocytes, erythroids, and megakaryocytes.
- CD38: Expressed in myeloid, monocytic cells, erythroid precursors, immature B and T cells, and plasma cells.
- CD41 and CD61: Both expressed in megakaryocytes.
- CD56: Expressed in NK cells.
- CD117: Expressed in immature neutrophils and mast cells.
- CD235a: Expressed in erythroid precursors.
- HLA-DR: Observed in myeloblasts, monocytes, B cells, and activated T cells.
- MPO: Observed in neutrophilic cells.

8.6   FLOW CYTOMETRY AND MYELODYSPLASTIC SYNDROME

Several features of myelodysplastic syndrome (MDS) can be identified by flow cytometry. First, granulocytes are hypogranular, and the normal population of granulocytes as seen in the side scatter versus CD45 is distorted. The hypogranular cells occupy a lower position in the scattergram compared to a normal population. Two other plots are also helpful: CD11b versus CD16 and CD13 versus CD16. Immature granulocytes start by expressing CD16 and subsequently express CD11b, the expression of which becomes increasingly brighter with maturation. Subsequently, CD16 expression becomes very high. Thus, mature neutrophils are positive for both CD11b and CD16. This is the normal pattern of expression. When the CD13 versus CD16 plot is considered, immature cells have a bright expression of CD13 but very little CD16 expression. With maturation, CD13 becomes dim (at the stage of metamyelocyte), but there are still low levels of CD16. Subsequently, both CD13 and CD16 become bright. Therefore, mature neutrophils are positive for both CD13 and CD16. In MDS, these normal patterns may be lost. Another feature of abnormal myeloid differentiation is gain of CD56 in the myeloid cells with lack of CD16. The normal distribution of CD11b versus CD16 is shown in Figure 8.1. Figure 8.2 shows the normal distribution of CD13 versus CD16.

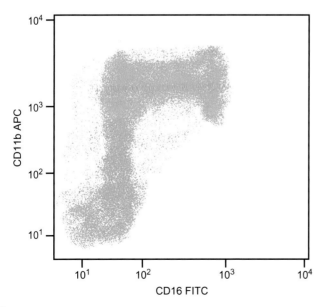

FIGURE 8.1
Normal distribution of CD11b versus CD16.

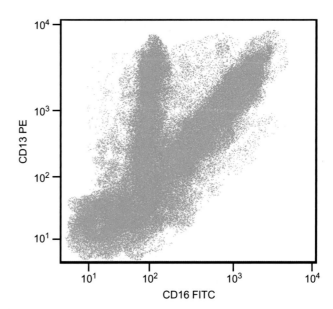

FIGURE 8.2
Normal distribution of CD13 versus CD16.

8.7 FLOW CYTOMETRY AND HEMATOGONES

Hematogones are maturing B cell precursors representing a normal component of bone marrow. Typically, these cells decrease with age and usually constitute 1% or fewer bone marrow cells. However, there may be hyperplasia of hematogones, especially in the setting of regeneration after chemotherapy or stem cell transplantation or in patients with congenital or immune cytopenias. Hematogones may be confused with neoplastic B cells because there is morphologic and immunophenotypic overlap. The earliest hematogones are positive for CD34, CD10, and CD22 but negative for CD20. Subsequently, CD10 is downregulated with progressive gain of CD20. Therefore, in hematogones, expression of CD20 and CD34 is mutually exclusive, and CD34 expression in hematogones is biphasic (either positive or negative).

KEY POINTS

- Mature B cell lymphoid neoplasms are distinguished from non-neoplastic cells by immunoglobulin light-chain class restriction and aberrant antigen expression.

- Light-chain restriction can be seen in mature B cell lymphoid neoplasms, rare reactive B cell populations such as in tonsils of children (which have been shown to exhibit λ light-chain restriction), multicentric Castleman disease, and some cases of florid follicular hyperplasia.
- $CD5^+$ B cell lymphomas can be observed in CLL/SLL, mantle cell lymphoma, and B-PLL (sometimes), DLBCL (rarely), MZL (rarely), and LPL (very rarely).
- $CD10^+$ B cell neoplasms include B-ALL, FL, DLBCL, Burkitt lymphoma, and HCL (uncommon—10% of cases).
- Mature T and NK cell lymphoid neoplasms are identified with flow cytometry by detection of aberrant antigen expression. This can take the form of loss of one or more pan T cell markers (e.g., loss of CD5 or CD7) and the presence of antigens not normally expressed (e.g., CD15, CD13, CD33, or even CD20; NK cells expressing CD5).
- NK cells by flow cytometry are negative for surface CD3 and positive for CD16, CD56, and CD57. NK cells are also positive for CD2, cCD3, CD7, CD8 (some), and CD26.
- Plasma cells are $CD38^+$ (bright) and $CD138^+$ (although less sensitive). Plasma cells in the blood and tonsils are $CD4^+$, whereas in the bone marrow, plasma cells may be either $CD45^+$ or $CD45^-$. Normal plasma cells are $CD19^+$ and $CD20^-$. Abnormal plasma cells are $CD19^-$ and $CD20^-$. These cells are typically also $CD56^+$.
- Myeloblasts will demonstrate markers of immaturity—CD34 (immature myeloid and lymphoid marker) and CD117 (immature myeloid marker); demonstrate pan myeloid markers (MPO, CD13, and CD33); and demonstrate lack of mature myeloid markers, such as CD11b, CD15, and CD16.
- Lymphoblasts will demonstrate markers of immaturity—CD34 and TdT. B lymphoblasts will also lack surface immunoglobulin and show lack of CD20 (a more mature B cell marker; more cells will be positive for CD19 than for CD20). These lymphoblasts are also positive for cCD22. T lymphoblasts will show CD1a (an immature T cell marker) and lack surface CD3. However, they are positive for cCD3.
- In AML, aberrant expression of lymphoid antigens may be present, and those that are frequently expressed include CD2, CD7, CD19, and CD56. Similarly, in ALL, aberrant expression of one or two myeloid antigens may be observed.
- In APL, the abnormal promyelocytes are $CD34^-$; completely or partially negative for HLA-DR, CD11b, and CD15; and positive for CD13 (heterogeneous), CD33 (homogeneous and bright), and CD117.

The abnormal promyelocytes lack CD10 and CD16, which are mature myeloid markers. It has been reported that the combination of the absence of CD11b, CD11c, and HLA-DR identifies 100% of APLs.

- Acute monocytic leukemia is positive for monocytic markers CD11b, CD14, CD15, CD16, CD36, and CD64.
- Acute erythroid leukemia is positive for CD235a (glycophorin) and for CD36 in the absence of CD64. CD36 is positive for monocytes, erythroid cells, and megakaryocytes. If the case is not megakaryocytic (i.e., CD41$^-$ and CD61$^-$) and CD64 is negative (i.e., not monocytic), then it favors the diagnosis of erythroid leukemia.
- In acute megakaryocytics leukemia, megakaryoblasts are positive for CD41 and CD61.
- Several features of MDS can be identified by flow cytometry. First, granulocytes are hypogranular, and the normal population of granulocytes as seen in the side scatter versus CD45 is distorted. The hypogranular cells occupy a lower position in the scattergram compared to a normal population. Two other plots are also helpful: CD11b versus CD16 and CD13 versus CD16. Immature granulocytes start by expressing CD16 and subsequently express CD11b, the expression of which becomes increasingly brighter with maturation. Subsequently, CD16 expression becomes very high. Thus, mature neutrophils are positive for both CD11b and CD16. This is the normal pattern of expression. When the CD13 versus CD16 plot is considered, immature cells have a bright expression of CD13 but very little CD16 expression. With maturation, CD13 becomes dim (at the stage of the metamyelocyte), but there are still low levels of CD16. Subsequently, both CD13 and CD16 become bright. Therefore, mature neutrophils are positive for both CD13 and CD16. In MDS, these normal patterns may be lost. Another feature of abnormal myeloid differentiation is gain of CD56 in the myeloid cells with lack of CD16.
- Hematogones are maturing B cell precursors representing a normal component of bone marrow. Typically, these cells decrease with age and usually constitute 1% or fewer bone marrow cells. However, there may be hyperplasia of hematogones, especially in the setting of regeneration after chemotherapy or stem cell transplantation or in patients with congenital or immune cytopenias. Hematogones may be confused with neoplastic B cells because there is morphologic and immunophenotypic overlap. The earliest hematogones are positive for CD34, CD10, and CD22 but negative for CD20. Subsequently, CD10 is downregulated with progressive gain of CD20. Therefore, in hematogones, expression of CD20 and CD34 is mutually exclusive, and CD34 expression in hematogones is biphasic (either positive or negative).

References

[1] Brown M, Wittwer C. Flow cytometry: principles and clinical applications in hematology. Clin Chem 2000;46:1221−9.

[2] Faure GC, Amsellem S, Arnoulet C, Bardet V, et al. Mutual benefits of B-ALL and HLDA/HCDM HLDA 9th Barcelona 2010. Immunol Lett 2011;134:145−9.

[3] Spencer J, Choy MY, MacDonald TT. T cell receptor Vβ expression by mucosal T cells. J Clin Pathol 1991;44:915−18.

[4] Barlogie B, Alexanian R, Jagannath S. Plasma cell dyscrasias. JAMA 268;2946−51.

[5] Dong HY, Kung JX, Bhardwaj V, McGill J. Flow cytometry rapidly identifies all acute promyelocytic leukemias with high specificity independent of underlying cytogenetic abnormalities. Am J Clin Pathol 2011;135:76−84.

CHAPTER 9

Cytogenetic Abnormalities and Hematologic Neoplasms

9.1 INTRODUCTION

Cancer is the second most common cause of death in developed countries and the most common cause of untimely deaths (>35% of deaths before age 65 years). Cancer has a strong genetic component, and prognosis sometimes depends on the abnormality. For example, leukemia with inv(3)(q21, q26) has a poor prognosis, but other types of defects that cause leukemia may have a good prognosis, such as Philadelphia chromosome-positive chronic myeloid leukemia [1]. Therefore, cytogenetics, a subdiscipline of genetics dealing with cytological and molecular analysis of chromosomes during cell division and the location of genes on chromosomes as well as movement of chromosomes during cell division, plays an important role in the diagnosis and the prognosis of various hematological disorders. An in-depth discussion of various cytogenetics techniques is beyond the scope of this book, but a brief description is provided in this chapter.

Various different specimens can be used for cytogenic analysis, such as peripheral blood (e.g., blood lymphocytes), bone marrow, amniotic fluid, cord blood, tissue specimens, and even urine (for fluorescence *in situ* hybridization (FISH) analysis in a suspected case of bladder cancer), but blood is the most common specimen used. The development of the chromosome banding technique was the first major advancement in cytogenetic analysis. The first banding technique, Q-banding (with quinacrine dihydrochloride and further examination with fluorescence microscopy), was later replaced mostly by G-banding (staining of chromosomes with Giemsa solution). The polymeric region of chromosomes can be visualized by C-banding, and NOR-banding stains the nuclear organizing region; both are useful banding methods. In 1986, FISH was developed, which was another breakthrough in molecular cytogenetics. This technique allows detection of specific nucleic acid sequences in morphologically preserved chromosomes and thus the visualization of small segments of DNA using specific probes. Later development of spectral karyotyping and the multicolor FISH

CONTENTS

A. Wahed and A. Dasgupta: Hematology and Coagulation. DOI: http://dx.doi.org/10.1016/B978-0-12-800241-4.00009-7

technique allowed visualization of all 24 different chromosomes using unique color combinations for each chromosome with the aid of various combinations and concentrations of fluorescent dyes. More recently, microarray-based methods have been developed using large insert genomic clones, cDNA, or oligonucleotides that can replace metaphase chromosome as DNA targets for analysis [2]. Common chromosomal abnormalities are numerical abnormalities (aneuploidy; deviation of normal 46 chromosomes) and structural rearrangements of chromosome (terminal and interstitial deletion abbreviated as "del," inversion abbreviated as "inv," and translocation abbreviated as "t"). In general, translocation means exchange between two or more chromosomes, deletion indicates loss of part of a chromosome, and inversion indicates rearrangement within an individual chromosome. Other abnormalities—duplications, ring chromosomes, and isochromosomes—are also considered as structural rearrangements of chromosomes. The short arm of a chromosome is denoted as "p" and the long arm as "q." For reporting purposes, numerical abnormalities precede structural abnormalities of chromosomes. For example, Philadelphia chromosome is denoted as t(9;22)(q34;q11), indicating that this chromosome is formed due to reciprocal translocation of chromosomes 9 and 22 involving region q34 (long arm) in chromosome 9 and region q11 (long arm) in chromosome 22.

9.2   CYTOGENETIC ABNORMALITIES IN CHRONIC MYELOID LEUKEMIA

Chronic myeloid leukemia (CML) is associated with an abnormal chromosome 22 known as the Philadelphia chromosome (see above), which was the first discovered genetic abnormality associated with human cancer. The Philadelphia chromosome, t(9;22)(q34;q11), results in the formation of a unique fusion gene product (BCR—ABL1: the ABL1 gene from chromosome 9 fuses with the BCR gene on chromosome 22; "BCR" is the abbreviation for break point cluster region, and ABL1 is an oncogene known as Abelson murine leukemia viral oncogene) that encodes and enhances tyrosine kinase activity. It is implicated in the development of CML and is the primary target for the treatment of this disorder using the tyrosine kinase inhibitor imatinib mesylate. The Philadelphia chromosome is present in hematopoietic cells from patients with CML but not in nonhematopoietic tissues, including bone marrow fibroblasts. The ABL1 gene is located on chromosome 9q34. This gene had been previously identified as the cellular homolog of the transforming gene of Abelson murine leukemia virus. The ABL1 gene encodes a non-receptor protein tyrosine kinase, c-ABL. ABL1 on chromosome 9 has 11 exons, with 2 alternative 5′ first exons, and a very large first intron of more than 250 kilobases (kb). The BCR gene is located on chromosome 22, which has 25 exons. The *BCR* gene product is a 160-kDa cytoplasmic phosphoprotein denoted BCR.

Formation of Philadelphia chromosome is due to a break in the first intron of the *ABL1* gene. Depending on the location of the breakpoint within the major BCR region, the consequence of the t(9;22) translocation in CML is to fuse the first 13 or 14 exons of the *BCR* gene upstream of the second exon of the *ABL1* gene. The two alternative fusion genes are traditionally described according to the original bcr exon nomenclature as *b2a2* and *b3a2* fusions or by the subsequent nomenclature as *e13a2* and *e14a2*, respectively. Transcription of the fusion gene followed by RNA splicing results in the generation of a novel 8.5-kb fusion *BCR−ABL1* mRNA that encodes a 210-kDa fusion protein designated p210 BCR−ABL1 or p210$^{BCR-ABL1}$. The protein product of the *e14a2* fusion is 25 amino acids longer than that of *e13a2*.

If the breakpoint on chromosome 22 is at a different location (minor or m-bcr), which leads to the first exon of BCR fused upstream of *ABL1* exon 2, then this is *e1a2*. This results in a 190-kDa protein, p190$^{BCR-ABL1}$. A third minor BCR region (μ-BCR) on chromosome 22 resulting in fusion of BCR exon 19 to *ABL1* exon 2 (*e19a2*) has been described in several patients, leading to the generation of a p230 form of BCR−ABL1. Thus, the three common variants of BCR−ABL1 are as follows:

- p210$^{BCR-ABL1}$: Created by the fusion of the *ABL1* gene at a2 with a breakpoint in the major *BCR* region at either e13 or e14 to produce an *e13a2* or *e14a2* transcript that is translated into a 210-kDa protein. This variant is present in most patients with CML and one-third of those with Philadelphia-positive B cell acute lymphoblastic leukemia (Philadelphia chromosome positive B-cell ALL).
- p190$^{BCR-ABL1}$: Created by the fusion of the *ABL1* gene at a2 with a breakpoint in the minor *BCR* region at e1 to produce an *e1a2* transcript that is translated into a 190-kDa protein. This variant is present in two-thirds of those with Philadelphia chromosome-positive B-cell ALL and a minority of patients with CML. The presence of p190 in chronic phase CML is correlated with monocytosis and a low neutrophil: monocyte ratio in the peripheral blood. Compared with other patients with CML, patients with the p190 fusion protein may have an inferior outcome when treated with tyrosine kinase inhibitors.
- p230$^{BCR-ABL1}$: Created by the fusion of the *ABL1* gene at a2 with a breakpoint in the μ-*BCR* region at e19 to produce an *e19a2* transcript that is translated into a 230-kDa protein. This variant is seen in some patients with chronic neutrophilic leukemia.

The World Health Organization (WHO) diagnostic criteria for CML require the detection of the Philadelphia chromosome or its products, the BCR−ABL1 fusion mRNA and the BCR−ABL1 protein. This can be accomplished through conventional cytogenetic analysis (karyotyping), FISH analysis, or reverse-transcription polymerase chain reaction (RT-PCR).

The cytogenetics technique requires *in vitro* culture and is time- as well as labor-intensive. In addition, this method can detect approximately 5% of Philadelphia-positive cells in a population of normal cells, and it can give false-negative results in cells with complex chromosomal rearrangements. However, FISH employs large DNA probes linked to fluorophores and permits direct detection of the chromosomal position of the BCR and ABL1 genes when employed with metaphase chromosome preparations. It can also be utilized on interphase cells from bone marrow or peripheral blood, in which physical colocalization of BCR and ABL probes is indicative of the presence of the BCR−ABL fusion gene [3]. The specificity of metaphase FISH is slightly higher than that of May−Grunwald−Giemsa banding for detection of the Philadelphia chromosome, and this technique allows easy identification of complex chromosomal rearrangements that mask the t(9;22) translocation. The specificity of interphase FISH is lower, by approximately 10%, due to false-positive results from coincidental colocalization of nonfused BCR and ABL1 genes in interphase nuclei.

RT-PCR is a highly sensitive technique that employs specific primers to amplify a DNA fragment from BCR−ABL1 mRNA transcripts. Depending on the combination of primers used, the method can detect the *e1a2*, *e13a2* (*b2a2*), *e14a2* (*b3a2*), and *e19a2* fusion genes. The use of nested primers and sequential PCR reactions renders the technique extremely sensitive, capable of routine detection of one Philadelphia-positive cell in 10^5 to 10^6 normal cells. Several features must be considered in the interpretation of RT-PCR data:

- There are many variables in the RT-PCR assay, including the internal standard to be used and the method used to compare results obtained in different laboratories. Although initial efforts to create an international standardized scale for quantitative PCR for BCR−ABL measurements appear to be promising, it is preferable to obtain serial samples from individual patients at the same laboratory, if possible.
- Patients with rare fusions, such as *e6a2* or *b2a3*, may not be detected with standard primer sets.
- Patients with 5′ m-bcr breakpoints can sometimes exhibit both *e13a2* (*b2a2*) and *e14a2* (*b3a2*) transcripts, whereas M-bcr (major breakpoint bcr) can also produce *e1a2* transcripts at lower levels, probably due to alternative splicing.

- *BCR−ABL1* fusion transcripts (M-bcr or m-bcr) can be detected at very low levels (one cell in 10^8 to 10^9) in hematopoietic cells from some normal individuals; this defines the limits of useful sensitivity of RT-PCR for diagnosis of leukemia.

9.3 CYTOGENETIC ABNORMALITIES IN MYELODYSPLASTIC SYNDROME

Patients with myelodysplastic syndrome (MDS) may have chromosomal abnormalities during diagnosis, or abnormal clones may appear later during the course of the disease. These abnormalities include a numerical change in chromosome numbers (i.e., monosomy or trisomy), a structural abnormality involving only one chromosome (e.g., inversion and interstitial deletion), or, less commonly, a balanced translocation involving two chromosomes. Approximately 10−15% of MDS patients exhibit complex karyotypes with multiple abnormalities. Additional chromosomal aberrations may evolve during the course of MDS, or an abnormal clone may emerge in a patient with a previously normal karyotype; these changes appear to accelerate progression of disease to acute leukemia. All of these common chromosome abnormalities observed in MDS are also commonly seen in other myeloid diseases (i.e., acute myeloid leukemia and myeloproliferative neoplasms).

Clonal chromosomal abnormalities can be detected in bone marrow cells in 40−70% of patients with primary MDS, as opposed to 70−80% detected in patients with acute myeloid leukemia (AML) *de novo*. The likelihood of chromosomal abnormalities is increased in patients with advanced MDS. Although rates vary depending on the technique used and the population studied, the most common chromosomal abnormality seen in MDS is del(5q) or (−5), which is deletion of the long arm of chromosome 5 (5q), occurring in approximately 15% of cases overall. There are two small commonly deleted regions: 5q33.1 (deletion of this locus is most commonly associated with the 5q− syndrome, with a relatively good prognosis) and 5q31 (more commonly seen with therapy-related MDS and associated with more aggressive disease).

Approximately 10% of patients with *de novo* MDS and up to 50% of patients with therapy-related MDS demonstrate −7 or del(7q), either alone or as part of a complex karyotype. Approximately 90% of cases have loss of a whole chromosome 7 (−7), and 10% are actual deletions, del(7q). Trisomy 8 is seen in less than 10% of patients with MDS and is considered an intermediate-risk finding.

Deletions of the long arm of chromosome 20, del(20q), occur in less than 5% of cases of MDS and are also seen in patients with AML and myeloproliferative disorders. When found as the sole chromosomal abnormality, del (20q) is associated with a favorable prognosis. However, loss of the Y chromosome is common in men without hematologic disorders and is not thought to play a role in the pathogenesis of MDS.

In addition, the presence of one of the following chromosomal abnormalities is presumptive evidence of MDS in patients with otherwise unexplained refractory cytopenia and no morphologic evidence of dysplasia: del(7q), del (5q), del(13q), del(11q), del(12p) or t(12p), del(9q), idic(X)(q13), t(17p) (unbalanced translocations) or i(17q) (i.e., loss of 17p), t(11;16)(q23; p13.3), t(3;21)(q26.2;q22.1), t(1;3)(p36.3;q21), t(2;11)(p21;q23), inv(3) (q21q26.2), and t(6;9)(p23;q34) (Table 9.1).

It is difficult to use cytogenetic analysis to predict the outcome for individual patients with MDS because many patients die from persistent and profound pancytopenia, regardless of whether or not progression to AML occurs. Patients with normal karyotype, isolated del(5q), isolated del(20q), and loss of Y have a low risk of developing AML, whereas patients with three or more abnormalities (in these situations, typically chromosomes 5 and 7 are

Table 9.1 Chromosomal Abnormalities in Patients with Myelodysplastic Syndrome (MDS)

Common Chromosomal Abnormalities in MDS	Chromosomal Abnormalities That May Be Presumptive Evidence of MDS in Patients with Unexplained Refractory Cytopenia but without Morphologic Evidence of Dysplasia
■ del(5q): Deletion of the long arm of chromosome 5 (5q) (most common). There are two small commonly deleted regions—5q33.1 (relative good prognosis; also seen in all patients with 5q deletion syndrome) and 5q31 (more commonly with therapy-related MDS causing more aggressive disease). ■ del(7q): Either alone or as part of a complex karyotype. Approximately 90% of cases have loss of a whole chromosome 7 (− 7), and 10% are actual deletions, del(7q). ■ Trisomy 8: Seen in less than 10% of patients with MDS and considered an intermediate-risk finding. ■ del(20q): Occurs in less than 5% of cases of MDS and also seen in patients with acute myeloid leukemia and myeloproliferative disorders. If the only abnormality, del(20q) is associated with a favorable prognosis. ■ Loss of the Y chromosome: May not play a role in the pathogenesis of MDS.	■ del(7q) ■ del(5q) ■ del(13q) ■ del(11q) ■ del(12p) or t(12p) ■ del(9q) ■ idic(X)(q13) (isodicentric X chromosome) ■ t(17p) (unbalanced translocations) or i(17q) (i.e., loss of 17p) ■ t(11;16)(q23;p13.3) ■ t(3;21)(q26.2;q22.1) ■ t(1;3)(p36.3;q21) ■ t(2;11)(p21;q23) ■ inv(3)(q21q26.2) ■ t(6;9)(p23;q34)

affected) or abnormalities of chromosome 7 have a high risk of progression to AML. All other abnormalities have an intermediate risk of progression to AML.

Abnormalities in certain genes have been identified in patients with MDS and AML with or without the presence of chromosomal abnormalities. These gene mutations have been found to affect DNA methylation, tumor suppressor genes, and oncogenes, but the prognostic significance of such mutations is not well characterized. Although many of these gene mutations are even more frequent in AML, particularly in AML with a normal karyotype, they are thought to provide insight into the pathobiology of MDS and its progression to AML:

- *TET2* mutations: Somatic mutations in *TET2* (ten−eleven translocation) family gene members occur in approximately 15% of patients with myeloid cancers, including MDS. Loss of function mutations of *TET2* result in increased methylation and silencing of genes that are normally expressed. When present in MDS, *TET2* mutations have been associated with a more favorable prognosis [4].
- *RUNX1* transcriptional core binding factor gene: *RUNX1* (runt-related transcription factor 1 family of genes; also known as acute myeloid leukemia 1 gene) gene mutations are seen in 7−15% of cases of *de novo* MDS, but they are more common in cases of therapy-related MDS and impart a poorer prognosis. In addition, *RUNX1* is a translocation partner for *RUNX1T1* (*ETO*) in cases of AML with t(8;21).
- *TP53* tumor suppressor gene: This tumor suppressor gene is located on 17p and mediates cell cycle arrest in response to a variety of cellular stressors. Approximately 5−15% of patients with MDS have known *TP53* mutations at the time of diagnosis. Abnormalities in p53 are more common in patients with MDS associated with prior exposure to alkylating agents of radiation (i.e., therapy-related MDS). Loss of wild-type *TP53* is associated with resistance to treatment and is a marker of poor prognosis independent of the International Prognostic Scoring System risk score.
- *RAS* oncogenes: Mutations of *RAS* (these genes encode RAS proteins, which were first isolated in rat sarcoma and are now abbreviated as RAS) have been identified in 10−35% of patients with MDS and in a subset of patients with AML. The majority of the mutations in MDS are in the *NRAS* gene, whereas *KRAS* and *HRAS* mutations occur less frequently. In both AML and MDS, *RAS* mutations have been reported more frequently in cases with a monocytic morphology (e.g., chronic myelomonocytic leukemia). The significance of *RAS* mutations in MDS remains unclear; however, *RAS* mutations are associated with MDS characterized by −7/del(7q).

BOX 9.1 PROGNOSES OF VARIOUS MUTATIONS ASSOCIATED WITH MDS

- *TET2* (ten—eleven translocation family of genes) mutation: Favorable prognosis
- *RUNX1* (runt-related transcription factor 1 family of genes) mutation: Poorer prognosis than that for *TET2* mutations
- *TP53* tumor suppressor gene mutations: Poor prognosis
- *RAS* oncogene mutations of RAS (these genes encode RAS proteins, which were first isolated in rat sarcoma and abbreviated as RAS): Significance in MDA unclear
- *IDH* mutations: Mutations in the isocitrate dehydrogenase oncogenes (i.e., *IDH1* and *IDH2*), associated with poor prognosis
- *FLT3* gene (human analog of the murine fetal liver tyrosine kinase gene) mutations: Although uncommon in MDS, if present, associated with the worst prognosis

BOX 9.2 MOST COMMON KARYOTYPIC ABNORMALITIES IN PATIENTS WITH ACUTE MYELOID LEUKEMIA

- t(15;17)(q24.1;q21.1)
- Trisomy 8
- t(8;21)(q22;q22),
- inv(16)(p13.1q22)/t(16;16)(p13.1;q22)
- 11q rearrangements

- *IDH* mutations: Mutations in the isocitrate dehydrogenase oncogenes (i.e., *IDH1* and *IDH2*) have been reported in some cases of MDS resulting in DNA hypermethylation and alteration of gene expression. The presence of *IDH* mutations is considered as portending a poor prognosis.
- *FLT3* gene mutations: Although mutations in the *FLT3* gene (human analog of the murine fetal liver tyrosine kinase gene) are uncommon in MDS, they have been associated with a worse prognosis.

The prognoses of various mutations associated with MDS are listed in Box 9.1.

9.4   CYTOGENETIC ABNORMALITIES IN PATIENTS WITH ACUTE MYELOID LEUKEMIA

Approximately 50—60% of patients with *de novo* AML (acute myeloid leukemia or acute myeloblastic leukemia) demonstrate abnormal karyotypes. Commonly observed abnormalities are listed in Box 9.2.

The following cytogenetic abnormalities, if found, result in the diagnosis of AML regardless of blast count:

- AML with t(8;21)(q22;q22); *RUNX1−RUNX1T1* (previously *AML1−ETO*): This is seen in approximately 5% of adults with newly diagnosed AML and is the most frequent abnormality in children with AML. The *RUNX1* (previously *AML1* or core binding factor α_2) gene on chromosome 21 and the *RUNX1T1* (previously *ETO* or *MTG8*) gene on chromosome 8 form a chimeric product that regulates the transcription of a number of genes that are critical to hematopoietic stem and progenitor cell growth, differentiation, and function. Cases of AML with t(8;21) have a morphologically distinct phenotype. It imparts a favorable prognosis in adults but a poor prognosis in children.
- AML with inv(16)(p13.1q22) or t(16;16)(p13.1;q22); *CBFB−MYH11* (inversion of chromosome 16 produces fusion transcript involving *CBFB*, the core binding factor gene, and *MYH11*, the smooth muscle myosin heavy-chain gene): AML with inv(16) represents approximately 5−8% of newly diagnosed AML cases and typically demonstrates abnormal eosinophils at all stages of maturation in the bone marrow. Patients with inv(16) generally have a good response to intensive chemotherapy.
- Acute promyelocytic leukemia (APL) with t(15;17)(q24.1;q21.1); *PML−RARA*: This translocation is highly specific for APL where the gene coding for the retinoic acid receptor α (*RARA*) normally located on chromosome 17 is disrupted by t(15;17) and is fused with the promyelocytic leukemia gene (*PML*) on chromosome 15 [5]. This disruption results in the PML−RARA fusion protein, which is thought to lead to persistent transcriptional repression, thereby preventing differentiation of promyelocytes.

In addition to the cytogenetic abnormalities described previously, the following AML subtypes, which require at least 20% myeloid blasts for diagnosis, have been identified as subcategories of AML with genetic abnormalities of prognostic significance:

- AML with t(9;11)(p22;q23); *MLLT3−MLL*: The mixed lineage leukemia gene (*MLL*) located at chromosome 11 encodes a transcription factor with pleiotropic effects on hematopoietic gene expression and chromatin remodeling. The *MLL* gene can fuse with one of more than 50 different known chromosomal patterns, but the most common translocation is *MLLT3*, which is seen more often in children and is associated with monocytic features [6]. These patients tend to have an intermediate response to standard therapy.

- AML with t(6;9)(p23;q34); *DEK–NUP214*: This abnormality is seen in approximately 1% of patients with newly diagnosed AML. Cases with this translocation are characterized by basophilia, pancytopenia, and dysplasia. Patients with this abnormality have a poor response to standard therapy, which may be a result of the lesion or its association with *FLT3–ITD*, which is also known to have a poor prognosis.
- AML with inv(3)(q21q26.2) or t(3;3)(q21;q26.2); *RPN1–EVI1*: The t(3;3) and inv(3) translocations account for approximately 1% of AML cases and are associated with a poor response to therapy. These cytogenetic changes lead to the activation of the *EVI1* (ectopic viral integration site 1) gene, which is thought to result in peripheral blood thrombocytosis and increased atypical megakaryocytes in the bone marrow characteristic of cases with this abnormality.
- AML (megakaryoblastic) with t(1;22)(p13;q13); *RBM15–MKL1*: This type of leukemia is also rare, but it is typically a megakaryoblastic process occurring in infants. It is not seen in patients with Down syndrome.

Characteristic chromosomal abnormalities have also been identified in patients who developed MDS or AML after chemotherapy and/or radiation therapy for a previous disorder. The most common type of therapy-related MDS/AML is due to damage from alkylating agents and is characterized by abnormalities in chromosomes 5 and/or 7. The second most common subtype of therapy-related AML occurs in patients who have been treated with chemotherapeutic drugs that inhibit DNA topoisomerase II. AML associated with these drugs is often characterized by balanced translocations involving the *MLL* gene at 11q23 or the *RUNX1* gene at 21q22.

In addition to the chromosomal abnormalities described previously, the following specific gene mutations also occur in AML:

- *FLT3* (fms-related tyrosine kinase 3): Seen most often in AML with t(6;9) and in APL and AML with normal karyotype. The two primary types of *FLT3* mutations are internal tandem duplications (*FLT3-ITD*) and mutations affecting codon 385 or codon 386 of the second tyrosine kinase domain (*FLT3-TKD*). *FLT3-ITD* mutations are associated with an adverse outcome.
- Nucleophosmin (*NPM1*).
- *CEBPA*: Mutations of *NPM1* and *CEBPA* are frequently observed in AML patients with a normal karyotype, and in the absence of *FLT3-ITD* they have a favorable prognosis.
- *KIT*: Observed in AML with t(8;21) and inv(16)/t(16;16). The presence of *KIT* mutations imparts poor prognosis.
- *MLL*, *WT1*, *NRAS*, and *KRAS*.

9.5 CYTOGENETIC ABNORMALITIES IN ACTUTE LYMPHOBLASTIC LEUKEMIA

Chromosomal abnormalities are seen in approximately 80% of B cell ALL. The t(4;11) translocation is seen in up to 60% of infants (up to 1 year of age) with ALL, but it is rarely observed in adult patients with ALL. This translocation results in a *MLL−AFF1* fusion gene and is associated with a poor prognosis. Individuals with ALL and t(4;11) have high leukocyte counts, an immature immunopheno-type, B cell lineage, and frequent coexpression of myeloid antigens.

The t(9;22) translocation that produces the Philadelphia chromosome is observed in approximately 2−5% of children and approximately 30% of adults. This translocation is associated with a poor prognosis. The t(1;19) translocation occurs in approximately 30% of patients with pre-B cell childhood ALL and less commonly in other B lineage.

The t(12;21) translocation, resulting in the fusion *ETV6−RUNX1* gene, is present in approximately 25% of children with B lineage ALL and 3% of adults with ALL. Patients with ALL and t(12;21) have a favorable prognosis. Patients who have hyperdiploidy with more than 50 chromosomes often have a good prognosis. The individual structural abnormalities do not appear to influence outcome in patients with hyperdiploidy except for t(9;22), which is associated with a poor prognosis.

Approximately 5 or 6% of ALL patients, independent of age, have clonal loss of various chromosomes, resulting in a hypodiploid clone with fewer than 46 chromosomes. ALL patients with hypodiploid clones generally have a poor prognosis, especially patients with near-haploid and low-hypodiploid clones. Deletion of 9p is an unfavorable risk factor associated with a high rate of relapse in precursor B ALL in children.

Among children with T cell ALL, approximately 60% have an abnormal kar-yotype. These patients have a distinct pattern of recurring karyotypic abnor-malities involving both T cell receptor (TCR) and non-TCR gene loci, including activating mutations of *NOTCH1* in more than 50%.

9.6 CYTOGENETIC ABNORMALITIES IN MULTIPLE MYELOMA

Conventional karyotyping detects cytogenetic abnormalities in 20−30% of patients with myeloma. In contrast, FISH detects abnormalities in almost all cases. The most common chromosomal translocations in multiple myeloma involve 14q32, the site of the immunoglobulin heavy-chain (IgH) locus, and are known as "primary IgH translocations." These include t(11;14)(q13;q32), t(4;14) (p16.3;q32.3), t(6;14)(p25;q32), t(8;14)(q24;q32), and t(14;16)(q32.3;q23).

Abnormalities that impart a standard risk for developing multiple myeloma include t(11;14) and t(6;14) and trisomies of odd-numbered chromosomes. Abnormalities that impart poor prognosis include t(4;14), t(14;16), t(14;20), and/or del(17p13). The poor prognosis of these high-risk factors may be abrogated by the presence of at least one trisomy.

Deletions of 17p—including 17p13, the p53 locus—are found in 10% of multiple myeloma patients and are associated with a shorter survival after both conventional chemotherapy and hematopoietic cell transplantation. There is a low rate of complete response, rapid disease progression, advanced disease stages, plasma cell leukemia, and central nervous system multiple myeloma.

Trisomy (hyperdiploidy) is associated with a favorable outcome in multiple myeloma. Hyperdiploidy occurs in myeloma typically due to trisomies of odd-numbered chromosomes. Trisomies occurring in high-risk patients (by FISH) can cause the risk to be reduced to standard risk.

9.7   CYTOGENETIC ABNORMALITIES IN B AND T CELL LYMPHOMAS

9.7.1   CLL/SLL

Trisomy 12 is seen in approximately 20% of cases of chronic lymphocytic leukemia (CLL) and is associated with poor prognosis, whereas del(13q14) is seen in approximately 50% of cases and is also associated with a favorable prognosis. Other deletions seen in CLL include those of 11q and 17p. The 11q deletions are the most common type of karyotypic evolution over time.

9.7.2   Follicular Lymphoma

In follicular lymphoma, the classic cytogenetic abnormality observed is t(14;18)(q32;q21). At the molecular level, this translocation juxtaposes the *bcl-2* proto-oncogene (band 18q21) with the Ig heavy-chain gene (band 14q32), resulting in deregulation of *bcl-2* gene expression and elevation of bcl-2 mRNA and protein. The classic abnormality is seen in 90% of cases of follicular lymphoma, grades I and II. In addition to the classic abnormality, other alterations are seen in 90% of follicular lymphoma cases. The number of additional chromosomal alterations increases with histologic grade and transformation. Abnormalities of 3q27 and/or BCL6 rearrangements are seen in 5−15% of cases of follicular lymphoma, mostly grade 3B.

9.7.3 Mantle Cell Lymphoma

Mantle cell lymphoma is characterized by the presence of a balanced chromosomal translocation, t(11;14)(q13;q32). This abnormality juxtaposes the *CCND1* gene (11q13) with the *IgH* (14q32) gene, resulting in cyclin D1 overexpression.

9.7.4 Marginal Zone Lymphoma

The following are the most common anomalies in extranodal marginal zone B cell lymphoma (MZBCL) of MALT (mucosa-associated lymphoid tissue) type:

- The translocation t(11;18)(q21;q21)/*API2−MLT* fusion with 20−50% incidence. The translocation is associated with low-grade MALT lymphoma of the stomach and the lung.
- The translocation t(14;18)(q32;q21)/*IgH−MLT1* fusion, leading to enhanced *MLT1* expression, may occur in 10−20% of all MALT lymphomas. It is associated with MALT lymphoma of the liver, skin, ocular adnexa, lung, and salivary gland.
- The translocation t(1;14)(p22;q32) and/or the corresponding deregulation or rearrangement of *BCL10* at 1p22 is another recurrent chromosome aberration in a minority of cases, and it appears to be more frequent in high-grade MALT than in low-grade MALT lymphoma.
- The translocation t(3;14)/*IgH−FOXP1* fusion may occur in 10% of all MALT lymphomas. It is associated with MALT lymphoma of the orbit, thyroid, and skin, whereas it is not found in MALT lymphoma of the stomach or salivary gland, and other forms of MZBCL.
- Trisomy 3 and trisomy 18 have been reported in low-grade as well as high-grade MALT lymphoma.

In splenic MZBCL, the 7q deletions are the most common abnormality observed. Other abnormalities include total or partial trisomy 3. However, nodal MZBCL does not have a distinct cytogenetic profile.

9.7.5 Diffuse Large B Cell Lymphoma

Different cytogenetic abnormalities are observed in diffuse large B cell lymphomas, including the following:

- The translocation t(3;v)(q27;v)/*Bcl6* rearrangement, seen in 30% of cases of diffuse large B cell lymphomas.
- The translocation t(14;18)(q32;q21)/*Bcl2* rearrangement, a feature of follicular lymphoma, seen in 15−25% of cases.

- *MYC* rearrangement, seen in approximately 10% of cases. The *MYC* partner is the *IG* gene, which is observed in 60% of cases, and the non-*IG* gene is observed in the remainder of cases.

9.7.6   Burkitt Lymphoma

Various cytogenetic abnormalities are observed in Burkitt lymphoma, including the following:

- The translocation t(8;14)(q24;q32), which is seen in the vast majority of cases: The *MYC* gene is on chromosome 8, and the *IgH* gene is on chromosome 14. In 2001, the WHO classification seemed to require a translocation of *MYC* to an immunoglobulin gene for diagnosis of Burkitt lymphoma, but in 2008, the classification allowed for a minor proportion of cases without demonstrable translocation of *MYC* to be diagnosed with Burkitt lymphoma [7].
- The translocation t(8;22) (q24;q11): The gene for κ light chain is on chromosome 22.
- The translocation t(2;8) (p12;q24): The gene for λ light chain is on chromosome 2.

9.7.7   Anaplastic Large Cell Lymphoma

In all cases of anaplastic large cell lymphoma (ALCL) and anaplastic large cell lymphoma (ALK), rearrangement involving the anaplastic lymphoma kinase (*ALK*) gene on chromosome 2p23 is observed. There are several translocations and inversions involving *ALK*, with the most common one being t(2;5), encoding a nuclear phosphoprotein (NPM)/ALK fusion protein (70−75% of cases).

KEY POINTS

- The Philadelphia chromosome, t(9;22)(q34;q11), results in the formation of a unique gene product (BCR−ABL1) encoding tyrosine kinase. There are three common variants of BCR−ABL1.
  - p210$^{BCR-ABL1}$: Created by the fusion of the *ABL1* gene at a2 with a breakpoint in the major *BCR* region at either e13 or e14 to produce an *e13a2* or *e14a2* transcript that is translated into a 210-kDa protein. This variant is present in most patients with CML and one-third of those with Philadelphia-positive B cell acute lymphoblastic leukemia (Philadelphia chromosome positive B-cell ALL).
  - p190$^{BCR-ABL1}$: Created by the fusion of the *ABL1* gene at a2 with a breakpoint in the minor *BCR* region at e1 to produce an *e1a2*

transcript that is translated into a 190-kDa protein. This variant is present in two-thirds of those with Philadelphia chromosome-positive B-cell ALL and a minority of patients with CML. The presence of p190 in chronic phase CML is correlated with monocytosis and a low neutrophil:monocyte ratio in the peripheral blood. Compared with other patients with CML, patients with the p190 fusion protein may have an inferior outcome when treated with tyrosine kinase inhibitors.

- p230$^{BCR-ABL1}$: Created by the fusion of the *ABL1* gene at a2 with a breakpoint in the μ-BCR region at e19 to produce an *e19a2* transcript that is translated into a 230-kDa protein. This variant is seen in some patients with chronic neutrophilic leukemia.

- Patients with normal karyotype, isolated del(5q), isolated del(20q), and loss of Y have a low risk of developing AML, whereas patients with three or more abnormalities (in these situations, typically chromosomes 5 and 7 are affected) or abnormalities of chromosome 7 have a high risk of progression to AML. All other abnormalities have an intermediate risk of progression to AML.

- The following cytogenetic abnormalities, if found, result in the diagnosis of AML regardless of blast count:
  - AML with t(8;21)(q22;q22); *RUNX1−RUNX1T1* (previously *AML1−ETO*). It imparts a favorable prognosis in adults but a poor prognosis in children.
  - AML with inv(16)(p13.1q22) or t(16;16)(p13.1;q22); *CBFB−MYH11*: AML with inv(16) represents approximately 5−8% of newly diagnosed AML and typically demonstrates abnormal eosinophils at all stages of maturation in the bone marrow. Patients with inv(16) generally have a good response to intensive chemotherapy.
  - APL with t(15;17)(q24.1;q21.1); *PML−RARA*: This translocation is highly specific for APL and results in a PML−RARA fusion protein that is thought to lead to persistent transcriptional repression, thereby preventing differentiation of promyelocytes.

- The following four AML subtypes, which require at least 20% myeloid blasts for diagnosis, have been identified as subcategories of AML with genetic abnormalities of prognostic significance:
  - AML with t(9;11)(p22;q23); *MLLT3−MLL*: This is seen more often in children and is associated with monocytic features.
  - AML with t(6;9)(p23;q34); *DEK−NUP214*: This abnormality is seen in approximately 1% of patients with newly diagnosed AML and cases with this translocation are characterized by basophilia, pancytopenia, and dysplasia. Patients with this abnormality have a poor response to standard therapy, which may be a result of the

lesion or its association with *FLT3−ITD*, which is also known to convey a poor prognosis.

- AML with inv(3)(q21q26.2) or t(3;3)(q21;q26.2); *RPN1−EVI1*: The t(3;3) and inv(3) translocations account for approximately 1% of AML cases and are associated with a poor response to therapy. These cytogenetic changes lead to activation of the *EVI1* gene, which is thought to result in the peripheral blood thrombocytosis and increased atypical megakaryocytes in the bone marrow characteristic of cases with this abnormality.

- AML (megakaryoblastic) with t(1;22)(p13;q13); *RBM15−MKL1*: This type of leukemia is also rare, but it is typically a megakaryoblastic process occurring in infants. It is not seen in patients with Down syndrome.

- In addition to the chromosomal abnormalities described previously, specific gene mutations also occur in AML:

  - *FLT3* (fms-related tyrosine kinase 3): Seen most often in AML with t(6;9) and in APL and AML with normal karyotype. The two primary types of *FLT3* mutations are internal tandem duplications (*FLT3-ITD*) and mutations affecting codon 385 or codon 386 of the second tyrosine kinase domain (*FLT3−TKD*). *FLT3−ITD* mutations are associated with an adverse outcome.

  - Nucleophosmin (*NPM1*).

  - *CEBPA*: mutations of *NPM1* and *CEBPA* are frequently observed in AML patients with a normal karyotype, and in the absence of *FLT3−ITD* they have a favorable prognosis.

- Chromosomal abnormalities are seen in approximately 80% of B cell ALL. t(4;11) is seen in up to 60% of infants (up to 1 year of age) with ALL, but it is rarely observed in adult patients with ALL. Individuals with ALL and t(4;11) have high leukocyte counts, an immature immunophenotype, B cell lineage, and frequent coexpression of myeloid antigens. The t(9;22) translocation that produces the Philadelphia chromosome is observed in approximately 2−5% of children and approximately 30% of adults. This translocation is associated with a poor prognosis. t(1;19) occurs in approximately 30% of patients with pre-B cell childhood ALL and less commonly in other B lineages. The t(12;21) translocation, resulting in the fusion *ETV6−RUNX1* gene, is present in approximately 25% of children with B lineage ALL and 3% of adults with ALL. Patients with ALL and t(12;21) have a favorable prognosis. Patients who have hyperdiploidy with more than 50 chromosomes often have a good prognosis. The individual structural abnormalities do not appear to influence outcome in patients with hyperdiploidy except for t(9;22), which is associated with a poor prognosis. Approximately 5 or 6% of ALL patients, independent of age, have clonal loss of various chromosomes, resulting in a hypodiploid clone with fewer than 46 chromosomes. ALL patients with hypodiploid clones generally have a poor prognosis, especially patients with near-haploid and low-hypodiploid clones. Deletion of 9p is an unfavorable risk factor associated with a high rate of relapse in precursor B ALL in children. Among children with T cell ALL, approximately 60% have an abnormal karyotype. These patients have a distinct pattern of recurring karyotypic abnormalities involving both T cell receptor (TCR) and non-TCR gene loci, including activating mutations of *NOTCH1* in more than 50%.

- Cytogenetic abnormalities are observed in multiple myeloma. The most common chromosomal translocations in multiple myeloma involve 14q32, the site of the immunoglobulin heavy-chain (IgH) locus, and are known as "primary IgH translocations."

- Abnormalities that impart a standard risk are t(11;14) and t(6;14) and trisomies of odd-numbered chromosomes. Abnormalities that impart poor prognosis include t(4;14), t(14;16), t(14;20), and/or del(17p13). The poor prognosis of these high-risk factors may be abrogated by the presence of at least one trisomy.

- Cytogenetic abnormalities are observed in patients with CLL. Trisomy 12 is seen in approximately 20% of cases of CLL and is associated with a poor prognosis. del(13q14) is seen in approximately 50% of cases and is associated with a favorable prognosis. Other deletions seen in CLL include those of 11q and 17p. 11q deletions are the most common type of karyotypic evolution over time.

- In follicular lymphoma, the classic cytogenetic abnormality is t(14;18)(q32;q21). In addition to the classic abnormality, other alterations are seen in 90% of follicular lymphoma cases. The number of additional chromosomal alterations increases with histologic grade and transformation.

- Mantle cell lymphoma is characterized by the presence of a balanced chromosomal translocation, the t(11;14)(q13;q32). This abnormality juxtaposes the *CCND1* gene (11q13) with the *IgH* (14q32) gene, resulting in cyclin D1 overexpression.

- The following are the most common anomalies in extranodal marginal zone B cell lymphoma (MZBCL) of MALT (mucosa-associated lymphoid tissue) type:
  - The translocation t(11;18)(q21;q21)/*API2−MLT* fusion with 20−50% incidence. The translocation is associated with low-grade MALT lymphoma of the stomach and the lung.
  - The translocation t(14;18)(q32;q21)/*IgH−MLT1* fusion, leading to enhanced MLT1 expression; it may occur in 10−20% of all MALT lymphomas. It is associated with MALT lymphoma of the liver, skin, ocular adnexa, lung, and salivary gland.

- The translocation t(1;14)(p22;q32) and/or the corresponding deregulation or rearrangement of *BCL10* at 1p22 is another recurrent chromosome aberration in a minority of cases, and it appears to be more frequent in high-grade MALT than in low-grade MALT lymphoma.
    - The translocation t(3;14)/*IgH−FOXP1* fusion may occur in 10% of all MALT lymphomas. It is associated with MALT lymphoma of the orbit, thyroid, and skin, whereas it is not found in MALT lymphoma of the stomach, salivary gland, and other forms of MZBCL.
    - In splenic MZBCL: 7q deletions are the most common abnormality observed. Other abnormalities include total or partial trisomy 3. However, nodal MZBCL does not have a distinct cytogenetic profile.
- Diffuse large B cell lymphoma (DLBCL): t(3;v)(q27;v)/*Bcl6* rearrangement is seen in 30% of cases of DLBCL. t(14;18)(q32;q21)/*Bcl2* rearrangement, a feature of follicular lymphoma, is seen in 15−25% of cases.
- Burkitt lymphoma: t(8;14)(q24;q32) is seen in the vast majority of cases. The *MYC* gene is on chromosome 8, and the *IgH* gene is on chromosome 14. t(8;22)(q24;q11): the gene for κ light chain is on chromosome 22. t(2;8) (p12;q24): the gene for λ light chain is on chromosome 2.
- All cases of ALK and ALCL have a rearrangement involving the anaplastic lymphoma kinase (*ALK*) gene on chromosome 2p23. There are several translocations and inversions involving *ALK*, with the most common one being t(2;5), encoding a nuclear phosphoprotein (NPM)/ALK fusion protein (70−75% of cases).

References

[1] Huret JL, Ahmad M, Arsaban M, Bernheim A, et al. Atlas of genetics and cytogenetics in oncology and hematology in 2013. Nucleic Acid Res 2013;41(Database issue):D920−4.

[2] Kannan TP, Zilfalil BA. Cytogenetics: past, present and future. Malaysian J Med Sci 2009;16:4−9.

[3] Primo D, Tabernero MD, Rasillo A, Sayagués JM, et al. Patterns of *BCR/ABL* gene rearrangements by interphase fluorescence *in situ* hybridization (FISH) in *BCR/ABL*+ leukemias: incidence and underlying genetic abnormalities. Leukemia 2003;17:1124−9.

[4] Abdel-Wahab O, Mullally A, Hedvat C, Garcia-Manero G, et al. Genetic characterization of TET1, TET2, and TET3 alterations in myeloid malignancies. Blood 2009;114:144−7.

[5] Chen Z, Chen SJ. *RARA* and *PML* genes in acute promyelocytic leukemia. Leuk Lymphoma 1992;8:253−60.

[6] Chandra P, Luthra R, Zuo Z, Yao H, et al. Acute myeloid leukemia with t(9;11)(p21−22;q23): common properties of dysregulated Ras pathway signaling and genomic progression characterize *de novo* and therapy related cases. Am J Clin Pathol 2010;133:686−93.

[7] Said J, Lones M, Yea S. Burkitt lymphoma and MYC: what else is new? Adv Anat Pathol 2014;21:160−5.

CHAPTER 10

Benign Lymph Nodes

10.1 INTRODUCTION

The lymph system is an important component of the body's immune system. Lymph nodes, which are small bean-shaped glands that are an integral part of the lymph system, are located in clusters in various parts of the body, including the neck, armpit, and groin, as well as inside the center of the chest and abdomen. A lymph node consists of cortex and medulla, encased in a fibrous capsule. Subcapsular sinus is located just underneath the capsule. The outer cortex consists mainly of B cells arranged as follicles, and T cells are found in deeper cortex. Paracortex is the portion of the lymph node immediately surrounding the medulla and contains a mixture of mature and immature T cells. The unstimulated follicles are primary follicles, which consist of small lymphocytes and follicular dendritic cells. The small lymphocytes are predominantly B lymphocytes that exhibit surface IgD and surface IgM and are positive for CD5. Follicular dendritic cells are stained through CD21, CD23, and CD35. Secondary follicles are derived from primary follicles following antigenic stimulation. Secondary follicles have an inner germinal center and an outer mantle zone. Within the germinal center in a reactive follicle, a light zone and a dark zone are present. This is referred to as polarity. The light zone consists of centrocytes and follicular dendritic cells. The mitotic rate is lower in this zone. There are fewer tingible body macrophages in this area. In contrast, the dark zone has centroblasts with a high mitotic rate and an increased number of tingible body macrophages. The germinal center of a reactive follicle stains through CD10 and Bcl-6 (B cell lymphoma 6 protein) and is negative for Bcl-2. Because the secondary follicle is a B cell area, B cell markers such as CD20 will highlight both the germinal center and the mantle zone. A few T cells are found in the germinal center. The Ki-67 staining (Ki-67 is a nuclear protein, the name being derived from Kiel, Germany; "67" denotes the clone number) of the germinal centers demonstrates a high proliferation index.

CONTENTS

A. Wahed and A. Dasgupta: Hematology and Coagulation. DOI: http://dx.doi.org/10.1016/B978-0-12-800241-4.00010-3

The outer mantle zone of the secondary follicle is composed of B lympho-cytes, which are positive for CD5 and Bcl-2. Outside the mantle zone is the marginal zone, which is still a B cell area, but the cells are more loosely packed. The area beyond the marginal zone is the paracortex, which is a pre-dominantly T cell area. The paracortex is thus easily stained via CD3, a pan T cell marker. T cells are also stained via Bcl-2. Thus, the Bcl-2 staining pat-tern of a reactive lymph node includes staining of the paracortex and the outer parts of the secondary follicle but not the germinal center.

The primary follicle consists of naive B cells that are as yet not stimulated. When stimulated by an antigen, the naive B cells move into the germinal center and transform into centroblasts, which in turn form centrocytes. From the centrocytes are derived plasma cells and memory cells. The plasma cells move into the medulla, an ideal location to release immunoglobulins into the circulation. Memory cells typically reside in the marginal zone, whereas naive B cells are found in the mantle zone layer.

The paracortex predominantly consists of T cells, interdigitating dendritic cells, and high endothelial venules. A few B cells are also present in the para-cortex. Once activated, cells may transform into immunoblasts. The immu-nohistochemical pattern of benign lymph nodes is summarized in Table 10.1.

10.2 REACTIVE LYMPHOID STATES

Reactive lymphoid hyperplasia may be due to numerous causes, and the typi-cal patterns of lymphoid hyperplasia are follicular hyperplasia, paracortical hyperplasia, sinusoidal expansion, granulomatous lymphadenitis, and mixed

Table 10.1 Immunohistochemical Pattern of Benign Lymph Nodes

Immunohistochemical Marker	Comments
Pan B cell markers (e.g., CD20, CD79a)	Enables staining of B cells of mantle zone, marginal zone, germinal center, and scattered B cells in paracortex
CD10	Enables staining of B cells in the germinal center
Bcl-6	Enables staining of B cells in the germinal center
Bcl-2	Does not enable staining of cells in the germinal center; enables staining of B cells in the mantle zone and marginal zone
CD3	Enables staining of T cells in the paracortex; enables staining of the few T cells in the germinal center
CD21, CD23, CD35	Enables staining of the follicular dendritic cells
S100	Enables staining of the interdigitating cells in the paracortex

BOX 10.1 PATTERNS OF LYMPHOID HYPERPLASIA

- Follicular hyperplasia (bacterial and viral infections, rheumatoid arthritis, Castleman disease)
- Paracortical hyperplasia (viral infections), drug (e.g., phenytoin)-related lymphadenopathy
- Sinusoidal expansion
- Mantle zone hyperplasia
- Marginal zone hyperplasia
- Nodular paracortical T cell hyperplasia
- Mixed pattern

Table 10.2 Features of Follicular Hyperplasia and Follicular Lymphoma

Follicular Hyperplasia	Follicular Lymphoma
Follicles of variable size	Back-to-back follicles; follicles uniform in size and may occupy the medulla
Distinct mantle zone present	Mantle zone indistinct
Germinal center with light and dark zone	Distinction of light and dark zone lost
Germinal center has TBM	Germinal center lacks TBM
Lymphocytes do not extend beyond capsule	Expansion beyond capsule
Ki-67 high	Ki-67 low; in grade III, Ki-67 may be high
Bcl-2 negative in germinal center	Bcl-2 positive in germinal center

TBM, tingible body macrophage.

pattern (Box 10.1). Infections are a common cause of both follicular and paracortical hyperplasia. Viral infections can result in both follicular and paracortical hyperplasia. However, it is important to differentiate follicular hyperplasia from follicular lymphoma. Follicular lymphoma is the most common indolent type of non-Hodgkin lymphoma and the second most common form of non-Hodgkin lymphoma involving B cells (see Chapter 11). Table 10.2 lists the features of follicular hyperplasia and follicular lymphoma.

10.2.1 Viral Lymphadenopathy

Often, viral infections lead to paracortical hyperplasia in which expansion of the paracortex with the presence of transformed lymphocytes is observed. There is also hyperplasia of the interdigitating dendritic cells, and these cells have pale cytoplasm that causes the paracortex to have a mottled appearance. Follicular hyperplasia may also be present. The large transformed cells are immunoblasts and may mimic large cell lymphoma. Occasionally, these cells may appear like Reed–Sternberg cells. The immunoblasts may be B cells (more often) or T cells. Sometimes, with the presence of B immunoblasts, CD20 expression is downregulated. These cells frequently express CD30, thus mimicking Reed–Sternberg cells. However, these are negative for CD15. In cases of cytomegalovirus (CMV) infections, intranuclear viral inclusions with a halo may be observed. In herpes simplex virus infections, multinucleated giant cells with ground glass nuclei may be seen. Like CMV, intranuclear inclusions may also be present. In measles, Warthin–Finkeldey giant cells may be seen.

In patients with HIV infections, lymph nodes typically exhibit follicular hyperplasia. Three patterns have been described. In pattern A (early pattern), there are reactive follicles with reduced mantle zone. There are folliculosis and interfollicular hemorrhage. Aggregates of monocytoid B cells are evident, and scattered Warthin–Finkeldey cells may also be present. In pattern C (late stage), the follicles are atrophic with vascular proliferation, which is highly similar to the "lollipop" appearance of follicles often seen in hyaline vascular Castleman disease. Pattern B is a transition from pattern A to pattern C and has features in between. Lymph node CD4$^+$ T cells are selectively depleted during HIV infection. However, except for end stage disease, lymph node changes do not correlate with those of peripheral blood; therefore, peripheral blood is a more sensitive parameter for monitoring immunologic changes in various stages of HIV infection [1].

10.2.2   Bacterial Infections and Lymphadenopathy

In most instances, bacterial infections involve the lymph node or the spleen as a secondary site of dissemination. A notable exception is cat scratch disease, which primarily affects the lymph nodes. Most of the bacterial infections produce nonspecific changes, although some infections do produce morphologic changes characteristic of a specific organism. Cat scratch disease is the most common cause of chronic lymphadenopathy among children and adolescents. It features subacute regional lymphadenitis with an associated inoculation site due to a cat scratch or bite, and the causative agent is *Bartonella henselae* [2]. Various bacterial infections and associated morphologic changes are described in Table 10.3.

10.2.3   *Toxoplasma gondii* and Lymphadenopathy

Toxoplasma gondii is an intracellular protozoan capable of infecting many species, including humans. The sexual form of the parasite is found in the intestinal epithelium of domestic cats (only known host), and oocytes are shed in the feces of cats. Contact with cat feces or eating the undercooked meat of an

Table 10.3 Bacterial Infections and Lymphadenopathy

Bacterial Infection	Comments
Staphylococcus and *Streptococcus*	Acute inflammatory cells with necrosis (central); abscess formation may be seen. Inflammation of the capsule is evident. Increased number of macrophages will be seen and subsequently granulation tissue and fibrosis.
Actinomyces	Similar to above; in addition, sulfur granules that are large bacterial colonies may be seen.
Bartonella henselae (cat scratch disease)	Follicular hyperplasia with necrosis and acute inflammation. Areas of necrosis are surrounded by palisading histiocytes. Giant cells may be seen. Warthin–Starry stain highlights the bacteria, similarly to immunostains.
Bartonella henselae (bacillary angiomatosis)	Bacillary angiomatosis is typically seen in immunocompromised individuals, especially HIV/AIDs patients. The lymph node shows nodules, composed of vessels. The endothelial cells of the vessels have vacuolated cytoplasm. In between the blood vessels are amphophilic materials in the form of clumps of bacteria. Also present are foamy macrophages between the blood vessels.
Tropheryma whipplei	Dilatation of the sinuses by large lipid vacuoles and multinucleated giant cells. Also found are macrophages with vacuolated cytoplasm. These macrophages contain the bacteria that are diastase resistant and PAS positive. Overall, the appearance resembles that of "Swiss cheese."
Treponema pallidum	Follicular hyperplasia with significant intertollicular hyperplasia is seen. There is also vascular proliferation with vasculitis. The capsule may become thickened. Granulomas may be seen.
Tuberculosis	Tuberculosis by *Mycobacterium tuberculosis* forms granulomatous lymphadenitis. Granulomas are a focal collection of epithelioid macrophages. Some may be converted to giant cells. There is a peripheral collar of fibroblasts, plasma cells, and lymphocytes. There is central coagulative necrosis, referred to as caseous necrosis.

animal infected with *T. gondii* may cause human infection. Toxoplasmosis is a very common human infection throughout the world; primary infection during pregnancy is of most importance because it may result in congenital toxoplasmosis, stillbirth, or even spontaneous abortion [3]. *Toxoplasma* lymphadenopathy is characterized by a triad of histologic findings: follicular hyperplasia, clusters of epithelioid macrophages within or in close proximity to germinal centers, and clusters of monocytoid B cells within sinuses.

10.2.4 Granulomatous Lymphadenopathy

Granulomatous lymphadenopathy is an important category of lymphadenopathy, and there are a wide variety of causes. Granuloma formation is a chronic inflammatory reaction in which macrophages and other inflammatory cells are involved. In granulomatous lymphadenopathy caused by cat scratch disease and tularemia, monocytoid B lymphocytes with T cells and macrophages contribute to the formation of granuloma. However, granulomatous lymphadenopathy may be due to a noninfective cause, such as sarcoidosis and sarcoid-like reaction [4]. Sarcoidosis classically presents with

> ### BOX 10.2 COMMON CAUSES OF GRANULOMATOUS LYMPHADENOPATHY
>
> - Infections (viral, fungal, mycobacterial, bacterial)
> - Systemic (sarcoidosis, rheumatoid arthritis, vasculitis)
> - Associated with tumors (lymphomas, carcinomas, melanoma)
> - Immunodeficiency states
> - Foreign body

hilar lymphadenopathy and rounded, nodular, non-necrotizing granulomas. Common causes of granulomatous lymphadenopathy are listed in Box 10.2.

10.2.5 Necrotizing Lymphadenopathy

In necrotizing lymphadenopathy, focal or geographic areas of necrosis occur that are associated with a cellular response. Both bacterial (e.g., cat scratch disease, lymphogranuloma venarum, and tularemia) and viral (e.g., Epstein–Barr virus and herpes simplex virus) infections are associated with necrotizing lymphadenopathy. An important differential diagnosis of necrotizing lymphadenopathy is Kikuchi–Fujimoto disease.

10.2.6 Progressive Transformation of Germinal Centers

In typical cases of progressive transformation of germinal centers (PTGC), large lymphoid nodules in a background of follicular hyperplasia are observed. These large nodules are progressively transformed germinal centers. It is thought that the cells of the mantle zone infiltrate and disrupt the germinal center, and during this process the germinal center becomes ill-defined. The follicular center cells are reduced. Tingible body macrophages are rare or absent. In the majority of cases, PTGC is seen to affect the lymph nodes focally. PTGC is considered as a benign, reactive change. However, it is associated with lymphomas, especially nodular lymphocyte predominant Hodgkin lymphoma (NLPHL).

10.2.7 Regressive Changes in Germinal Centers

This condition is characterized by small, compact B cell nodules in which the germinal center lymphocytes are lost. The stromal components are prominent, including the follicular dendritic cells. These features are commonly associated with hyaline vascular Castleman disease. Other conditions associated with similar findings are Epstein–Barr virus infection, autoimmune diseases, post radiation or post chemotherapy, HIV/AIDS, and syphilis.

10.3 SPECIFIC CLINICAL ENTITIES WITH LYMPHADENOPATHY

Lymphadenopathy may have many causes, and in this section the association of lymphadenopathy with specific diseases is addressed. Lymphadenitis, which refers to inflammation of lymph nodes, is often associated with lymphadenopathy. It is usually difficult to distinguish between these two diseases and the terms are often used interchangeably.

10.3.1 Kikuchi—Fujimoto Disease

The etiology of Kikuchi disease remains elusive, and various infective organisms have been implicated, including viral infection. This disease was first reported in Japan in 1972, and it is observed predominantly in Japan and other Asian countries. It is a rare disease that is typically self-limiting within 1—4 months, but a recurrence rate of 3 or 4% has been reported. The disease presents as painless cervical lymphadenopathy in young adults; it occurs in both males and females, but females are four times more likely to be affected. Diagnosis is based on histology of the affected lymph node specimen obtained by excisional biopsy [5]. Morphologically, areas of necrosis in the paracortex are observed along with the presence of abundant karyorrhectic debris. Interestingly, neutrophils are absent. The necrotic areas contain bright eosinophilic fibrinoid deposits and are surrounded by activated large T lymphocytes and numerous histiocytes. The histiocytes have abundant cytoplasm. Plasma cells and plasmacytoid monocytes are also present. Due to the areas of necrosis and the surrounding numerous histiocytes, Kikuchi disease is synonymous with necrotizing histiocytic lymphadenitis.

10.3.2 Kimura Disease

The etiology of Kimura disease is also unknown, and like Kikuchi disease, it is seen more often in Asians. It is a chronic inflammatory disorder of the subcutaneous tissue and affected regional lymph nodes. The cervical area is the most common site of involvement. Affected lymph nodes exhibit follicular hyperplasia, intense eosinophilia with eosinophilic microabscess, and infiltration of the germinal centers. Warthin—Finkeldey polykaryocytes may be seen in the germinal centers. Hyperplasia of capillaries is also a feature. This disease is associated with peripheral blood eosinophilia and elevated levels of IgE.

10.3.3 Kawasaki Disease

This is also known as a mucocutaneous lymph node syndrome. It was first reported in Japan, and although it is seen most often in the Japanese, it is also seen in Western countries, affecting mostly children. The etiology, however, remains unclear. Morphologically, expansion of interfollicular areas with loss of normal architecture is observed. There are areas of necrosis with the presence of neutrophils and nuclear debris. Vessels may be increased in number.

10.3.4   Dermatopathic Lymphadenitis

This is seen in lymph nodes draining areas of skin involved with irritation, infection, or inflammation. Langerhans cells present in skin migrate to regional lymph nodes upon antigenic stimulation. Subsequently, T cell activation occurs, and as a result, T cell hyperplasia in the lymph node is observed. Therefore, the paracortical area is expanded. Macrophages with brown pigments are found within the paracortex. The pigment is typically melanin. There are also interspersed interdigitating dendritic cells and Langerhans cells.

10.3.5   Lymphadenopathy in Autoimmune Diseases

Autoimmune diseases such as systemic lupus erythematosus (SLE) and rheumatoid arthritis are important causes of lymphadenopathy. In SLE, necrotic areas within the paracortex are seen with eosinophilic debris within the necrotic areas. In time, histiocytes and lymphocytes surround the necrotic areas. Multiple hematoxylin bodies, which are ill-defined structures representing degenerated nuclei, are seen. In cases of lymphadenopathy due to rheumatoid arthritis, the lymph nodes exhibit florid follicular hyperplasia. Plasmacytosis is seen in the interfollicular areas. Deposition of hyaline material within the lymph nodes also occurs.

10.3.6   Rosai—Dorfman Disease

This disorder is also known as sinus histiocytosis with massive lymphadenopathy. The disease presents as bilateral painless cervical lymphadenopathy. Typically, it is a benign disorder that resolves spontaneously. Morphologically, the lymph nodes demonstrate dilatation of the sinuses by large macrophages with round vesicular nuclei, nucleoli, and foamy cytoplasm. Emperipolesis (which is the presence of intact lymphocytes within histiocytes) is present and is the hallmark of the disease. Eosinophils are not prominent, which is a feature of Langerhans cell histiocytosis (LCH). The macrophages are positive for S100 and also for typical macrophage/histiocytic markers (e.g., CD68). However, macrophages are negative for CD1a. In contrast, CD1a is positive in LCH.

10.3.7   Langerhans Cell Histiocytosis

LCH is a group of disorders, ranging from focal lesions (typically of bone) to systemic disease. Lymph nodes and spleen may be sites of primary disease or of secondary involvement. Affected lymph nodes show sinusoidal dilation by Langerhans cells. These cells have elongated irregular nuclei with a linear groove and inconspicuous cytoplasm. Multinucleated cells as well as eosinophils are also seen. Langerhans cells are positive for S100, histiocytic markers, and also CD1a.

10.3.8   Castleman Disease

Castleman disease, also known as angiofollicular hyperplasia, can be broadly classified into two categories—localized and systemic (also known as multi-centric). The types of localized disease include hyaline vascular–Castleman disease (HV-CD) and plasma cell–Castleman disease (PC-CD). The etiology of Castleman disease is still not fully elucidated, but one known etiologic agent is human herpes viris-8 (HHV-8). HHV-8 encodes for a homolog of interleukin-6 (IL-6), which is a B cell growth factor. High levels of IL-6 may also derive from endogenous sources. The HV-CD variant is seen in 80–90% of all cases of unicentric Castleman disease. The mediastinum is a common site of involvement and typically forms a mass. There occurs follicular hyperplasia, and the follicles may show two or more germinal centers. The germinal centers are depleted of small lymphocytes. They are occupied by numerous follicular dendritic cells. Hyaline deposits are found within the germinal canters. The mantle zone of the follicles is expanded, and in some follicles, the mantle zone is composed of concentric rings (onion skin appearance). Sclerotic blood vessels are seen to radially penetrate the follicles (lollipop appearance). In the interfollicular areas, there are numerous high endothelial venules, with plasma cells, small lymphocytes, immunoblasts, eosinophils, and plasmacytoid monocytes.

In PC-CD, the interfollicular area has a large number of plasma cells. Vascular proliferation in the interfollicular areas may also be present. The follicles in PC-CD are usually hyperplastic. The histologic findings in multi-centric Castleman disease (M-CD) are similar to those of PC-CD.

Castleman disease may predispose to a variety of neoplasms. Patients with HV-CD are at risk of developing follicular dendritic cell sarcoma. Malignant lymphoma may develop in patients with PC-CD.

KEY POINTS

- Primary follicles consist of small lymphocytes and follicular dendritic cells. The small lymphocytes are predominantly B lymphocytes that exhibit surface IgD and surface IgM and are positive for CD5. Follicular dendritic cells are stained via CD21, CD23, and CD35.

- Secondary follicles are derived from primary follicles following antigenic stimulation. Secondary follicles have an inner germinal center and an outer mantle zone. Within the germinal center in a reactive follicle, there is a light zone and a dark zone. This is referred to as polarity. The light zone consists of centrocytes and follicular dendritic cells. The mitotic rate is lower in this zone. There are fewer tingible body macrophages in this area. In contrast, the dark zone has centroblasts with a high mitotic rate and increased number of tangible body macrophages. The germinal center of a reactive follicle stains through CD10 and Bcl-6 and is negative for Bcl-2.
- The outer mantle zone of the secondary follicle is composed of B lymphocytes, which are positive for CD5 and Bcl-2. Outside the mantle zone is the marginal zone, which is still a B cell area, but the cells are more loosely packed. The area beyond the marginal zone is the paracortex and is a predominantly T cell area. The paracortex is thus easily stained via CD3, a pan T cell marker. Bcl-2 also enables staining of T cells. Thus, the Bcl-2 staining pattern of a reactive lymph node includes staining of the paracortex and the outer parts of the secondary follicle but not the germinal center.
- Often, viral infections lead to paracortical hyperplasia, which results in expansion of the paracortex with the presence of transformed lymphocytes. There is also hyperplasia of the interdigitating dendritic cells, and these cells have pale cytoplasm. This causes the paracortex to have a mottled appearance. Follicular hyperplasia may also be present. The large transformed cells are immunoblasts and may mimic large cell lymphoma. Occasionally, these cells may appear like Reed–Sternberg cells.
- In cases of CMV infections, intranuclear viral inclusions with a halo may be evident. In HSV infections, multinucleated giant cells with ground glass nuclei may be seen. Like CMV, intranuclear inclusions may also be present. In measles, Warthin–Finkeldey giant cells may be seen.
- In patients with HIV infections, lymph nodes typically exhibit follicular hyperplasia. Three patterns are described. In pattern A (early pattern), there are reactive follicles with reduced mantle zone. There is folliculosis and interfollicular hemorrhage. Aggregates of monocytoid B cells are evident, and scattered Warthin–Finkeldey cells may also be present. In pattern C (late stage), the follicles are atrophic with vascular proliferation. Pattern B is a transition from pattern A to pattern C and has features in between.
- *Toxoplasma* lymphadenopathy is characterized by a triad of histologic findings: follicular hyperplasia, clusters of epithelioid macrophages, within or in close proximity to germinal centers, and clusters of monocytoid B cells within sinuses.

- In typical cases of PTGC, large lymphoid nodules in a background of follicular hyperplasia are observed. These large nodules are progressively transformed germinal centers. It is thought that the cells of the mantle zone infiltrate and disrupt the germinal center, making the germinal center ill-defined. The follicular center cells are reduced. Tingible body macrophages are rare or absent. In the majority of cases, PTGC is seen to affect the lymph nodes focally. PTGC is considered as a benign, reactive change. However, it is associated with lymphomas, especially NLPHL.
- Kikuchi−Fujimoto disease is seen predominantly in Asians. The disease presents as painless cervical lymphadenopathy in young adults. Morphologically, areas of necrosis in the paracortex are observed along with the presence of abundant karyorrhectic debris. Interestingly, neutrophils are absent. The necrotic areas contain bright eosinophilic fibrinoid deposits. The necrotic areas are surrounded by activated large T lymphocytes and numerous histiocytes that have abundant cytoplasm. Plasma cells and plasmacytoid monocytes are also present. Due to the areas of necrosis and the surrounding numerous histiocytes, Kikuchi disease is synonymous with necrotizing histiocytic lymphadenitis.
- In Kimura disease, affected lymph nodes exhibit follicular hyperplasia, intense eosinophilia with eosinophilic microabscess, and infiltration of the germinal centers. Warthin−Finkeldey polykaryocytes may be seen in the germinal centers. Hyperplasia of capillaries is also a feature. This disease is associated with peripheral blood eosinophilia and elevated levels of IgE.
- Kawasaki disease is also known as mucocutaneous lymph node syndrome. Morphologically, there occurs expansion of interfollicular areas with loss of normal architecture. There are areas of necrosis with the presence of neutrophils and nuclear debris. Vessels may be increased.
- Dermatopathic lymphadenitis is seen in lymph nodes draining areas of skin involved with irritation, infection, or inflammation. Langerhans cells present in skin migrate to regional lymph nodes upon antigenic stimulation. Subsequently, there is T cell activation and T cell hyperplasia in the lymph node. Thus, the paracortical area is expanded. Macrophages with brown pigments are found within the paracortex. The pigment is typically melanin. There are also interspersed interdigitating dendritic cells and Langerhans cells.
- Rosai−Dorfman disease is also known as sinus histiocytosis with massive lymphadenopathy. The disease presents as bilateral painless cervical lymphadenopathy. Typically, it is a benign disorder that spontaneously resolves. Morphologically, the lymph nodes demonstrate dilatation of the sinuses by large macrophages with round vesicular nuclei, nucleoli, and foamy cytoplasm. Emperipolesis is present and is

the hallmark of the disease. Eosinophils are not prominent, which is a feature of LCH. The macrophages are positive for S100 and typical macrophage/histiocytic markers (e.g., CD68). They are negative for CD1a. In contrast, CD1a is positive in LCH.

- Langerhans cell histiocytosis is a group of disorders, ranging from focal lesions (typically of bone) to systemic disease. Lymph nodes and spleen may be sites of primary disease or secondary involvement. Affected lymph nodes show sinusoidal dilation by Langerhans cells. These cells have elongated irregular nuclei with a linear groove and inconspicuous cytoplasm. Multinucleated cells as well as eosinophils are also seen. Langerhans cells are positive for S100, histiocytic markers, and also CD1a.

- Castleman disease (also known as angiofollicular hyperplasia) has essentially two forms—localized and systemic (also known as multicentric). Localized disease includes the hyaline vascular (HV-CD) type and the plasma cell (PC-CD) type. The HV-CD variant is seen in 80−90% of all cases of unicentric CD. Follicular hyperplasia occurs, and the follicles may show two or more germinal centers. The germinal centers are depleted of small lymphocytes. They are occupied by numerous follicular dendritic cells. Hyaline deposits are found within the germinal centers. The mantle zone of the follicles is expanded, and in some, the mantle zone is composed of concentric rings (onion skin appearance). Sclerotic blood vessels are seen to radially penetrate the follicles (lollipop appearance). In the interfollicular areas, there are numerous high endothelial venules, with plasma cells, small lymphocytes, immunoblasts, eosinophils, and plasmacytoid monocytes. In PC-CD, the interfollicular area has a large number of plasma cells. Vascular proliferation in the interfollicular areas may also be present. The follicles in PC-CD are usually hyperplastic. The histologic findings in multicentric M-CD are similar to those of PC-CD. CD may predispose to a variety of neoplasms. Patients with HV-CD are at risk of developing follicular dendritic cell sarcoma. Malignant lymphoma may develop in patients with PC-CD.

References

[1] Wood GS. The immunohistology of lymph nodes in HIV infection. Prog AIDS Pathol 1990;2:25−32.

[2] Lamps LW, Scott MA. Cat-scratch disease: historical, clinical and pathological perspectives. Am J Clin Pathol 2004;121(Suppl.):S71−80.

[3] Pappas G, Roussos N, Falagas ME. Toxoplasmosis, snapshots: global status of *Toxoplasma gondii* seroprevalence and implications for pregnancy and congenital toxoplasmosis. Int J Parasitol 2009;39:1385−94.

[4] Asano S. Granulomatous lymphadenitis. J Clin Exp Hematop 2012;52:1−16.

[5] Dalton J, Shaw R, Democratis J. Kikuchi−Fujimoto disease. Lancet 2014;383(9922):1098.

CHAPTER 11

B Cell Lymphomas

11.1 INTRODUCTION

Cancer is one of the leading causes of death worldwide (according to the World Health Organization (WHO), it caused 7.6 million deaths in 2008 which represents 13% of all deaths), and lymphomas are cancers that arise from clonal proliferations of lymphoid cells at various stages of differentiation [1]. Broadly, lymphomas can be classified as Hodgkin lymphoma and non-Hodgkin lymphoma. The American Cancer Society estimated that in the USA 2014, there would be approximately 70,800 new cases of non-Hodgkin lymphoma and that an estimated 18,990 people might die from these lymphomas; it also estimated that there would be 9190 new cases of Hodgkin lymphoma and that an estimated 1180 deaths might occur from Hodgkin lymphoma. Currently, non-Hodgkin lymphoma is the fifth most common type of cancer in the United States.

The WHO classification of lymphomas is commonly used in clinical practice. WHO recognizes approximately 70 different types of lymphomas and classifies them into four broad groups: mature B cell neoplasm, mature T cell and natural killer cell neoplasm, Hodgkin lymphoma, and immunodeficiency-associated lymphoproliferative disorders. Although Hodgkin lymphoma is a B cell malignancy, it is characterized by Reed–Sternberg cells, which are not found in any other B cell lymphomas. This chapter focuses on B cell lymphomas.

11.2 FOLLICULAR LYMPHOMA

Follicular lymphoma is defined as a proliferation of malignant germinal center B cells that are admixed with other nonmalignant cells, such as T cells, follicular dendritic cells (FDCs), and macrophages. Germinal B cell proliferation at least partially must follow a follicular pattern of growth in lymph

CONTENTS

177

A. Wahed and A. Dasgupta: Hematology and Coagulation. DOI: http://dx.doi.org/10.1016/B978-0-12-800241-4.00011-5

nodes. The neoplastic cells are centrocytes (cleaved follicle center cells) and centroblasts (noncleaved follicle center B cells), and the relative portion of centrocytes to centroblast determines the grading scheme of this lymphoma. Follicular lymphoma is predominantly observed in adults (median age of onset is 60 years) and is the second most common lymphoma diagnosed in the United States and Western Europe, accounting for 35% of all non-Hodgkin lymphoma and 70% of indolent lymphomas. Most patients have widespread disease at diagnosis (bone marrow is involved in 40−50% of cases), even though most are usually asymptomatic, except for lymph node enlargement.

Morphologically, lymph nodes show neoplastic follicles (follicular architecture), which are poorly defined and closely packed and devoid of mantle zone as well as polarization. The absence of tingible body macrophages is another feature. The architectural pattern of this disease is subdivided into follicular (>75% follicular), follicular and diffuse (25−75% follicular), minimally follicular (<25% follicular), and diffuse (a diffuse area is defined as an area of tissue completely lacking follicles defined by $CD21^+/CD23^+$ FDCs). Grading is from 1 to 3, depending on the number of centroblasts per high-power field (hpf): grade 1, 0−5 centroblasts/hpf; grade 2, 6−15 centroblasts/hpf; and grade 3, >15 centroblasts/hpf. Grade 3 is further subdivided into 3a (some centrocytes present) and 3b (solid sheets of centroblasts). FDCs, which are large cells with a central nucleolus, should not be counted as centroblasts. Centroblasts have multiple nucleoli located adjacent to the nuclear membrane. Morphological features and grading of follicular lymphoma are summarized in Table 11.1. If involved, bone marrow typically shows paratrabecular involvement.

Table 11.1 Morphological Features and Grading of Follicular Lymphoma

Morphological Features of Neoplastic Follicles	Patterns	Grading
Poorly defined and closely packed Devoid of mantle zone Devoid of polarization (centroblasts and centrocytes occupy different zones in a non-neoplastic follicle) Absence of tingible body macrophages	Follicular: >75% follicular Follicular and diffuse: 25−75% follicular Minimally follicular: <25% follicular Diffuse: A diffuse area is defined as an area of tissue completely lacking follicles defined by $CD21^+/CD23^+$ follicular dendritic cells	Grade 1: 0−5 centroblasts/hpf Grade 2: 6−15 centroblasts/hpf Grade 3: >15 centroblasts/hpf 3a: Some centrocytes present 3b: Solid sheets of centroblasts

11.2.1 Immunophenotyping in Follicular Lymphoma

In the diagnosis of follicular lymphoma, it is important to demonstrate that the tumor cells are B cells and not T cells. Commonly used B cell markers include CD19 (cluster of differentiation 19), CD20, CD79a, and PAX5 (paired box protein 5 encoded by the *PAX5* gene). Typically, at least two markers are selected, and these are most often CD20 and PAX5. CD3 is most frequently used to document that the tumor cells are not T cells. Ki-67 is a marker for proliferation index. This can be used in all lymphomas. Light-chain restriction can be demonstrated by immunostains for κ and λ. Follicular lymphoma is a tumor of germinal center B cells. Markers for germinal center include CD10, Bcl-2 (B cell lymphoma 2 protein encoded by the *Bcl* gene), and Bcl-6. Bcl-2 should enable staining of the tumor cells in the germinal center of the neoplastic follicles of follicular lymphoma. Reactive follicles will be negative. Therefore, markers in follicular lymphoma include the following:

- CD20, PAX5, CD3, Ki-67
- CD10, Bcl-2, and Bcl-6 (markers for germinal center).

However, grade 3 follicular lymphoma and the cutaneous type may be negative for Bcl-2.

11.2.2 Genetics of Follicular Lymphoma

Follicular lymphoma is a malignant counterpart to normal germinal B cells, and the majority of patients with this disorder show chromosomal translocation t(14;18)(q32;q21), which creates a derivative of chromosome 14 on which the *BCL-2* gene is juxtaposed to immunoglobulin heavy-chain gene (*IgH*) sequence. This abnormality results in overproduction of Bcl-2 protein, a family of proteins that block apoptosis. Therefore, as expected, overproduction of Bcl-2 protein in these patients prevents cells from undergoing apoptosis. The following two points concern the genetic aspect of follicular lymphoma:

- Rearrangement of the *BCL-2* gene is the molecular consequence of the t(14;18)(q32;q21) chromosomal translocation that is present in 75–85% of cases, and cytogenetics is the best way to detect karyotypic changes in the tumor specimen [2]. However, this abnormality is not associated with prognosis.
- Proper function of the *BCL-2* gene confers a survival advantage on B cells but at the same time failure to switch off *BCL-2* during blast transformation may contribute to development of lymphoma by preventing apoptosis.

Grade 1 and grade 2 follicular lymphoma are indolent. However, grade 3 follicular lymphoma is aggressive and is usually treated as diffuse large B cell lymphoma (DLBCL). Usually, 25–33% of cases of follicular lymphoma progress to DLBCL.

11.3   CHRONIC LYMPHOCYTIC LEUKEMIA/SMALL LYMPHOCYTIC LYMPHOMA

Chronic lymphocytic leukemia (CLL) and small lymphocytic lymphoma (SLL) are neoplasms of monomorphic small, round B lymphocytes in blood, bone marrow, and lymph nodes. CLL usually originates in the bone marrow and spills over into the blood. SLL starts in the lymphoid tissue. In fact, CLL and SLL are considered as the same entity, with SLL restricted to tissue cases, without a leukemic phase.

CLL is considered the most common leukemia in Western countries and also has the highest genetic predisposition. The majority of patients diagnosed with CLL are above age 50 years, and unlike other leukemias, radiation exposure does not increase the risk of CLL. Most patients are asymptomatic, although some may have autoimmune hemolytic anemia. The absolute lymphocyte count is greater than 5000/mm^3 of blood and persists for more than 3 months. Monoclonal B cell lymphocytosis is a condition seen in 3.5% of individuals above age 40 years, and it is uncertain whether this is a forerunner of CLL.

The morphology of lymph nodes in CLL/SLL includes the following:

- Effacement of lymph node architecture; predominant cells are small lymphocytes.
- Pseudofollicles are present, which resemble vague nodules and are sometimes described as having a cloudy sky appearance.
- Pseudofollicles have proliferation centers that contain prolymphocytes (medium-sized cells with dispersed chromatin and small nucleoli) and para-immunoblasts (large cells with dispersed chromatin and central nucleoli).
- Plasmacytoid differentiation may be observed.

Bone marrow involvement may be interstitial or diffuse or nodular.

11.3.1   Immunophenotyping for CLL/SLL

CLL/SLL is a clonal B cell lymphoproliferative disorder, and flow cytometry is useful in phenotyping. As expected in CLL/SLL, B cell markers such as CD19 and CD20 should be positive. To demonstrate clonality, these B cells will show light-chain restriction. The markers mentioned so far show dim expression. CLL cells exhibit aberrant expression of CD5 and CD23. Therefore, coexpression of CD5 and CD23 should be observed in CD19$^+$ or CD20$^+$ cells. FMC7 is typically negative in CLL/SLL.

Two additional markers, CD38 and ZAP-70 (ζ-chain associated protein kinase 70 kDa molecular weight), should also be considered because their presence indicates a poor prognosis. Immunostaining that may be considered for SLL includes that via B cell markers (should be positive), T cell markers (e.g., CD3, which should be negative), and CD5 and CD23 (both should be positive).

Cytogenetic studies demonstrate association of CLL with del(13q14), trisomy 12, del(11q22−q23), and del(17p13) [3]. However, the following are two of the most common abnormalities associated with CLL:

- Del(13q14.3) (seen in 50−60% of cases), the most frequently observed chromosomal abnormality associated with CLL; but individuals with this abnormality usually have a long survival time
- Trisomy 12 (seen in approximately 15% cases), which has an atypical morphology and aggressive clinical course (intermediate prognosis).

Deletion of 17p and 11q is associated with a poor prognosis.

In recent years, new molecular prognostic factors, such as the mutation status of the immunoglobulin variable heavy-chain gene (*IgVH* gene), CD38, and ZAP-70, have emerged with significantly improved prediction of prognosis of CLL. The mutated *IgVH* gene from a postgerminal center or memory-type "B" cell is associated with stable disease and long survival because such cells do not express ZAP-70. Important aspects of the prognosis of CLL include the following:

- If no somatic mutation of the *IgVH* gene (nonmutated *IgVH*) is present, then these cells express ZAP-70, indicating a worse prognosis.
- The presence of somatic mutations consistent with derivation from postgerminal center B cells, these cells not expressing the tyrosine kinase ZAP-70. This imparts a good prognosis.

In summary, prognosis is worse if there is diffuse marrow involvement, the presence of ZAP-70, CD38 positivity, and the presence of trisomy 12, del(17p), and del(11q). If the absolute lymphocyte doubling time is less than 1 year, this also implies poor prognosis.

CLL may transform into DLBCL (Richter transformation, 3.5% cases) and may also transform into Hodgkin lymphoma (0.5% cases).

11.4 B CELL PROLYMPHOCYTIC LEUKEMIA

B cell prolymphocytic leukemia (PLL) is a B cell lymphoproliferative disorder with prolymphocytic morphology. Most patients are above age 60 years at diagnosis, and they often have marked splenomegaly and rapidly rising lymphocyte count. Leukemic cells derived from mantle cell lymphoma, which may have similar morphology, are excluded from the diagnosis.

This leukemia is characterized by the presence of prolymphocytes (>55% but often >90%) in blood. Important features of these lymphoid cells include the following:

- Medium size (lymphocyte × 2)
- Round nucleus with central nucleolus
- Condensed chromatin
- Small amount of faintly basophilic cytoplasm.

Bone marrow shows diffuse infiltrate of prolymphocytes. Spleen shows extensive red and white involvement. Pseudofollicles are not seen in lymph nodes. Immunophenotypic characteristics of B cell prolymphocytic leukemia include the following:

- B cell markers positive with light-chain restriction
- Surface immunoglobulins (sIg; usually IgM$^+$ or IgM$^+$/IgD$^+$)
- CD5: Positive in approximately 33% of cases
- CD23: Negative
- FMC 7: Positive.

11.5 MANTLE CELL LYMPHOMA

Mantle cell lymphoma is a B cell neoplasm of monomorphous, small to medium-sized cells that resemble centrocytes and constitute approximately 6% of all non-Hodgkin lymphomas. Median age of patients at diagnosis is 60 years, and a male predominance is observed. In mantle cell lymphoma, pseudofollicles (proliferation centers) or transformed cells (centroblast/para-immunoblasts) are not observed. In addition, there is no transformation to large cell lymphoma.

Most patients with mantle cell lymphoma present with lymphadenopathy and hepatosplenomegaly (stage III or IV). Bone marrow involvement at the time of diagnosis is observed in 50–60% patients. Involvement of the gastro-intestinal tract is usually 30% (large gut: multiple lymphomatous polyposes and Waldeyer ring). The following are important morphological features of mantle cell lymphoma:

- Monomorphic proliferation of small to medium-sized lymphoid cells that resemble centrocytes
- Vague nodular/diffuse/mantle zone growth pattern
- Increase in hyalinized small blood vessels.

Immunophenotypic characteristics of mantle cell lymphoma include the following:

- B cell markers positive with light-chain restriction
- Strong expression of immunoglobulins (IgM^+ or IgM^+/IgD^+)
- $CD5^+$ but $CD23^-$
- Positive CD43, Bcl-2, and cyclin D1 (positive cyclin D1 is also seen in a small percentage of patients with of CLL, PLL, and multiple myeloma)
- Negative CD10 and Bcl-6.

Mantle cell lymphoma is a distinct subtype of malignant lymphoma characterized by chromosomal translocation t(11;14)(q13;q32) involving the cyclin D1 gene present in chromosome 11 and the Ig heavy-chain gene present in chromosome 14. This chromosomal translocation results in overexpression of cyclin D1 and cell cycle dysregulation in almost all cases. Clinically, this disease displays an aggressive course, with a continuous relapse pattern and a median survival of only 3−7 years. However, emerging strategies of including proteasome inhibitors, immune modulatory drugs, and mammalian target of rapamycin (mTOR) inhibitors along with the use of other chemotherapeutic agents may improve survival [4].

Mantle cell lymphoma variants include the following:

- Blastoid variant: Cells resemble lymphoblasts with dispersed chromatin; high mitotic rate (>10/10 high-power field)
- Pleomorphic
- Small
- Marginal zone like.

11.6 MARGINAL ZONE B CELL LYMPHOMA

Marginal zone lymphoma represents 5−17% of all non-Hodgkin lymphoma in adults. The WHO classification groups marginal zone lymphoma into three categories: splenic marginal zone lymphoma, nodal marginal zone lymphoma, and extranodal marginal zone B cell lymphoma involving mucosal tissues. Although occurring in diverse anatomic lesions, lymphoma of mucosa-associated lymphoid tissues (MALT) is a distinct clinical pathological entity, originally described by Isaacson and Wright in 1983. Key features of marginal zone lymphoma include the following:

- There is a heterogeneous population of cells that include centrocyte-like cells, small lymphocytes, centroblasts, immunoblasts, and monocytoid cells (cells with abundant cytoplasm).
- Plasmacytic differentiation is manifest.

- There is tumor infiltrate in the marginal zone that extends into the interfollicular area.
- Under low power, the tumor appears "pink" due to the monocytoid B cells.
- In epithelial tissue, the tumor cells infiltrate the epithelium, forming lymphoepithelial lesions (three or more tumor cells with distortion or destruction of the epithelium).
- There is possible transformation to DLBCL, especially with high-grade MALT.
- The gastrointestinal tract is the most common site of MALT lymphomas, and the stomach is the most common location within the gastrointestinal tract.
- Bone involvement is observed in 10−20% of cases (bone marrow involvement in nodal marginal is higher than MALT, whereas bone marrow involvement in splenic marginal is highest).

Immunophenotyping for marginal zone lymphoma includes the use of B cell markers. Positive CD43 (aberrant expression) is seen in up to 50% of cases. CD5, CD10, and cyclin D1 are negative. However, there is no specific marker for MALT lymphoma.

11.6.1 MALT Lymphoma

MALT lymphomas can occur in a variety of organs, including the orbit, conjunctiva, salivary glands, skin, thyroid glands, lungs, stomach, and intestine. These tumors are often localized and have indolent clinical behavior. Pathological evaluation of tumors, including immunohistochemical and cytogenetic studies, is useful for diagnosis. The discovery of an association between *Helicobacter pylori* infection and MALT lymphoma and cure of this disease with eradication of *H. pylori* indicates that underlying antigenic stimulation due to *H. pylori* infection may be related to this disorder. Hashimoto thyroiditis is a risk factor for thyroid MALT lymphoma, and Sjögren syndrome is a risk factor for salivary gland MALT lymphoma. For treatment of MALT lymphoma of nongastric locations, radiotherapy, chemotherapy, or a combination of both may be used, depending on the grading of tumor [5].

Chromosome translocation t(11;18)(q21;q21) is frequently associated with gastric MALT lymphoma and may be found in 30−50% of all cases, but this translocation is rare in MALT lymphoma involving other sites. The chromosome translocation t(11;18)(q21;q21) leads to a fusion of the apoptosis inhibitor gene *API2* on chromosome 11 mucosa associated lymphoid tissue lymphoma translocation 1 (*MLT/MALT1* gene) on chromosome 18. The presence of this mutation may cause gastric MALT lymphoma to be less responsive to *H. pylori* eradication therapy using multiple antibiotics. Trisomy 3 may be present in up to 60% of cases of MALT lymphomas.

11.6.2 Splenic Marginal Zone Lymphoma

Splenic marginal zone lymphoma is a rare tumor in which tumor cells surround the white pulp germinal centers with effacement of the mantle zone. The tumor cells expand into the outer marginal zone, where cells with abundant pale cytoplasm are seen. Transformed cells are also seen in the outer area. The red pulp is infiltrated with small and larger tumor cells. Splenic hilar nodes and bone marrow are often involved. Tumor cells may be seen in the peripheral blood, in which these cells appear as lymphocytes with polar villi (in hairy cell leukemia, the villi are circumferential, not polar).

11.6.3 Nodal Marginal Zone Lymphoma

In nodal marginal zone lymphoma, there is no evidence of extranodal or splenic marginal zone lymphoma. This lymphoma is in general indolent and uncommon. Morphology is similar to that of MALT lymphoma, and immunophenotyping characteristics include the following:

- The presence of B cell markers
- $CD43^+$ (aberrant expression, seen in up to 50% of cases)
- $Bcl-2^+$ but $Bcl-6^-$
- CD5, CD10, and cyclin D1 negative.

11.7 BURKITT LYMPHOMA

This is a high-grade, highly aggressive B cell tumor that was first recognized by British surgeon Denis Burkitt among children when he was working in Africa. Burkitt lymphoma is an aggressive form of disease that if not treated in a timely manner may be fatal. In Africa, Burkitt lymphoma is common among children who also may be infected with malaria and Epstein–Barr virus. Outside Africa, this type of lymphoma is relatively rare. The types of Burkitt lymphoma are as follows:

- Endemic: This type is observed mainly in Africa among children between the ages of 4 and 7 years, and this disease is more common in boys than in girls. Most often, jaws and abdomen are involved. Epstein–Barr virus can be detected in most cases.
- Sporadic: This type is observed in both children and young adults (median age, 30 years). Epstein–Barr virus may be detected in up to 30% of cases, and both HIV and Epstein–Barr virus infection may be

detected in up to 25% cases. In this type, involvement of kidneys, ovaries, and breasts is observed.

■ Immunodeficiency associated: Individuals may be positive for HIV; Epstein–Barr virus is detected in 25% of cases.

Key morphologic features of Burkitt lymphoma include diffuse growth pattern with starry sky appearance; medium-sized cells with squared-off, well-defined borders; clumped chromatin; prominent nucleoli; and high mitotic rate.

Immunophenotyping demonstrates positivity for B cell markers. CD10 and Bcl-6 are positive, but Bcl-2 and TdT (terminal deoxynucleotidyl transferase, a DNA polymerase) are negative. Ki-67 is approximately 100%.

Three specific chromosomal translocations—t(8;14), t(2;8), and t(8;22)—are commonly present in cases involving Burkitt lymphoma. These translocations have been shown to be related to DNA molecular rearrangements of the immunoglobulin genes and c-*myc* genes (cellular homolog of retroviral myelocytomatosis oncogene). In translocation t(8;14), the c-*myc* gene on chromosome 8 and the Ig heavy gene on chromosome 14 are involved in rearrangement. In translocation t(2;8), the κ gene on chromosome 2 is involved, whereas in t(8;22), the λ gene on chromosome 22 is involved.

There are two variants of Burkitt lymphoma: Burkitt lymphoma with plasmacytoid differentiation and atypical Burkitt/Burkitt-like, in which there is more pleomorphism in nuclear size and shape.

11.8   LYMPHOBLASTIC LEUKEMIA/LYMPHOBLASTIC LYMPHOMA

Lymphoblastic leukemia arises in the bone marrow, whereas lymphoblastic lymphoma arises in the lymphoid tissue. Lymphoblastic leukemia is most often B cell, and lymphoblastic lymphoma is most often T cell. Lymphoblastic lymphoma is a high-grade neoplasm that produces a mass lesion with 25% or fewer lymphoblasts in the bone marrow. There is diffuse effacement of architecture with starry sky pattern. The lymphoblasts are medium-sized cells with nuclei having condensed to dispersed chromatin and inconspicuous nucleoli. B cell lymphoblastic leukemia affects skin, bone, and lymph nodes, whereas T cell lymphoblastic lymphoma usually arises as a mediastinal mass.

Immunophenotyping involves positive B cell markers. Usually, TdT, CD10, and Bcl-2 are positive.

11.9 LYMPHOPLASMACYTIC LYMPHOMA/ WALDENSTRÖM MACROGLOBULINEMIA

Lymphoplasmacytic lymphoma (LPL) is a low-grade lymphoma with a familial predisposition (in 20% cases) and associated with hepatitis C infection. Key features include the following:

- There is involvement of bone marrow and, sometimes, lymph node and spleen.
- A spectrum of cells are seen, including small B lymphocytes, plasmacytoid lymphocytes, and plasma cells. There is an increase in mast cells as well as an increase in epithelioid histiocytes.
- Serum monoclonal protein with hyperviscosity or cryoglobulinemia may be present.
- If the monoclonal IgM concentration is greater than 3 g/dL, then this disease is called Waldenström macroglobulinemia (WM).

Other B cell disorders may also have monoclonal IgM concentrations greater than 3 g/dL, and these disorders are also classified as WM. Therefore, LPL is not synonymous with WM.

Variants of LPL include lymphoplasmacytoid (small lymphocytes with occasional plasma cells), lymphoplasmacytic (small lymphocytes, plasmacytoid lymphocytes, and plasma cells), and polymorphous (increased large cells). This disease can transform into DLBCL.

Immunophenotype involves observation of positive B cell markers. CD38 is also positive, but CD5, CD10, and CD23 are usually negative. Surface and cytoplasmic immunoglobulin may be present in some cells. Common cytogenetic abnormality is chromosomal translocation t(9;14). Rearrangement of the PAX5 gene is observed in 50% of cases.

11.10 DIFFUSE LARGE B CELL LYMPHOMA

Diffuse large B cell lymphoma (DLBCL) is a high-grade lymphoma with diffuse proliferation of large neoplastic B cells. It is the most common lymphoma (follicular lymphoma is the second most common) and accounts for 30−40% of adult non-Hodgkin lymphomas. It presents with a rapidly enlarging, often symptomatic mass at a single nodal/extranodal site. In 40% of cases, it is initially confined to an extranodal site. Bone marrow involvement is low (10% of cases) at diagnosis. Bone marrow may demonstrate a lower grade of lymphoma (referred to as discordant lymphoma).

DLBCL may occur as primary or *de novo*, but it also may be secondary to progression/transformation from other types of lymphoma, including SLL

(Richter transformation, 3.5% of cases), follicular lymphoma (25−33% of cases), marginal zone lymphoma, nodular lymphocyte predominant Hodgkin lymphoma (NLPHL), and LPL. Morphologic variants include the following:

- Centroblastic: Round, vesicular nuclei with two to four nucleoli
- Immunoblastic: Central nucleolus, basophilic cytoplasm
- Anaplastic: Very large, polygonal cells with bizarre pleomorphic nuclei; may grow in cohesive pattern, mimicking carcinoma; tend to be CD30$^+$.

Immunophenotyping usually shows the presence of positive B cell markers such as CD20, CD22, and CD79a. Surface immunoglobulins (IgM > IgG > IgA) are present in 50−75% of cases. Bcl-2 marker is positive in approximately 30% of cases, but positive Bcl-6 is observed in a much higher percentage of cases. Approximately 30% of DLBCL cases show positive CD30. CD10 positivity implies transformation from follicular lymphoma. Immunohistochemical subgroups include the following:

- CD5$^+$ DLBCL
- Germinal center B cell-like (GCB): CD10$^+$ (>30% of cells) or CD10$^-$ but Bcl-6$^+$. However, MUM1 is negative, and activation of nuclear transcription factor-κB (NF-κB) is absent.
- Nongerminal center B cell-like (also known as ABC type, which stands for activated B cell): Here, activation of NF-κB is positive.

The various subtypes of DLBCL are summarized in Box 11.1. Testing for Epstein−Barr virus is useful in patients suspected of having DLBCL upon review of tissue biopsy. In biopsy tissue, molecular detection of Epstein−Barr virus-encoded RNA transcript (EBER) by *in situ* hybridization is the gold standard for proving that a histopathological lesion is indeed Epstein−Barr virus related. In addition, latent membrane protein 1 (LMP1) immunostaining may also be performed to detect latent Epstein−Barr virus in affected tissue [6].

11.10.1   B Cell Lymphoma, Unclassifiable with Features Intermediate Between DLBCL and Burkitt Lymphoma (Gray Zone Lymphoma)

These lymphomas have morphologic and genetic features of both DLBCL and Burkitt lymphoma. The cells are typically a mixture of large and smaller cells. Starry sky may be seen. Ki-67 proliferation index is usually very high—approximately 90%. Some of these lymphomas may show evidence of "double-hit" genetic abnormalities, both Bcl-2 and MYC translocations.

BOX 11.1 DLBCL VARIANTS

- Primary central nervous system lymphoma of brain: Diffuse growth pattern with tumor cells in the perivascular area.
- Primary cutaneous DLBCL, leg type.
- Primary mediastinal lymphoma: Most likely thymus in origin with interstitial fibrosis and clear cell appearance of tumor cells. CD30 is positive in 80% of cases, but cells lack surface immunoglobulins.
- Intravascular large B cell lymphoma.
- DLBCL associated with chronic inflammation: Prototype is pyothorax-associated lymphoma, which may be associated with Epstein—Barr virus.
- Lymphomatoid granulomatosis: Angiocentric and angiodestructive tumor in which Epstein—Barr virus-induced immunodeficiency is associated with increased risk. Lungs are the most common site, and there are three grades according to distribution of inflammatory cells and large cells. In grade 1, large cells are minimal, whereas in grade 3, large cells are the most numerous cells.
- T cell/histiocyte-rich DLBCL.
- Anaplastic lymphoma kinase-positive DLBCL: These are CD 30$^-$.
- Epstein—Barr virus-positive DLBCL of elderly.
- Primary effusion lymphoma: Patients are usually positive for HIV, Epstein—Barr virus, and HHV-8 virus (also known as Kaposi sarcoma-associated herpes virus). This disorder may show plasmablastic morphology.
- Plasmablastic lymphoma: Involvement of oral cavity, and patients are positive for HIV.
- Large cell lymphoma arising in HHV-8-associated Castleman disease: Patients are usually HIV$^+$ and exhibit large cells with plasmablastic morphology.

11.11 HAIRY CELL LEUKEMIA

Hairy cell leukemia (HCL) is a rare neoplasm of peripheral B cells that was first described by Bouroncle and colleagues as leukemia reticuloendotheliosis and later renamed hairy cell leukemia because of typical cytoplasmic projections (like hair) in tumor cells. The incidence of HCL is less than 1 in 100,000 people, and it comprises less than 2% of lymphoid leukemia, affecting mainly males (male-to-female ratio, 4:1). The median age of diagnosis is 55 years. Most patients present with enlarged but rarely symptomatic splenomegaly and nonspecific symptoms such as fatigue and weakness. HCL is an indolent lymphoma involving bone marrow and splenic red pulp. Hepatomegaly is observed in nearly half of patients, and lymph nodes are rarely enlarged [7]. Until 1984, splenectomy was considered the treatment of choice, associated with good therapeutic response. Currently, however, HCL is treated with the purine analogs pentostatin and cladribine, which can result in complete remission in 76—98% of cases, and today HCL is considered a treatable disease [8].

In the peripheral blood, atypical lymphocytes with hairy projections are also observed. Typical features include pancytopenia and monocytopenia. The tumor cells are small to medium-sized cells with oval-indented nuclei (heavy chromatin, absent nucleoli) and abundant pale cytoplasm. In the bone marrow biopsy, increased reticulin and cells with pale cytoplasm (fried egg appearance) are observed. Red pulp disease with white pulp atrophy is observed in the spleen.

Immunophenotyping shows positive CD11c, CD25, CD103, CD123, and annexin A1. In addition, the tartrate resistant acid phosphatase (TRAP) test in blood cells or bone marrow is also positive, which is characteristic of HCL.

Hairy cell leukemia variant (HCL-V) is a rare B cell lymphoproliferative disease that shares many clinical, morphological, and immunophenotyping features with classic HCL, including splenomegaly, neoplastic lymphocytes with cytoplasmic projections, or hair and bone marrow involvement. However, in contrast to classic HCL, patients with HCL-V have significantly elevated white blood cell count and easy to aspirate bone marrow (sometimes it is difficult to aspirate bone marrow in HCL patients), and neoplastic cells infrequently show a positive TRAP test. In addition, patients with HCL-V may have anemia or thrombocytopenia but usually no monocytopenia, neutropenia, or pancytopenia. The HCL-V tumor cells exhibit prominent nucleoli—a feature typically absent in conventional HCL. Most patients with HCL-V show diminished response or even lack of response to conventional therapy for HCL, including therapy with interferon-α and cladribine [9].

11.11.1 Approach to the Diagnosis of Lymphoma

The following are useful points to consider in the evaluation of lymphomas:

- Lymphomas with the presence of follicles: The presence of follicles may indicate reactive follicles or follicular lymphoma. Sometimes there may be neoplastic follicles, which represent the follicular growth pattern of mantle cell lymphoma. In marginal zone lymphoma, benign follicles may be present with tumor cells in the interfollicular area. In LPL, there may be retention of normal lymph node architecture, or sometimes there may be a follicular growth pattern. Peripheral T cell lymphoma not otherwise specified has a variant called follicular pattern. Angioimmunoblastic T cell lymphoma (AITL) is characterized by partial effacement of lymph node architecture.
- Lymphomas with a nodular pattern (even vague nodules): Include SLL (vague nodules), NLPHL, nodular sclerosis, and lymphocyte-rich classical Hodgkin lymphoma and mantle cell lymphoma.
- Lymphomas with increased vessels: This category includes AITL, peripheral T cell lymphoma, and mantle cell lymphoma.
- Lymphomas with a starry sky pattern: Include Burkitt lymphoma, blastoid mantle, lymphoblastic lymphoma, aggressive DLBCL, and gray zone lymphoma.
- Lymphoma with a monotonous population of cells: Mantle cell lymphoma.
- Lymphomas with the presence of plasmacytoid and plasma cells: Include LPL, SLL, and marginal zone lymphoma.

- Lymphomas with the presence of intrasinusoidal tumor cells: Include ALCL and LPL.
- Lymphomas with pale or pink areas (which implies the presence of cells with abundant cytoplasm): Include marginal zone lymphoma. T cell lymphomas (especially AITL) also may have cells with pale cytoplasm.
- Lymphomas with characteristic cell sizes include: Small cells, SLL; intermediate cells, lymphoblastic lymphoma and Burkitt lymphoma; and large cells, DLBCL and ALCL.
- B cell lymphomas with CD10$^+$ tumor cells: Include follicular lymphoma, DLBCL, lymphoblastic lymphoma, and Burkitt lymphoma (typically Bcl-2$^-$ and Bcl-6$^+$).
- B cell lymphomas that are typically Bcl-2$^+$: Include follicular lymphoma, DLBCL, marginal zone lymphoma (typically Bcl-6$^-$), and mantle cell lymphoma (typically Bcl-6$^-$).

Characteristics features of various B cell lymphomas are listed in Table 11.2.

KEY POINTS

- Neoplastic follicles in follicular lymphoma are poorly defined and closely packed, devoid of mantle zone, devoid of polarization (centroblasts and centrocytes occupy different zones in a non-neoplastic follicle), and lack tingible body macrophages.
- Patterns of follicular lymphoma include follicular ($>75\%$ follicular), follicular and diffuse (25−75% follicular), minimally follicular (<25% follicular), and diffuse.
- Grading of follicular lymphoma is as follows: Grade 1, 0−5 centroblasts/hpf; grade 2, 6−15 centroblasts/hpf); and grade 3, >15 centroblasts/hpf (3a, some centrocytes present; 3b, solid sheets of centroblasts). Bone marrow, if involved in follicular lymphoma, typically shows paratrabecular involvement.
- Follicular lymphoma is a tumor of germinal center B cells. Markers for germinal center include CD10, Bcl-2, and Bcl-6. Bcl-2 will allow staining of the tumor cells in the germinal center of the neoplastic follicles of follicular lymphoma. Reactive follicles will be negative. Grade 3 and cutaneous type may be negative for Bcl-2. In general, 25−33% of cases of follicular lymphoma may progress to DLBCL.
- Chronic lymphocytic leukemia (CLL) is the most common leukemia in Western countries, with the highest genetic predisposition. Radiation exposure does not increase the risk of development of CLL, unlike with other leukemias.
- Lymph nodes with small lymphocytic lymphoma (SLL)/CLL will show the presence of pseudofollicles (proliferation centers) that contain

Table 11.2 Characteristics of Various B Cell Lymphomas

Lymphoma	Low Power	High Power	Immunohistochemistry	Genetics	Comments

Follicular lymphoma
Differential diagnosis
Reactive follicles vs. FL
Follicular growth pattern of MCL
MZL: Tumor cells in interfollicular around benign follicles
LPL: May retain normal LN architecture or have follicular growth pattern
	Back to back follicles
	Follicular pattern >75% follicular
	Follicular and diffuse: 25–75% follicular
	Minimally follicular: <25% follicular
	Diffuse
	Diffuse area completely lacking follicles
	Diffuse
	Diffuse area defined by CD21$^+$/CD23$^+$ FDC (i.e., CD21$^-$/CD23$^-$)
		No polarization; no mantle zone; no tingible body macrophages
		Grading
		  Grade 1: 0–5 centroblasts/hpf
		  Grade 2: 6–15 centroblasts/hpf
		  Grade 3: >15 centroblasts/hpf
		    3a: Some centrocytes present
		    3b: Solid sheets of centroblasts
		FDC should not be counted as centroblasts.
			B cell markers are positive.
			CD10$^+$
			Bcl-2$^+$ (in the germinal centers)
			Ki-67 low
			Grade 3 and cutaneous FL may be Bcl2$^-$ and CD10$^-$.
				t(14;18)
				Bcl-2 in chromosome 18
					Second most common lymphoma/most common low-grade lymphoma
					Bone marrow involvement, classically paratrabecular
					Can transform to DLBCL
					Report architecture and grade
					Grade 3 aggressive and treated as DLBCL
					Widespread disease at diagnosis but patient may be asymptomatic

SLL	Vague nodules (cloudy sky)	Proliferation centers (pseudofollicle): Small lymphs, prolymphocytes, and paraimmunoblast Small cells	B cell markers positive These B cells are CD5$^+$, CD23$^+$. Dim sIg (by flow)	del(13q14) Trisomy 12	Worse prognosis: CD38$^+$, Zap-70$^+$, trisomy 12, 17p$^-$, 11q$^-$, diffuse BM involvement and rapid doubling time (<1 year) Can transform to HL and DLBCL
CLL	Absolute lymph count: >5000 for 3 months Majority of patients >50 years Most patients are asymptomatic. Some patients have autoimmune hemolytic anemia.	Most common leukemia in Western countries Highest genetic disposition Radiation not a risk Monoclonal B cell lymphocytosis: >40 years; 3.5%	CLL patients have prolymphocytes (large cells with blue cytoplasm and prominent nucleoli). CLL has up to 10% prolymphocytes. CLL/PLL has up to 55% PLL has >55% prolymphocytes	PLL: Bright surface immunoglobulin (sIg), FMC7$^+$, CD23$^-$		

Entity / Epidemiology	Histology	Cytology / Differential	Immunophenotype	Genetics	Notes
MCL Median age: 60 years Male predominance	Diffuse/nodular/mantle No pseudofollicle/"proliferation centers"	Differential diagnosis of prolymphocytes: Leukemic phase of blastoid mantle, hairy cell variant Monotonous population No transformed cells (centroblasts/paraimmunoblasts)	B cell markers positive These B cells are CD5$^+$ and CD23$^-$ Bcl-2$^+$, cyclin D1$^+$ CD10$^-$ and Bcl-6$^-$	t(11; 14) Cyclin D1 on chromosome 11	Does not transform to DLBCL Variants Blastoid mantle: Higher grade; cells resemble lymphoblasts with dispersed chromatin; high mitotic rate (>10/10 hpf) Pleomorphic Small Marginal zonelike (pale cytoplasm)
MZL MALT lymphoma (most common GIT) Nodal MZL Splenic MZL	Pink (including cytoplasm in monocytoid B cells)	Heterogeneous population; monocytoid B cells; some large cells, plasmacytoid differentiation, small lymphs, neutrophils	B cell markers positive, CD5$^-$, CD10$^-$, cyclin D1$^-$, Bcl-2$^+$, Bcl-6$^-$, CD43$^+$ (50% of cases)	t(11;18) Trisomy 3	Diagnosis of exclusion Transforms to DLBCL
Burkitt lymphoma	Starry sky (this is a feature of high-grade lymphomas such as Burkitt, lymphoblastic lymphoma). The lymphocytes impart the blue background (sky), and the tingible body macrophages represent the stars.	Intermediate-size cells Prominent nucleoli Cytology: cell with deep blue cytoplasm and vacuoles. The vacuoles are stained with oil red O (due to lipid content).	B cell markers positive, CD10$^+$, Bcl-2$^-$, Bcl-6$^+$, TdT$^-$, Ki-67 ~100% Burkitt lymphoma was previously designated as ALL L3 (FAB). By flow surface immunoglobulin show light-chain restriction.	t(8;14) t(2;8) t(8;22) Chr 8: myc Chr 14: IgH Chr 2: κ Chr 22: λ	Variants Atypical BL with plasmacytic differentiation Types Endemic: Africa; children; jaw and abdominal involvement; EBV$^+$ in majority Sporadic: Kids and young adults (median age, 30 years); EBV 30% In both endemic and sporadic, involvement of ovaries, kidneys, and breasts Immunodeficiency associated: HIV$^+$, EBV 25%

Continued

Table 11.2 Characteristics of Various B Cell Lymphomas *Continued*

Lymphoma	Low Power	High Power	Immunohistochemistry	Genetics	Comments
Lymphoblastic lymphoma/ lymphoblastic leukemia	Starry sky	Intermediate-size cells Inconspicuous nucleoli	B cell markers positive CD10$^+$, TdT$^+$, Bcl-2$^+$		Lymphoblastic lymphoma most often T cell Lymphoblastic leukemia most often B cell
LPL	Retention of architecture	Spectrum of cells: Small lymphs, plasmacytoid, plasma cells. Increase in mast cells; increase epithelioid histiocytes	B cell markers positive CD38$^+$	t(9;14) Rearrangement of *PAX-5* gene (50%)	Can transform to DLBCL Other B cell disorders may have IgM > 3 g/dL (also WM), so WM not synonymous with LPL Variants Lymphoplasmacytoid— small lymphocytes, plasmacytoid lymphocytes, and plasma cells Lymphoplasmacytic— small lymphocytes, plasmacytoid lymphocytes, and increased plasma cells Polymorphous— increased large cells
		May have serum monoclonal protein with hyperviscosity or cryoglobulinemia If IgM > 3 g/dL, is Waldenström macroglobulinemia			
DLBCL	Diffuse growth pattern	Large cells Centroblastic: Round vesicular nuclei with 2–4 nucleoli Immunoblastic: Central nucleolus, basophilic cytoplasm Anaplastic: Very large, polygonal cells with bizarre pleomorphic nuclei; may grow in cohesive pattern, mimicking carcinoma; CD30$^+$	High Ki-67 (40–90%) IHC subgroups CD5$^+$ DLBCL GCB: CD10$^+$ (>30% cells) *or* CD10$^-$, Bcl-6$^+$, MUM1$^-$ NF-κB activation$^-$ Non-GCB (ABC): NF-κB$^+$		Most common lymphoma Primary or from FL, SLL, MZL, LPL, NLPHL
DLBCL variants See Box 11.1.					

ABC, activated B cell; DLBCL, diffuse large B cell lymphoma; EBV, Epstein–Barr virus; GCB, germinal center B cell-like; HHV-8, human herpesvirus-8;

prolymphocyte (medium-sized cells with dispersed chromatin and small nucleoli) and para-immunoblasts (large cells with dispersed chromatin and central nucleoli). Plasmacytoid differentiation may also be observed.

- CLL cells exhibit aberrant expression of CD5 and CD23. Therefore, coexpression of CD5 and CD23 should also be observed in $CD19^+$ or $CD20^+$ cells. FMC7 is typically negative in CLL/SLL.
- Features of poor prognosis in CLL include diffuse marrow involvement, the presence of trisomy 12, no somatic mutations (i.e., $ZAP-70^+$), CD38 positivity, and an absolute lymphocyte count doubling time of less than 1 year.
- SLL/CLL may transform into DLBCL (Richter transformation, 3.5% of cases). In addition, SLL/CLL may also transform into Hodgkin lymphoma (0.5% of cases).
- Mantle cell lymphoma (MCL) is a B cell neoplasm of monomorphous small to medium-sized cells that resemble centrocytes. It is characterized by the absence of pseudofollicle/proliferation centers. In addition, transformed cells (centroblasts/para-immunoblasts) are also absent, and no transformation to large cell lymphoma is observed.
- Mantle cell lymphoma aberrantly expresses CD5 but not CD23. The blastoid variant of MCL is an aggressive high-grade lymphoma.
- Extranodal marginal zone B cell lymphoma of mucosa-associated lymphoid tissue (MALT lymphoma) demonstrates a heterogeneous population of cells that include centrocyte-like cells, small lymphocytes, centroblasts, immunoblasts, and monocytoid cells (cells with abundant cytoplasm). Plasmacytic differentiation is also present. Marginal zone lymphoma does not have any classical markers. It is a diagnosis of exclusion. In approximately 50% of cases, there may be aberrant expression of CD43.
- Burkitt lymphoma is a high-grade, highly aggressive B cell tumor. Key morphologic features include diffuse growth pattern with starry sky appearance; medium-sized cells with squared off, well-defined borders; clumped chromatin; and prominent nucleoli with a high mitotic rate (Ki-67 ~ 100%).
- Lymphoblastic leukemia arises in the bone marrow, whereas lymphoblastic lymphoma arises in the lymphoid tissue. Lymphoblastic leukemia is most often B cell, and lymphoblastic lymphoma is most often T cell. Lymphoblastic lymphoma is a high-grade neoplasm that produces a mass lesion with 25% or fewer lymphoblasts in the bone marrow. There is diffuse effacement of architecture with a starry sky pattern. The lymphoblasts are medium-sized cells with nuclei having condensed to dispersed chromatin and inconspicuous nucleoli.
- Lymphoplasmacytic lymphoma (LPL) is a low-grade lymphoma with a familial predisposition (in 20%) and is associated with hepatitis C

infection. Key features include involvement of bone marrow and, sometimes, lymph node and spleen. In LPL, a spectrum of cells are seen, including small B lymphocytes, plasmacytoid lymphocytes, and plasma cells. There is an increase in mast cells as well as epithelioid histiocytes.

- Serum monoclonal protein with hyperviscosity or cryoglobulinemia may be present. If monoclonal IgM is greater than 3 g/dL, then this is called Waldenström macroglobulinemia (WM). Other B cell disorders may also be associated with IgM greater than 3 g/dL (these are also classed as WM). Therefore, LPL is not synonymous with WM.

- Diffuse large B cell lymphoma (DLBCL) is a high-grade lymphoma with diffuse proliferation of large neoplastic B cells. It is the most common lymphoma (follicular lymphoma is second most common) and accounts for 30–40% of adult non-Hodgkin lymphomas. It presents with a rapidly enlarging, often symptomatic mass at a single nodal/extranodal site. In 40% of cases, it is initially confined to an extranodal site.

- DLBCL may occur as primary or *de novo* or as secondary to progression/ transformation from SLL (Richter transformation, ~3.5% of cases), follicular lymphoma (25–33% of cases), marginal zone lymphoma, NLPHL, and LPL.

- Morphologic variants of DLBCL include centroblastic (round, vesicular nuclei with two to four nucleoli), immunoblastic (central nucleolus and basophilic cytoplasm), and anaplastic (very large polygonal cells with bizarre pleomorphic nuclei; may grow in a cohesive pattern, mimicking carcinoma; tend to be $CD30^+$).

- Immunohistochemical subgroups of DLBCL include $CD5^+$ DLBCL, GCB, $CD10^+$ ($>30\%$ of cells), $CD10^-$ but $Bcl\text{-}6^+$ ($MUM1^-$ and NF-κB activation also negative), and nongerminal center B cell-like ($NF\text{-}\kappa B^+$).

- Typical features of hairy cell leukemia include pancytopenia and monocytopenia. The tumor cells are small to medium-sized cells with oval-indented nuclei (heavy chromatin and absent nucleoli) and abundant pale cytoplasm. In the bone marrow biopsy, there is increased reticulin and cells with pale cytoplasm (fried egg appearance). Hairy cell leukemia cells are positive for CD11c, CD25, CD103, CD123, and annexin A1.

References

[1] Novelli S, Briones J, Sierra J. Epidemiology of lymphoid malignancies: last decade update. Springerplus 2013;2:70.

[2] Freedman A. Follicular lymphoma: 2014 update on diagnosis and management. Am J Hematol 2014;89:429–36.

[3] Gaidano G, Foa R, Dalla-Favera R. Molecular pathogenesis of chronic lymphocytic leukemia. J Clin Invest 2012;122:3432−8.

[4] Dreyling M, Kluin-Nelemans HC, Bea S, Klapper W, et al. Update on the molecular pathogenesis and clinical treatment of mantle cell lymphoma: report of the 11th annual conference of the European Mantle Cell Lymphoma Network. Leuk Lymphoma 2013;54:699−707.

[5] Tsang RW, Gospodarowica MK, Pintile M, Bezjak A. Stage I and II MALT lymphoma: results of treatment with radiotherapy. Int J Radiation Oncol Biol Phys 2001;50:1258−64.

[6] Gulley M. Molecular diagnosis of Epstein−Barr virus related disease. J Mol Diag 2001;3:1−10.

[7] Forconi F. Hairy cell leukemia: biological and clinical overview from immunogenetic insight. Hematol Oncol 2011;29:55−6.

[8] Maevis V, Mey U, Schmidt-Wolf G, Schmidt-Wolf IGH. Hairy cell leukemia: short review, today's recommendations and outlook. Blood Cancer J 2014;4:e184.

[9] Cessna MH, Hartung L, Tripp S, Perkins SL. Hairy cell leukemia variant: fact or fiction. Am J Clin Pathol 2005;113:132−8.

CHAPTER 12

T Cell and Natural Killer Cell Lymphomas

12.1 INTRODUCTION

Peripheral T cell lymphomas comprise a variety of rare malignancies derived from mature (post-thymic) T cells and natural killer (NK) cells. These malignancies are less common than B cell lymphomas and account for 5−10% of all cases of non-Hodgkin lymphoma in North America and Europe; in Asia, this percentage may be as high as 24%. T cell lymphomas in general carry a poorer prognosis than B cell lymphomas [1]. These lymphomas represent a heterogeneous group of diseases differing in histology, tumor site, and cell origin. In addition, many subtypes are present in the 2008 World Health Organization classification for which clinical, morphological, molecular, and phenotypic data are necessary. For example, human T cell leukemia virus type 1 (HTLV-1) provirus is necessary for the diagnosis of adult T cell leukemia/lymphoma. However, proper diagnosis is hampered by several difficulties, including a significant morphological and immunophenotypic overlap across different entities and lack of characteristic genetic alterations in most of them [2]. These lymphomas can be broadly classified as nodal, extranodal, cutaneous, and leukemic or disseminated types.

12.2 NODAL T CELL LYMPHOMAS

Nodal T cell lymphomas are relatively rare, and diagnostic difficulties arise from their wide range of histological patterns because most mature T cell lymphomas retain some functional characteristics of non-neoplastic T cells, such as the capacity to secrete cytokines and to co-stimulate immune cell growth, thus obscuring non-neoplastic immune cells [3].

CONTENTS

A. Wahed and A. Dasgupta: Hematology and Coagulation. DOI: http://dx.doi.org/10.1016/B978-0-12-800241-4.00012-7

12.2.1   Angioimmunoblastic T Cell Lymphoma

Angioimmunoblastic T cell lymphoma (AITL) is a peripheral T cell lymphoma characterized by systemic disease in which the primary site of disease is the lymph node. It accounts for 1 or 2% of all non-Hodgkin lymphomas and 15−20% of all T cell lymphomas. Epstein−Barr virus (EBV) has been proposed as a possible infective agent involved in the pathogenesis of AITL, but tumor cells may lack the presence of Epstein−Barr-coded RNA; or occasionally such RNA may be present in cells in some cases.

The morphology of nodes includes polymorphous infiltrate of small to medium-sized lymphocytes. These cells have clear cytoplasm and clear cytoplasmic border. The tumor cells may be admixed with other cells, such as small lymphocytes, plasma cells, eosinophils, and histiocytes.

There is also an increase in high endothelial venules (HEVs). In addition, proliferation of follicular dendritic cells (CD21⁺), typically surrounding HEVs, is observed. Three histologic patterns (I−III) are recognized. In pattern I, the lymph node architecture is partially preserved, with many hyperplastic lymphoid follicles present. In pattern III, the lymph node architecture is lost, with very few follicles.

Immunophenotyping shows positive T cell markers (e.g., CD2, CD3, and CD5). AITL is thought to arise from the CD4⁺ T cells within the germinal centers; therefore, CD4 and CD10 are also positive. Moreover, Bcl-6 and CXCL13 (chemokine: C-X-C-motif ligand 13; also known as B lymphocyte chemoattractant) are also positive.

12.2.2   Peripheral T Cell Lymphoma

Peripheral T cell lymphoma is a T cell lymphoma with a broad cytological spectrum involving both nodal and extranodal distribution (bone marrow, peripheral blood, liver, spleen, and skin).

Cells are most often medium or large, but clear cells may also be seen. Reed−Sternberg-like cells may also be present, and HEVs may be increased. Inflammatory, polymorphous background is often seen. The following three variants are known:

- Lennert lymphoma (small lymphocytes with cluster of epithelioid histiocytes)
- Follicular
- T zone (perifollicular growth pattern).

Immunophenotyping shows positive T cell markers with frequent downregulation of CD5 and CD7. In addition, most often CD4 is positive but CD8 is negative. Unlike in AITL, in peripheral T cell lymphoma, CD10, Bcl-6, and CXCL13 expressions are negative.

12.2.3 Anaplastic Large Cell Lymphoma

Anaplastic large cell lymphoma (ALCL) is a T cell neoplasm characterized by large cells with abundant cytoplasm with expression of anaplastic lymphoma kinase (ALK) protein and CD30. ALCL frequently involves nodes and extranodal sites. Morphological features include diffuse involvement with the presence of hallmark cells (typically large cells with horseshoe/reniform nucleus). Some of these cells appear to have pseudo-inclusions (doughnut cells).

Sinusoidal involvement by tumor cells is typically seen in ALCL. Variants of ALCL include the following:

- Common variant
- Lymphohistiocytic variant (tumor cells admixed with a large number of histiocytes)
- Small cell variant (here, ALK staining is nuclear)
- Hodgkin-like pattern (mimics nodular sclerosis (NS), a subtype of classical Hodgkin lymphoma)
- Composite pattern (more than one pattern).

Immunophenotyping shows positive T cell markers, but B cell markers are negative. CD30 is positive, and ALK staining is also positive. If T cell markers are negative, then this is the "null" phenotype. The chromosomal translocation t(2;5)(p23;q35) is a recurrent abnormality present in ALCL, and 46% of patients with ALCL bear this signature translocation. This translocation results in a fusion gene involving the nucleophosmin (*NPM*) gene and a receptor tyrosine kinase gene known as the anaplastic lymphoma kinase (*ALK*) gene that encodes ALK protein. The *NPM*−*ALK* chimeric gene encodes a constitutively activated tyrosine kinase. Several t(2,5) variants have been described [4]. As expected, staining for ALK protein is both cytoplasmic and nuclear. With variant translocations, staining for ALK is cytoplasmic or membranous. The majority of cases are also positive for epithelial membrane antigen (EMA) and CD4. However, CD8 is negative. In addition, ALCL is consistently negative for EBV.

This is a T cell neoplasm that is indistinguishable from ALCL on morphological grounds; however, it lacks ALK, but it is CD30 positive.

This tumor is seen more often in adults (ALK-positive ALCL is seen more often in children and young adults), and like ALK-positive ALCL, most cases present with advanced disease stage (III or IV). The tumor cells in ALK-negative ALCL tend to be large and more pleomorphic than in ALK-positive ALCL.

12.3 EXTRANODAL NK/T CELL LYMPHOMAS

Extranodal NK/T cell lymphomas are not exceptional in Western countries but predominantly affect middle-aged men in Asia, Mexico, and South

America. These lymphomas present as tumors or destructive lesions in the nasal cavity, maxillary sinuses, or palate, and despite a localized presentation, prognosis is poor in most patients. These lymphomas are also morphologically heterogeneous.

12.3.1 Extranodal NK/T Cell Lymphoma, Nasal Type

Extranodal NK/T cell lymphoma, nasal type is an extranodal lymphoma in which vascular damage with consequent necrosis commonly occurs, and there is a very strong association of this lymphoma with EBV. The nasal cavity as well as adjacent areas (nasopharynx and paranasal sinuses) are the most frequent sites of involvement. Other extranodal sites include the skin and soft tissue. This lymphoma is more prevalent in Asians and Native Americans.

Morphological features include diffuse infiltrate with an angiocentric and angiodestructive pattern in which cells may be small, medium, or even large. Non-neoplastic inflammatory cells (plasma cells, small lymphocytes, eosinophils, and histiocytes) may accompany neoplastic cells. Immunophenotyping shows positive CD2 and CD56. Although surface CD3 is negative, cytoplasmic CD3E is positive. Cytotoxic granules (e.g., granzyme, perforin, and T cell-restricted intracellular antigen (TIA1) are present. Because extranodal NK/T cell lymphoma, nasal type is closely associated with EBV, positive expression of EBV-encoded protein such as latent membrane protein 1 (LMP1) can be demonstrated using immunohistochemistry.

12.3.2 Enteropathy-associated T Cell Lymphoma

There are two forms of enteropathy-associated T cell lymphoma (EATL): EATL type I, in which the tumor cells are medium to large cells with or without pleomorphism and admixed with inflammatory cells, and EATL type II (also known as the monomorphic variant), which has medium-sized tumor cells with no inflammatory component. EATL type I is associated with celiac disease, whereas EATL II is not. In both types, intraepithelial lesions are typically seen.

Immunophenotyping shows positive T cell markers (e.g., CD3, CD5, and CD7). CD4 is negative, whereas CD8 may be positive or negative. CD30 may be positive in a varying proportion of cells. However, EATL type II cells, in addition to being positive for T cell markers, are also typically positive for CD4, CD8, and CD56.

12.3.3 Hepatosplenic T Cell Lymphoma

Hepatosplenic T cell lymphoma (HSTL) is a T cell neoplasm derived from cytotoxic T cells of the $\gamma\delta$ T cell receptor type. Typically, the tumor cells are medium in size and exhibit sinusoidal infiltration of liver, spleen, and bone marrow. Individuals who are chronically immunosuppressed (e.g., solid organ transplant patients and patients on azathioprine and infliximab for Crohn's disease) are at risk of developing this type of lymphoma.

12.3.4 Subcutaneous Panniculitis-like T Cell Lymphoma

Subcutaneous panniculitis-like T cell lymphoma (SPTCL) is a T cell lymphoma that infiltrates the subcutaneous tissue. The dermis and epidermis are not involved. Cases expressing γδ T cell receptor are excluded. The tumor cells exhibit a range of sizes, but in any particular case, tumor cell size tends to be constant. Tumor cells rim individual fat cells, and the former have a rim of pale staining cytoplasm. Admixed with the tumor cells are histiocytes. However, other inflammatory cells, such as plasma cells, are typically absent.

12.4 CUTANEOUS T CELL LYMPHOMA

Cutaneous T cell lymphoma is composed of a group of rare lymphoproliferative disorders that are characterized by localization of neoplastic T lymphocytes in the skin. The most common types of cutaneous T cell lymphomas are mycosis fungoides and its leukemic variant, Sézary syndrome, and collectively cutaneous T cell lymphomas are classified under non-Hodgkin lymphomas. The incidence of cutaneous T cell lymphoma in the United States is 7.7 cases per 1 million people per year, whereas the combined incidence of mycosis fungoides and Sézary syndrome is 6.4 cases per 1 million people per year. Pruritus is a common manifestation of these diseases, which may not respond well to therapy, and most patients present with limited plaque stage disease [5].

12.4.1 Mycosis Fungoides

Mycosis fungoides (MF) accounts for 50% of all primary cutaneous lymphomas. MF is typically confined to skin. Skin lesions include patch, papules, and plaques. The disease is characterized by epidermal and, later, dermal infiltrate of small to medium-sized T cells with cerebriform nuclei. In the early lesions, the basal layer of the epidermis is typically involved. Later lesions exhibit Pautrier microabscesses (intraepidermal collection of atypical lymphocytes). Late lesions (tumor stage) will exhibit dermal infiltrate, and epidermotropism may not be apparent. In late lesions, the tumor cells may also be large; these large cells may also be CD30 positive.

Immunophenotyping shows positive CD2, CD3, CD4, CD5, and CD8. However, CD7 is frequently negative, and CD8 is also negative.

12.4.2   Sézary Syndrome

This is a triad of erythroderma, generalized lymphadenopathy, and Sézary cells (neoplastic T cells with cerebriform nuclei) in skin, lymph node, and peripheral blood. Smaller abnormal cells are referred to as Lutzner cells. The criteria for Sézary syndrome include the following:

- 1000 Sézary cells/mm^3 (per microliter)
- CD4/CD8 >10
- Loss of one or more T cell antigens.

12.4.3   Primary Cutaneous CD30$^+$ T Cell Lymphoproliferative Disease

This category accounts for the second most common cutaneous T cell lymphomas, and this group includes the following:

- Primary cutaneous ALCL
- Lymphomatoid papulosis; papules; spontaneous regression; can progress to lymphoma
- Borderline lesion.

Primary cutaneous ALCL is actually large cell lymphoma, which is T cell (T cell phenotype with variable loss of CD2, CD5, or CD3), CD30$^+$, cutaneous lymphocyte antigen (CLA) positive (unlike systemic ALCL, which is negative for CLA), and ALK and EMA negative (unlike systemic ALCL). Primary cutaneous ALCL is an indolent lymphoproliferative disease.

Lymphomatoid papulosis (LyP) is a chronic, recurring disease in which the cells are composed of large cells and may appear anaplastic, immunoblastic, or Hodgkin-like. There is a marked accompaniment of inflammatory cells. Lymphomatoid papulosis may be followed by mycosis fungoides, cutaneous ALCL, and Hodgkin lymphoma. There are three histolgy subtypes:

- Type A: Large multinucleated or Reed–Sternberg-like cells present (CD30$^+$); numerous inflammatory cells (e.g., histiocytes, small lymphocytes, neutrophils, and eosinophils)
- Type B: Cerebriform cells in epidermis (like MF); few inflammatory cells; tumor cells CD30$^-$
- Type C: Monotonous population of large CD30$^+$ T cells with few inflammatory cells.

12.5   LEUKEMIA/DISSEMINATED

Mature T cell leukemias are a clonal proliferation of post-thymic T cells that often exhibit systematic manifestation or involvement of extramedullary sites in conjunction with hematological abnormalities. Some types of mature T cell lymphomas, such as adult T cell leukemia/lymphoma, demonstrate particular epidemiological features, such as association with HTLV-1.

12.5.1 T Cell Prolymphocytic Leukemia

T cell prolymphocytic leukemia (T-PLL) is an aggressive T cell leukemia with leukemic cells in peripheral blood, bone marrow, liver, spleen, and, sometimes, skin. The cells are typically small to medium, with basophilic cytoplasm and a nucleolus. They typically also have cytoplasmic protrusions or blebs.

Immunophenotyping indicates positive CD2, CD3, CD7, and CD52 (can be used as a target for therapy), but terminal deoxynucleotide transferase (TdT) and CD1a are negative. However, the majority are $CD4^+$ and $CD8^-$; some are $CD4^+$ and $CD8^+$.

12.5.2 T Cell Large Granular Lymphocyte Leukemia

T cell large granular lymphocyte leukemia (LGL) is an indolent leukemia characterized by greater than 2000 large granular lymphocyte cells/mm^3 (per microliter) of blood for 6 months or more. It may be accompanied by cytopenia (most often neutropenia) and is associated with autoimmune diseases.

Immunophenotyping is positive for CD3 and CD8 along with cytotoxic effector proteins (TIA1, cytotoxic granule-associated RNA binding protein, and granzymes B and M).

12.5.3 Chronic Lymphoproliferative Disorders of NK Cells

This is an indolent leukemia characterized by greater than 2000 NK cells/mm^3 of blood for 6 months or more. Unlike aggressive NK cell leukemia, this is not EBV driven and does not exhibit racial or genetic predisposition. Patients may be asymptomatic or exhibit features of cytopenia or organomegaly (lymphadenopathy, hepatomegaly, and splenomegaly).

Immunophenotyping indicates $CD3^-$ (surface), $CD16^+$, and $CD56^+$, and positivity for cytotoxic effector proteins (TIA1 and granzymes B and M).

12.5.4 Aggressive NK Cell Leukemia

This is an aggressive leukemia seen more often in young Asians and associated with EBV infection. It has the same immunophenotype as extranodal NK/T cell lymphoma, nasal type.

Immunotyping shows CD2$^+$, CD3$^-$, CD3E$^+$, and CD56$^+$, as well as the presence of cytotoxic molecules. However, in aggressive NK cell leukemia, CD16 is frequently positive.

12.5.5   Adult T Cell Leukemia/Lymphoma

Adult T cell leukemia/lymphoma (ATCL) is a T cell neoplasm caused by HTLV-1 and characterized by the presence of highly pleomorphic cells. This disease is endemic in southwestern Japan, the Caribbean basin, and areas of central Africa. There are several variants of this disease:

- Acute: Leukocytosis, hypercalcemia, skin rash, and lymphadenopathy
- Lymphomatous: Lymphadenopathy without leukemic cells in blood
- Chronic: Skin rash, leukemic cells in peripheral blood, and without hypercalcemia
- Indolent.

The leukemic cells are medium to large cells with irregular nuclei and basophilic cytoplasm. There may be many nuclear convolutions and lobules. These are referred to as flower cells. Various T cell and NK cell lymphomas are summarized in Table 12.1.

KEY POINTS

- Angioimmunoblastic T cell lymphoma (AITL) is strongly associated with EBV; however, the tumor cells are EBV$^-$. There is a polymorphous infiltrate of small to medium-sized lymphocytes. These cells have clear cytoplasm and a clear cytoplasmic border. The tumor cells may be admixed with other cells, such as small lymphocytes, plasma cells, eosinophils, and histiocytes. There is an increase in HEVs, and there is also proliferation of follicular dendritic cells (CD21$^+$), typically surrounding the HEVs. By immunohistochemistry, T cell markers are positive (e.g., CD2, CD3, and CD5). AITL is thought to arise from the CD4$^+$ T cells within the germinal centers. Therefore, CD4 and CD10 are positive. Bcl-6 and CXCL13 are also positive.
- Peripheral T cell lymphoma is T cell lymphoma with a broad cytological spectrum, with nodal and extranodal distribution (bone marrow, peripheral blood, liver, spleen, and skin). Cells are most often medium or large. Clear cells may be seen. Reed–Sternberg-like cells may also be seen. HEVs may be increased. Inflammatory, polymorphous background is often seen. The variants are Lennert lymphoma (small lymphocytes with a cluster of epithelioid histiocytes) and follicular and T zone lymphoma (perifollicular growth pattern). By immunohistochemistry, T cell markers are positive with frequent

Table 12.1 Characteristics of Various T Cell and NK Cell Lymphomas

Lymphoma	Patterns	Morphology	Phenotype	Comments and Variants
Angioimmunoblastic T cell lymphoma (AITL)	Typically diffuse with increased vessels	Polymorphous cells: Neoplastic lymphocytes with reactive cells (plasma cells, eosinophil, histiocytes) Neoplastic cells with clear cytoplasm and clear cytoplasmic border; proliferation FDCs (CD21$^+$)	Arises from follicular T helper cells (CD4$^+$, CD10$^+$, Bcl-6$^+$, CXCL13$^+$)	EBV infection may be associated with this tumor but tumor cells are EBV$^-$. Three histologic patterns (I–III): From I to III, follicles decrease and FDCs increase. Pattern II is most common.
Peripheral T cell lymphoma (PTCL)	Diffuse; if with fibrosis nodular; increased vessels	Heterogeneous population: Neoplastic cells small/medium/large plus reactive cells as above. Can have clear cytoplasm.	Mature T cell phenotype: CD4$^+$; loss of pan T antigen (e.g., CD5, CD7); immature T markers negative (CD1a, CD99)	Lymphoepithelioid variant (Lennert's lymphoma); T zone variant; follicular variant; anaplastic variant (CD30$^+$)
Anaplastic large cell lymphoma (ALCL)	Diffuse or partial; sinus involvement; fibrous bands and nodules	Classical/common variant (70–80%); Hallmark cells (horseshoe/reniform nucleus with high power filed), some with nuclear pseudo-inclusion	CD30$^+$; ALK$^+$; t(2,5) (NPM–ALK fusion gene): Cytoplasmic and nuclear. With other translocations, cytoplasm or membranous ALK. If T cell markers are negative, then it is "null phenotype." EMA$^+$ indicates poor prognosis. Cytotoxic (granzyme B, TIA-1, perforin) molecules$^+$	Classical/common; lymphohistiocytic; small (ALK nuclear; small cells CD30$^-$); Hodgkin-like pattern; composite pattern; sarcomatoid; giant cell rich; eosinophil rich; neutrophil rich
ALK negative ALCL	As above	As above	ALK$^-$; CD30$^+$	
Extranodal NK/T-cell lymphoma, Nasal type	Angiocentric and angiodestructive, necrosis Diffuse	Small/med/large	CD2$^+$, CD56$^+$, surface CD3$^-$, cytoplasmic CD3E$^+$; EBV$^+$; cytotoxic granules$^+$	More frequent in Asians and Native Americans
Enteropathy-associated T cell lymphoma (EATL)	Intraepithelial infiltrate	Infiltrate shows varying degree of transformation. EATL, type I: Tumor cells and inflammatory cells EATL, type II: Monomorphic cells but no inflammatory cells	CD3$^+$, CD7$^+$, CD5$^-$, CD4$^-$; a varying proportion show CD30$^+$; EATL, type II, positive for CD56	EATL, type I: Associated with celiac disease

Continued

Table 12.1 Characteristics of Various T Cell and NK Cell Lymphomas *Continued*

Lymphoma	Patterns	Morphology	Phenotype	Comments and Variants
Hepatosplenic T cell lymphoma	Hepatic sinusoids; splenic red pulp	Monotonous cells with rim of pale cytoplasm	CD3$^+$, CD7$^+$, CD56$^+$, CD4$^-$, CD5$^-$, CD8$^-$	Setting of chronic immunosuppression; also seen in Crohn's patients on azathioprine
Subcutaneous panniculitis-like T cell lymphoma	Epidermis and dermis spared	Monotonous, pale cytoplasm; neoplastic cells rim fat spaces; admixed histiocytes; other inflammatory cells absent	Usually CD8$^+$	
Mycosis fungoides	Patch: Basal layer of epidermis infiltrated by T cells (atypical haloed cells) Plaque: Pautrier microabscesses; medium cells with cerebriform nuclei Tumor stage: Dermal infiltrate; may see large cells that are CD30$^+$		CD2$^+$, CD3$^+$, CD4$^+$, CD8$^-$, CD5$^+$, CD7$^-$	
Sézary syndrome	Erythroderma, generalized lymphadenopathy, and Sézary cells	Sézary cells are neoplastic T cells with cerebriform nuclei.		Diagnostic criteria: 1. 1000 Sézary cells/mm^3 2. CD4/CD8 > 10 3. Loss of one or more T cell antigens
Primary cutaneous CD30$^+$ T cell lymphoproliferative disease	Primary cut anaplastic large cell lymphoma (primary cALCL); lymphomatoid papulosis; borderline lesion	Primary cALCL: Large cell lymphoma composed of T cells (T cell phenotype with variable loss of CD2, CD5, or CD3), CD30$^+$, cutaneous lymphocyte antigen (CLA) positive (unlike systemic ALCL, which is negative for CLA) and ALK and EMA negative (unlike systemic ALCL)	LyP type A: Large multinucleated or RS-like cells present (CD30$^+$); numerous inflammatory cells (e.g., histiocytes, small lymphocytes, neutrophils, eosinophils) LyP type B: Cerebriform cells in epidermis (like MF); few inflammatory cells; tumor cells CD30$^-$ Type C LyP: Monotonous population of large CD30$^+$	

		T cells with few inflammatory cells
	LyP: A chronic, recurring disease in which the cells are composed of large cells and may appear anaplastic, immunoblastic, or Hodgkin-like. There is a marked accompaniment of inflammatory cells. LyP may be followed by MF, cutaneous ALCL, and Hodgkin lymphoma.	
T cell prolymphocytic leukemia (PLL)	Aggressive T cell leukemia with the presence of leukemic cells in peripheral blood, bone marrow, liver, spleen, and sometimes skin	The cells are typically small to medium, with basophilic cytoplasm and a nucleolus. They also typically have cytoplasmic protrusions or blebs. TdT^-, $CD1a^-$, $CD2^+$, $CD3^+$, $CD7^+$, and $CD52^+$ (can be used as a target for therapy); majority are $CD4^+$ and $CD8^-$; some are $CD4^-$ and $CD8^+$
T-cell large granular lymphocyte leukemia (LGL)	This is an indolent leukemia characterized by >2000 LGL cells/mm^3 of blood for 6 months or more. It may be accompanied by cytopenia (most often neutropenia) and is associated with autoimmune diseases.	$CD3^+$, $CD8^+$, and positive for cytotoxic effector proteins (TIA1, granzymes B and M)
Chronic lymphoproliferative disorders of NK cells	This is an indolent leukemia characterized by >2000 NK cells/mm^3 of blood for 6 months or more. Unlike aggressive NK cell leukemia, this is not EBV driven and does not exhibit racial or genetic predisposition. Patients may be asymptomatic, or they may exhibit features of	$CD3^-$ (surface), $CD16^+$, $CD56^+$, positive for cytotoxic effector proteins TIA1, granzymes B and M)

Continued

Table 12.1 Characteristics of Various T Cell and NK Cell Lymphomas *Continued*

Lymphoma	Patterns	Morphology	Phenotype	Comments and Variants
	cytopenia or organomegaly (lymphadenopathy, hepatomegaly, splenomegaly).			
Aggressive NK cell leukemia	This is an aggressive leukemia seen more often in young Asians and associated with EBV infection.		Same immunophenotype as extranodal NK/T cell lymphoma, nasal type $CD2^+$, $CD3^-$, $CD3E^+$, $CD56^+$, cytotoxic molecules However, here CD16 is frequently positive.	Asians associated with EBV infection
Adult T cell leukemia	This is a T cell neoplasm caused by HTLV-1 and characterized by the presence of highly pleomorphic cells. This disease is endemic in southwestern Japan, the Caribbean basin, and areas of central Africa.	• Acute: leukocytosis, hypercalcemia, skin rash, and lymphadenopathy • Lymphomatous: Lymphadenopathy without leukemic cells in blood • Chronic: Skin rash, leukemic cells in peripheral blood and without hypercalcemia • Indolent The leukemic cells are medium to large cells with irregular nuclei and basophilic cytoplasm. There may be many nuclear convolutions and lobules. These are referred to as flower cells.	$CD2^+$, $CD3^+$, $CD5^+$, $CD7^-$, $CD4^+$, large cells may be positive for CD30	Japan, Caribbean associated with HTLV-1 infection

EBV, Epstein–Barr virus; FDC, follicular dendritic cells; LyP, lymphomatoid papulosis; MF, mycosis fungoides; RS, Reed–Sternberg.

downregulation of CD5 and CD7, but most often CD4 is positive and CD8 is negative. Unlike AITL, it is negative for CD10, Bcl-6, and CXCL13.

- Anaplastic large cell lymphoma (ALCL) is a T cell neoplasm characterized by large cells with abundant cytoplasm with expression of ALK protein and CD30. ALCL frequently involves nodes and extranodal sites. There occurs diffuse involvement with the presence of hallmark cells (typically large cells with horseshoe/reniform nucleus). Some of these cells appear to have pseudo-inclusions (doughnut cells). Sinusoidal involvement by tumor cells is typically seen. The variants of ALCL are common variant, lymphohistiocytic variant (tumor cells admixed with a large number of histiocytes), small cell variant (ALK staining is nuclear), Hodgkin-like pattern (mimics NS subtype of classical Hodgkin lymphoma), and composite pattern (more than one pattern). Immunohistochemistry analysis shows that cell markers are positive, B cell markers are negative, CD30 is positive, and ALK is positive. If T cell markers are negative, then this is the "null" phenotype. With t(2;5) (a fusion gene involving NPM and ALK), staining for ALK protein is both cytoplasmic and nuclear. With variant translocations, staining for ALK is cytoplasmic or membranous. The majority of cases are EMA$^+$, CD4$^+$ (CD8$^-$). ALCL is consistently negative for EBV.
- ALK-negative ALCL is also a T cell neoplasm that is indistinguishable from ALCL on morphological grounds. However, it lacks ALK. It is CD30$^+$. This tumor is seen more often in adults (ALK-positive ALCL is seen more often in children and young adults), and like ALK-positive ALCL, most cases present with advanced disease stage (III or IV). The tumor cells in ALK-negative ALCL tend to be large and more pleomorphic than those in ALK-positive ALCL.
- Extranodal NK/T cell lymphoma, nasal type is an extranodal lymphoma in which vascular damage occurs along with consequent necrosis. This disorder has a very strong association with EBV. Nasal cavities as well as adjacent areas (nasopharynx and paranasal sinuses) are the most frequent sites of involvement. This lymphoma is more prevalent in Asians and Native Americans. In this disorder, a diffuse infiltrate with an angiocentric and angiodestructive pattern is usually observed. Cells may be small, medium, or even large. Non-neoplastic inflammatory cells (plasma cells, small lymphocytes, eosinophils, and histiocytes) may accompany the tumor cells. Immunohistochemistry analysis shows CD2$^+$, CD56$^+$, but negative surface CD3, whereas cytoplasmic CD3E is positive, In addition, cytotoxic granules (e.g., granzyme, perforin, and TIA1) are present along with positive indication for EBV.

- There are two forms of enteropathy-associated T cell lymphoma (EATL): EATL type I, in which the tumor cells are medium to large cells with or without pleomorphism and admixed with inflammatory cells. EATL type II, also known as the monomorphic variant, has medium-sized tumor cells with no inflammatory component. EATL type I is associated with celiac disease, whereas EATL type II is not. In both, intraepithelial lesions are typically seen.
- Hepatosplenic T cell lymphoma (HSTL) is a T cell neoplasm derived from cytotoxic T cells of the $\gamma\delta$ T cell receptor type. Typically, the tumor cells are medium sized and exhibit sinusoidal infiltration of liver, spleen, and bone marrow. Individuals who are chronically immunosuppressed (e.g., solid organ transplant patients and patients on azathioprine and infliximab for Crohn's disease) are at risk of developing this disorder.
- Subcutaneous panniculitis-like T cell lymphoma (SPTCL) is a T cell lymphoma that infiltrates the subcutaneous tissue. The dermis and epidermis are not involved. Cases expressing $\gamma\delta$ T cell receptor are excluded. The tumor cells exhibit a range of sizes, but in any particular case, size tends to be constant. Tumor cells rim individual fat cells, and the cells have a rim of pale-staining cytoplasm. Admixed with the tumor cells are histiocytes. However, other inflammatory cells, such as plasma cells, are typically absent.
- Mycosis fungoides (MF) accounts for 50% of all primary cutaneous lymphomas. MF is typically confined to skin, and skin lesions include patch, papules, and plaques. The disease is characterized by epidermal and, later, dermal infiltrate of small to medium-sized T cells with cerebriform nuclei. In the early lesions, the basal layer of the epidermis is typically involved. Later lesions exhibit Pautrier microabscesses (intraepidermal collection of atypical lymphocytes). Late lesions (tumor stage) will exhibit dermal infiltrate, and epidermotropism may not be apparent. In the late lesions, the tumor cells may also be large, and these large cells may also be CD30$^+$. Immunohistochemistry analysis show CD2$^+$, CD3$^+$, CD5$^+$, CD7$^-$ (frequently), CD4$^+$, and CD8$^-$.
- Sézary syndrome is a triad of erythroderma, generalized lymphadenopathy, and Sézary cells (neoplastic T cells with cerebriform nuclei) in skin, lymph node, and peripheral blood. Smaller abnormal cells are referred to as Lutzner cells.
- Primary cutaneous CD30$^+$ T cell lymphoproliferative disease: This category accounts for the second most common cutaneous T cell lymphomas, and this group includes primary cutaneous ALCL, lymphomatoid papulosis, and borderline lesion.

- Primary cutaneous ALCL: Large cell lymphoma composed of T cells (T cell phenotype with variable loss of CD2, CD5, or CD3), CD30$^+$, CLA$^+$ (unlike systemic ALCL, which is CLA$^-$), and ALK and EMA negative (unlike systemic ALCL).
- Lymphomatoid papulosis (LyP) is a chronic, recurring disease in which the cells are large and may appear anaplastic, immunoblastic, or Hodgkin-like. There is a marked accompaniment of inflammatory cells. LyP may be followed by MF, cutaneous ALCL, and Hodgkin lymphoma.
- T cell prolymphocytic leukemia (T-PLL) is an aggressive T cell leukemia with the presence of leukemic cells in peripheral blood, bone marrow, liver, spleen, and, sometimes, skin. The cells are typically small to medium, with basophilic cytoplasm and a nucleolus. They typically also have cytoplasmic protrusions or blebs. Immunophenotype includes TdT$^-$, CD1a$^-$, CD2$^+$, CD3$^+$, CD7$^+$, and CD52$^+$ (can be used as a target for therapy). The majority cases are CD4$^+$ and CD8$^-$; some are CD4$^+$ and CD8$^+$.
- T cell large granular lymphocyte leukemia (LGL) is an indolent leukemia characterized by greater than 2000 large granular lymphocyte cells/mm^3 of blood for 6 months or more. It may be accompanied by cytopenia (most often neutropenia) and is associated with autoimmune diseases. Immunophenotype: CD3$^+$, CD8$^+$, and positive for cytotoxic effector proteins (TIA1 and granzymes B and M).
- Chronic lymphoproliferative disorders of NK cells is an indolent leukemia characterized by greater than 2000 NK cells/mm^3 of blood for 6 months or more. Unlike aggressive NK cell leukemia, this is not EBV driven and does not exhibit racial or genetic predisposition. Patients may be asymptomatic or exhibit features of cytopenia or organomegaly (lymphadenopathy, hepatomegaly, and splenomegaly). Immunophenotype: CD3$^-$ (surface), CD16$^+$, CD56$^+$, and positive for cytotoxic effector proteins (TIA1 and granzymes B and M).
- Aggressive NK cell leukemia is an aggressive leukemia seen more often in young Asians and associated with EBV infection. It has the same immunophenotype as extranodal NK/T-cell lymphoma, nasal type, as evidenced by CD2$^+$, CD3$^-$, CD3E$^+$, CD56$^+$, with the presence of cytotoxic molecules. However, CD16 is frequently positive.
- Adult T cell leukemia/lymphoma (ATCL) is a T cell neoplasm caused by HTLV-1 and characterized by the presence of highly pleomorphic cells. This disease is endemic in southwestern Japan, the Caribbean basin, and areas of central Africa. The leukemic cells are medium to large cells with irregular nuclei and basophilic cytoplasm. There may be many nuclear convolutions and lobules. These are referred to as flower cells.

References

[1] Tang T, Tav K, Quek R, Tao M, et al. Peripheral T-cell lymphoma: review and updates of current management strategies. Adv Hematol 2010;2010:624040.

[2] De Leval L, Gaulard P. Pathology and biology of peripheral T-cell lymphomas. Histopathology 2011;58:49−68.

[3] Warnke RA, Jones D, His ED. Morphological and immunophenotypic variants on nodal T-cell lymphomas and T-cell lymphoma mimics. Am J Clin Pathol 2007;127:511−27.

[4] Drexler HG, Gignac SM, von Wasielewski R, Werner M, et al. Pathobiology of NPM-A:K and variant fusion genes in anaplastic large cell lymphoma and other lymphoma. Leukemia 2000;14:1533−59.

[5] Vij A, Duvic M. Prevalence and severity of pruritus in cutaneous T-cell lymphoma. Int J Dermatol 2012;51:930−4.

CHAPTER 13

Hodgkin Lymphoma

13.1 INTRODUCTION

Hodgkin-type lymphoma was first recognized in 1832 by Thomas Hodgkin when he described postmortem findings of 7 patients with enlarged lymph nodes and spleen. In 1865, Samuel Wilks confirmed Hodgkin's findings in 15 additional patients and called this lymphoma Hodgkin disease [1]. Later, this lymphoma was classified as Hodgkin lymphoma, a hematolymphoid neoplasm primarily of B cell linage with unique histological, immunophenotypic, and clinical features. Symptoms include painless enlargement of lymph nodes, spleen, or other immune tissues. Other nonspecific symptoms, such as fever, night sweats, weight loss, low appetite, itchy skin, and fatigue, may also be present. The American Cancer Society estimates that 9190 new cases of Hodgkin lymphoma will be diagnosed in 2014, and an estimated 1180 deaths may occur. Again, this is a rare disease: The expected number of new cases per year is 2.7 per 100,000 people, and the death rate is 0.4 per 100,000 people. The age distribution of this disease is bimodal, with the first peak occurring between the ages of 15 and 30 years and the second peak in the sixth decade of life. Treatment is based on disease stage and prognostic factors. Therapy is usually multiagent chemotherapy, most often using doxorubicin, bleomycin, vinblastine, and dacarbazine, or radiotherapy or a combination of chemotherapy and radiotherapy.

13.2 OVERVIEW OF HODGKIN LYMPHOMA

Hodgkin lymphoma comprises approximately 30% of all lymphomas. Throughout the years, the absolute incidence has remained unchanged. In general, Hodgkin lymphoma arises in lymph nodes, most often the cervical region, spreading to contiguous lymph nodes. Young adults are most often affected, although it may also occur in the sixth decade of life. There is a childhood form of Hodgkin lymphoma (0–14 years) that is seen more

CONTENTS

215

A. Wahed and A. Dasgupta: Hematology and Coagulation. DOI: http://dx.doi.org/10.1016/B978-0-12-800241-4.00013-9

often in developing countries. There is a male preponderance in Hodgkin lymphoma (male-to-female ratio, 1.5:1); however, this is not seen in the nodular sclerosis subtype.

Neoplastic tissues usually contain a small number of tumor cells. Hodgkin lymphoma is characterized by a small number of scattered tumor cells residing in an abundant heterogeneous admixture of non-neoplastic inflammatory and accessory cells. The tumor cells produce cytokines, which are responsible for the presence of the background cells, lymphocytes, histiocytes, plasma cells, eosinophils, and neutrophils.

13.3   CLASSIFICATION OF HODGKIN LYMPHOMA

The Rye classification (1966) was the first classification of Hodgkin lymphoma. This classification includes nodular sclerosis, lymphocyte predominant, mixed cellularity, and lymphocyte-depleted forms. However, the current classification is based on World Health Organization (WHO) guidelines, in which Hodgkin lymphoma is classified into two major types: nodular lymphocyte predominant Hodgkin lymphoma ($\sim$5% of all cases) and classical Hodgkin lymphoma ($\sim$95% of cases; this can be further subclassified into nodular sclerosis, mixed cellularity, lymphocyte-rich, and lymphocyte-depleted types). Both classifications are summarized in Table 13.1.

Various neoplastic cells are observed in Hodgkin lymphoma. The neoplastic cells seen in classical Hodgkin lymphoma are as follows:

■ Reed—Sternberg (RS) cell: This is a large cell (20—50 μm) with abundant cytoplasm and two mirror-image nuclei, each with an eosinophilic nucleolus. The nuclear membrane is thick, with chromatin being distributed close to the nuclear membrane. Carl Sternberg (from Austria) provided the first detailed description of these cells in 1898, and in 1902 Dorothy Reed Mendenhall (from the USA) independently described these cells.

Table 13.1 Rye and WHO Classifications of Hodgkin Lymphoma

Rye Classification (1966)	WHO Classification
Nodular sclerosis Lymphocytes predominant Mixed cellularity Lymphocytes depleted	Nodular lymphocyte predominant Hodgkin lymphoma (5% of Hodgkin lymphomas) Classical Hodgkin lymphoma (95% of Hodgkin lymphomas), which can be subclassified into four groups: ■ Nodular sclerosis (70% of cases) ■ Mixed cellularity (20% of cases) ■ Lymphocyte rich (5% of cases) ■ Lymphocyte depleted (5% of cases)

- Mononuclear Hodgkin cell: This cell has the same features as the RS cell but has only one nucleus.
- Lacunar cell: This cell has cytoplasm that is retracted around the nucleus, creating an empty space. The nucleus is single and hyperlobulated.
- Mummy cell: This cell contains basophilic cytoplasm and a compact nucleus but without the presence of a nucleolus.

In nodular lymphocyte predominant Hodgkin lymphoma, "popcorn" cells (LP cells, formerly known as lymphohistiocytic cells or L&H cells) are observed [2]. These are large cells with a hyperlobulated nucleus.

In Hodgkin lymphoma, microscopically, nodules may be seen in the following:

- Nodular lymphocyte predominant Hodgkin lymphoma
- Nodular sclerosis classical Hodgkin lymphoma
- Lymphocyte-rich classical Hodgkin lymphoma.

13.3.1 Nodular Lymphocyte Predominant Hodgkin Lymphoma

Nodular lymphocyte predominant Hodgkin lymphoma (NLPHL) accounts for 5% of all reported cases of Hodgkin lymphoma. This disease is a predominantly male disease, with a 3:1 male-to-female ratio in Caucasians and a 1.2:1 ratio in African individuals. In adults, the median age of onset is 30−35 years [2]. Most patients present with localized peripheral lymphadenopathy that develops slowly and is responsive to therapy. NLPHL tends to spare the mediastinum, spleen, and bone marrow. This lymphoma was considered to be analogous to "low-grade" B cell lymphoma, but disseminated disease is not usually observed. In addition, NLPHL is typically negative for Epstein−Barr virus (EBV). Progressively transformed germinal centers are seen in association with NLPHL, but it is uncertain whether or not these lesions are preneoplastic. However, most patients with reactive hyperplasia and progressive transformation of germinal centers (PTGC) do not develop Hodgkin lymphoma. NLPHL may progress to large B cell lymphoma in 2 or 3% of cases.

The architecture of NLPHL may be nodular or nodular and diffuse. The neoplastic cells (also known as popcorn cells) and the background cells are CD20 positive. CD20 also highlights the nodularity. The neoplastic cells are positive for CD45 and negative for CD15 and CD30.

13.3.2 Classical Hodgkin Lymphoma

Classical Hodgkin lymphoma accounts for 95% of Hodgkin lymphomas, and it has a bimodal age distribution. EBV has been postulated to play a role in classical Hodgkin lymphoma. The prevalence of EBV in RS cells varies according to the histological subtype, with the highest prevalence in mixed cellularity (75%) and the lowest in nodular sclerosis (10–40%). Cervical lymph nodes are the most common area of involvement, and 60% of patients have mediastinal involvement. However, bone marrow involvement is rare, observed in only 5% of patients. The disease is characterized by the presence of RS cells in the appropriate cellular background. The neoplastic cells are usually not $CD20^+$, and the background lymphocytes are T cells ($CD20^-$).

The most common subtype of classical Hodgkin lymphoma is nodular sclerosis classical Hodgkin lymphoma (NSCHL). In contrast to other forms of Hodgkin lymphoma, nodular sclerosis occurs more often in young adults than in the elderly, and this type is more frequently seen in developed countries (resource-rich countries) in patients who belong to high-socioeconomic status groups; it is also less frequently associated with EBV [3]. NSCHL is characterized by the presence of nodules and broad bands of sclerosis, and the lymph node capsule may also be thickened. The collagen bands surround at least one nodule. This disease has a cellular phase and a fibrotic phase. Lacunar cells are seen more often in this subtype, and they may form aggregates. These aggregates may be associated with necrosis and histiocytes. This may resemble necrotizing granulomas. There is also a syncytial variant, which is an extreme form of the cellular phase. NSCHL can be further subdivided into two types: NS1 (nodular sclerosis type I), in which 75% or more of the nodules contain scattered RS cells in a lymphocyte-rich, mixed cellular, or fibrohistiocytic background; and NS2, in which at least 25% of the nodules contain increased numbers of RS cells.

The mixed cellularity subtype of classical Hodgkin lymphoma is more frequently seen in patients with HIV infection and also in developing countries. A bimodal age distribution is not seen in this type of lymphoma. This subtype demonstrates an interfollicular growth pattern with the presence of typical RS cells within an inflammatory background.

The lymphocyte-rich subtype of classical Hodgkin lymphoma is characterized by nodular (common) and diffuse architecture. The nodules (which represent expanded mantle zones) are composed of small lymphocytes that may harbor germinal centers. The RS cells are found within the nodules but not in the germinal centers. Some of the Hodgkin and Reed–Sternberg (HRS) cells may resemble LP cells, and with the nodularity they may resemble NLPHL. Intact germinal centers are infrequent in NLPHL. Thus, the presence of nodules with germinal centers and the difference in immunophenotype will help to distinguish between the diagnosis of lymphocyte-rich classical Hodgkin lymphoma and that of NLPHL.

Table 13.2 Major Characteristics of Subtypes of Classical Hodgkin Lymphoma

Nodular Sclerosis	Mixed Cellularity	Lymphocyte Rich	Lymphocyte Depleted
Most common subtype and observed in developed countries.	More frequent in patients with HIV infection and in developing countries.	Nodular (common) and diffuse architecture.	Rarest subtype seen; more frequent in patients with HIV infection and in developing countries.
Capsular fibrosis and broad collagen bands; often, lacunar and mummified cells are present.	Demonstrates an interfollicular growth pattern with the presence of typical Reed–Sternberg cells within an inflammatory background.	The Reed–Sternberg cells are found within the nodules but not in the germinal centers. Some of the HRS cells may resemble LP cells, and immunophenotyping will help to resolve the diagnosis.	Diffuse fibrosis (with a few HRS cells) and reticular subtypes (increased number of HRS cells) may be observed.
Lacunar cells may form aggregates that may be associated with necrosis and histiocytes. There are two subtypes: NS1 and NS2.	Sites are peripheral lymph node and spleen, and prognosis is intermediate.	Sites are often peripheral lymph nodes, and this type has a good prognosis.	Sites are often retroperitoneal and abdominal (in advanced stage), and this lymphoma has an aggressive course.
Sites are often cervical, axillary, and mediastinal, and this type has intermediate prognosis.			

Lymphocyte-depleted classical Hodgkin lymphoma is the rarest subtype. It is observed more frequently in patients with HIV infection and also in developing countries. As the name implies, few background lymphocytes are seen. Two subtypes are described: diffuse fibrosis (with a few HRS cells) and reticular (with an increased number of HRS cells). The major characteristics of these subtypes of classical Hodgkin lymphoma are listed in Table 13.2.

The differential diagnosis of Hodgkin lymphoma includes the following:

- Non-Hodgkin lymphoma
- NS2, reticular variant of lymphocyte-depleted Hodgkin lymphoma, anaplastic large cell lymphoma (ALCL), and T cell-rich diffuse large B cell lymphoma (DLBCL) may appear histologically similar to Hodgkin lymphoma.

13.4 IMMUNOSTAINS FOR DIAGNOSIS OF HODGKIN LYMPHOMA

Immunological and molecular studies have shown that most HRS cells of classical Hodgkin lymphoma are derived from germinal center B cells with rearranged immunoglobulin genes bearing crippling mutations. Immunohistological studies have detected B cell markers in HRS cells, including CD20 and CD79a (although less often expressed). CD20 is a transmembrane protein involved in B cell growth and differentiation. Rassidakis *et al.* reported that CD20 was expressed by HRS cells in 22% of patients with classical Hodgkin disease [4]. Eberle *et al.* reviewed the histopathology of Hodgkin lymphoma and immunostaining available for diagnosis in clinical settings [5]. The following are important points regarding immunostains in the diagnosis of Hodgkin lymphoma:

- NLPHL is a B cell neoplasm, and thus the LP cells are CD20$^+$. The predominant cell population of the nodules is also B cells. In classical Hodgkin lymphoma (in most cases also derived from B cells), CD20 may be detectable in 30−40% of cases, but it is usually of varied intensity and usually present in a minority of cases. In NLPHL, the LP cells are also positive for CD79a (a B cell marker) in most cases. CD79a is less often expressed in classical Hodgkin lymphoma. Another B cell marker, PAX5 (paired box family of transcription factor 5), is weakly expressed by HRS cells.
- LP cells (popcorn cells) are CD15$^-$ (in nearly all cases) and CD30$^-$ (in nearly all cases). These cells are positive for CD45 (in nearly all cases) and positive for epithelial membrane antigen (EMA) in 50% of cases. HRS cells are positive for CD15 (in ∼80% of cases) and positive for CD30 (in nearly all cases). These cells are usually negative for CD45 and EMA.
- In both types of Hodgkin lymphoma, the tumor cells are ringed by T cells in a rosette-like manner. Most LP cells are ringed by T cells and, less often, by CD57$^+$ T cells.
- Immunostain by BOB.1 (B lymphocyte specific coactivator of octamer binding transcription factors OCT-1 and OCT-2) and OCT.2 typically fails to stain HRS cells but typically and consistently stains LP cells. The plasma cell transcription factor IRF4/MUM1 (multiple myeloma 1/interferon regulatory factor 4) is consistently seen in classical Hodgkin lymphoma.
- EBV has been postulated to play a role in the pathogenesis of classical Hodgkin lymphoma. The presence of EBER (Epstein−Barr virus encoded RNA) is indicative of classical Hodgkin lymphoma.
- Bcl-6 is positive in nearly all cases of NLPHL, and CD68 is typically negative in classical Hodgkin lymphoma.
- The LP cells express IgD in 9−27% of cases.
- Most tumor cells of NLPHL and classical Hodgkin lymphoma express Ki-67, a protein associated with cellular proliferation.
- The nodules of NLPHL contain an expanded meshwork of follicular dendritic cells that are stained by CD21 and CD23.

In recent years, flow cytometry has been applied for immunophenotyping of classical Hodgkin lymphoma. Fromm and Wood demonstrated that six-color flow cytometry has acceptable sensitivity and specificity for clinical application, allowing immunophenotyping by this method [6]. Therefore, in the near future, clinical laboratories may use flow cytometry for the diagnosis of Hodgkin lymphoma.

13.5 STAGING OF HODGKIN LYMPHOMA

Hodgkin lymphoma is a potentially curable malignancy with a 5-year survival of 81% of patients. However, this disease may relapse in up to 30% of patients. The most common sites of disease are cervical, supraclavicular, and mediastinal lymph nodes, whereas subdiaphragmatic presentation with bone marrow and hepatic involvement are less common. Splenic involvement is usually associated with liver disease. The staging of disease is important and is usually achieved by workup including physical examination, chest X-ray, chest and abdominal CT scan, and bone marrow biopsy. Recently, it has been shown that 18-fluorodeoxyglucose positron emission tomography is useful for staging of Hodgkin lymphoma [7]. Staging of Hodgkin lymphoma includes the following:

- Stage I: Involvement of a single lymph node region or lymphoid structure (e.g., Waldeyer ring, thymus, and spleen)
- Stage II: Involvement of two or more lymph node regions on the same side of the diaphragm
- Stage III: Involvement of lymph node regions or structures, both sides of the diaphragm
- Stage IV: Diffuse or disseminated involvement of one or more extralymphatic organs, including any involvement of liver or bone marrow.

Letters are often associated with staging: "A" indicates absence, and "B" represents the presence of symptoms such as fever, night sweats, and weight loss (10% or more). In addition, "E" indicates whether the disease is extranodal or has spread from lymph nodes to adjacent tissue, and "X" denotes bulky disease if the largest deposit is greater than 10 cm or if the mediastinum is wider than one-third of the chest on X-ray. "S" indicates spleen involvement. The treatment and prognosis of classic Hodgkin lymphoma typically depends on the stage of the disease rather than the histologic classification.

KEY POINTS

- Hodgkin lymphoma comprises approximately 30% of all lymphomas. Young adults are most often affected, and there is a bimodal age distribution (15−40 years and >55 years, most commonly in the sixth decade of life).

- Hodgkin lymphoma is characterized by a small number of scattered tumor cells residing in an abundant heterogeneous admixture of non-neoplastic inflammatory and accessory cells. The tumor cells produce cytokines, which are responsible for the presence of the background cells, lymphocytes, histiocytes, plasma cells, eosinophils, and neutrophils.

- The WHO classification of Hodgkin lymphoma includes nodular lymphocyte predominant Hodgkin lymphoma (5% of cases of Hodgkin lymphoma) and classical Hodgkin lymphoma (95% of cases of Hodgkin lymphoma), which can be further subdivided into nodular sclerosis (70%), mixed cellularity (20%), lymphocyte rich (5%), and lymphocyte depleted (5%).

- Neoplastic cells are seen in Hodgkin lymphoma. In classical Hodgkin lymphoma, RS cells are seen, which are large cells (20−50 μm) with abundant cytoplasm and two mirror-image nuclei, each with an eosinophilic nucleolus. The nuclear membrane is thick, with chromatin distributed close to the nuclear membrane. Mononuclear Hodgkin cells have the same features as RS cells but with one nucleus. Lacunar cells have cytoplasm that is retracted around the nucleus, creating an empty space. The nucleus is single and hyperlobulated. Mummy cells have a basophilic cytoplasm and a compact nucleus with no nucleolus.

- In nodular lymphocyte predominant Hodgkin lymphoma, popcorn cells (LP cells, formerly known as lymphohistiocytic cells or L&H cells) are observed. These are large cells with a hyperlobulated nucleus.

- Classical Hodgkin lymphoma represents 95% of Hodgkin lymphomas, with a bimodal age distribution. EBV has been postulated to play a role in this disease. The prevalence of EBV in RS cells varies according to the histological subtype, with the highest prevalence in mixed cellularity (75%) and the lowest in nodular sclerosis (10−40%). Cervical lymph nodes are the most common area of involvement, and 60% have mediastinal involvement. Bone marrow involvement is rare (5%).

- NLPHL is typically negative for EBV. Progressively transformed germinal centers are seen in association with NLPHL. It is uncertain whether or not these lesions are preneoplastic. However, most patients with reactive hyperplasia and PTGC do not develop Hodgkin lymphoma. NLPHL may progress to large B cell lymphoma in 2 or 3% of cases.

- NLPHL is a B cell neoplasm, and thus the LP cells are $CD20^+$. The predominant cell population of the nodules is also B cells. In classical Hodgkin lymphoma (in most cases also derived from B cells), CD20 may be detectable in 30−40% of cases, but it is usually of varied intensity and usually present in a minority of cases. In NLPHL, the LP cells are also positive for CD79a (a B cell marker) in most cases. CD79a is less often expressed in classical Hodgkin lymphoma. Another B cell marker, PAX5, is weakly expressed by HRS cells.

- LP cells are CD15$^-$ (in nearly all cases) and CD30$^-$ (in nearly all cases). They are positive for CD45 (in nearly all cases) and positive for EMA in 50% of cases. HRS cells are positive for CD15 (in ∼80% of cases) and positive for CD30 (nearly all cases). They are usually negative for CD45 and negative for EMA.
- In both types of Hodgkin lymphoma, the tumor cells are ringed by T cells in a rosette-like manner. Most LP cells are ringed by T cells and, less often, by CD57$^+$ T cells.
- The treatment and prognosis of classic Hodgkin lymphoma typically depends on the stage of the disease rather than the histologic classification.

References

[1] Tamaru J. Pathological diagnosis of Hodgkin lymphoma. Nihon Rinsho 2014;72:450−5.

[2] Goel A, Fan W, Patel AA, Devabhaktuni M. Nodular lymphocyte predominant Hodgkin lymphoma: biology, diagnosis and treatment. Clin Lymphoma Myeloma Leuk 2014;14:261−70.

[3] Mani H, Jaffe ES. Hodgkin lymphoma: an update on its biology with new insights into classification. Clin Lymphoma Myeloma 2009;9:206−16.

[4] Rassidakis GZ, Medeiros LJ, Viviani S, Bonfante V, et al. CD 20 expression in Hodgkin and Reed−Sternberg cells of classical Hodgkin's disease: association with presenting features and clinical outcome. J Clin Oncol 2002;20:1278−87.

[5] Eberle FC, Mani H, Jaffe ES. Histopathology of Hodgkin's lymphoma. Can J 2009;15: 129−37.

[6] Fromm JR, Wood BL. A six color flow cytometry assay for immunophenotyping classical Hodgkin lymphoma in lymph node. Am J Clin Pathol 2014;141:388−96.

[7] Gobbi PG, Ferreri AJ, Ponzoni M, Levis A. Hodgkin lymphoma. Crit Rev Oncol Hematol 2013;85:216−37.

CHAPTER 14

Lymphoproliferative Disorders Associated with Immune Deficiencies and Histiocytic and Dendritic Cell Neoplasms

14.1 INTRODUCTION

The immune deficiency state predisposes a patient not only to infectious diseases but also to cancer, particularly cancer of the immune system [1]. Immune deficiencies are associated with a range of lymphoproliferative disorders, from benign reactive hyperplasia to atypical hyperplasias and frank lymphomas. Of the lymphomas, non-Hodgkin lymphomas are the most common type seen in such situations. These lymphomas are known to be aggressive and resistant to therapy. Patients infected with HIV are at a higher risk of developing lymphoproliferative disorders. Histiocytic and dendritic cell tumors are rare diseases, and their pathogenesis is still under investigation [2].

14.2 LYMPHOPROLIFERATIVE DISORDERS ASSOCIATED WITH IMMUNE DEFICIENCY

As stated previously, immune deficiencies are associated with a range of lymphoproliferative disorders, from benign reactive hyperplasia to atypical hyperplasias and frank lymphomas. Of the lymphomas, non-Hodgkin lymphomas are the most common type seen in such situations. These lymphomas are known to be aggressive and resistant to therapy. Leukemias and Hodgkin lymphomas are also seen with increased frequency. This topic is discussed in three sections on lymphoproliferative disorders associated with primary immune deficiency, lymphoproliferative disorders associated with HIV infection, and post-transplant lymphoproliferative disorders.

CONTENTS

A. Wahed and A. Dasgupta: Hematology and Coagulation. DOI: http://dx.doi.org/10.1016/B978-0-12-800241-4.00014-0

14.2.1   Lymphoproliferative Disorders Associated with Primary Immune Deficiency

The risk of developing lymphoma in individuals with primary immune deficiency is 10- to 200-fold higher than that of individuals without primary immune deficiency, and the risk and frequency of lymphomas depend on the specific type of immune deficiency [3]. Usually, there is a latency period before the lymphoma develops. The primary immune deficiencies most often associated with lymphomas are ataxia−telangiectasia, Wiskott−Aldrich syndrome, common variable immunodeficiency, severe combined immunodeficiency, X-linked lymphoproliferative disorder, Nijmegen breakage syndrome, hyper-IgM syndrome, and autoimmune lymphoproliferative disorder. Epstein−Barr virus (EBV) is involved in the majority of cases. The lymphomas often present in extranodal sites. The gastrointestinal (GI) tract, lungs, and central nervous system (CNS) are most often involved. Overall, diffuse large B cell lymphomas (DLBCL) are most common. Burkitt lymphoma, Hodgkin lymphomas, and T cell leukemias and lymphomas are also seen. In ataxia−telangiectasia, T cell leukemias and T cell lymphomas are more common than B cell neoplasms [4].

14.2.2   Lymphoproliferative Disorders Associated with HIV Infection

HIV-positive patients are at increased risk for all types of non-Hodgkin lymphoma. The most common HIV-associated lymphomas are Burkitt lymphoma, DLBCL (often of the CNS), primary effusion lymphoma, and plasmablastic lymphoma. With the use of highly active antiretroviral therapy (HAART), the incidence of non-Hodgkin lymphoma has decreased; however, the incidence of Hodgkin lymphoma has increased [5].

14.2.3   Post-transplant Lymphoproliferative Disorders

Post-transplant lymphoproliferative disorders (PTLDs) are seen in individuals who have undergone solid organ or bone marrow transplantations, and they develop as a consequence of immunosuppression. These disorders range from EBV-driven polyclonal proliferations (infectious mononucleosis type) to frank lymphomas. The lymphomas may or may not be positive for EBV. The majority of PTLDs in solid organ transplants are of host origin. In the majority of bone marrow transplant cases, most PTLDs are of donor origin [6].

The broad categories of PTLDs are as follows:

- Early lesions: Plasmacytic hyperplasias and infectious mononucleosis-like lesions.
- Polymorphic PTLD: A full range of B cell maturation is observed. B cells may or may not exhibit light-chain restriction.

- Monomorphic PTLD: Various B and T cell lymphomas.
- Classical Hodgkin lymphoma type PTLD.

14.3 HISTIOCYTIC AND DENDRITIC CELL NEOPLASMS

Neoplasms of this category are very rare, representing less than 1% of tumors presenting in lymph nodes or soft tissues. Histiocytes are derived from bone marrow-derived monocytes. Dendritic cells may be myeloid derived or derived from mesenchymal stem cells.

14.3.1 Histiocytic Sarcoma

As defined by the World Health Organization (WHO) classification, this category of malignant tumors includes those with morphologic and immunophenotypic features of mature tissue histiocytes. Histiocytic sarcomas are aggressive tumors. Most cases occur at extranodal sites. The most common sites of involvement are the GI tract, skin, and soft tissue. Presentation may include a mass with systemic symptoms (e.g., fever and weight loss). The tumor typically has a diffuse proliferation of large cells with a background of reactive cells. The reactive cells may be small lymphocytes, plasma cells, eosinophils, and benign histiocytes. The tumor cells are typically large with abundant eosinophilic cytoplasm. The chromatin is vesicular, and atypia may be mild to marked. A spindle cell pattern may be seen, As may multinucleated giant cells. In addition, erythrophagocytosis and emperipolesis may be seen [7,8]. The tumor cells are positive for CD45, CD4, CD11c, CD68, and lysozyme. Myeloid markers such as CD13, CD33, and MPO will be negative. Markers for follicular dendritic cells such as CD21, CD23, and CD35 are negative. CD1a, which stains Langerhans cells, is negative.

14.3.2 Dendritic Cell Neoplasms

Dendritic cell neoplasms are derived from antigen presenting cells and include tumors derived from follicular dendritic cells, interdigitating dendritic cells, and Langerhans cells.

14.3.2.1 *Follicular Dendritic Cell Sarcoma*

Follicular dendritic cell (FDC) sarcoma is a neoplastic proliferation of cells, typically spindle to ovoid cells with immunophenotypic features similar to those of non-neoplastic FDCs. FDCs are mesenchymal in origin and are located within the lymphoid follicle; they are most numerous in the light zone of the germinal center. Most tumors present as a mass, with lymphadenopathy being a frequent finding. The tumor cells are spindle to ovoid and form fascicles, whorls, or a storiform pattern. Diffuse sheets may also be seen. The cytoplasm is eosinophilic. Binucleated and multinucleated cells may be observed. The tumor cells are positive for CD21, CD23, and CD35. There is variable positivity for S100, CD68, and epithelial membrane antigen (EMA). FDC sarcomas are typically indolent tumors.

14.3.2.2 Interdigitating Dendritic Cell Sarcoma

Interdigitating dendritic cell (IDC) sarcoma is a neoplastic proliferation of cells, typically spindle to ovoid cells with immunophenotypic features of interdigitating dendritic cells. IDCs are myeloid derived. Patients present with a mass. When affected, lymph nodes demonstrate a tumor in the paracortical area; residual follicles are seen. The tumor cells are spindle to ovoid and demonstrate whorls, fascicles, or a storiform pattern. Features are thus very similar to those of FDC sarcoma. The tumor cells are positive for S100 and vimentin. Markers for FDC sarcoma, CD21, CD23, and CD35 are negative.

14.3.2.3 Langerhans Cell Histiocytosis and Langerhans Cell Sarcoma

Langerhans cell histiocytosis (LCH) is a neoplastic proliferation of Langerhans cells. Langerhans cell are also myeloid derived. Most cases of LCH occur in children. Also known as histiocytosis X, the disease can have various forms. Eosinophilic granuloma is the term used if the disease is solitary and localized to the bone. The term Hand—Schuller—Christian disease is used when there are multiple sites of involvement. Letterer—Siwe disease is the term used if the disease is disseminated or with visceral involvement. The lesions demonstrate the presence of Langerhans cells. LCH cells are oval cells distributed among eosinophils, histiocytes, neutrophils, and small lymphocytes. The nuclei of LCH cells are grooved, folded, indented, or lobulated. Electron microscopy demonstrates Birbeck granules. These cells may also be identified by langerin expression. LCH cells are positive for CD1a, langerin, and S100. The cells may also be positive for vimentin and CD68. Immunophenotype characteristics of histiocytic and dendritic cell neoplasms are summarized in Table 14.1.

Table 14.1 Histiocytic and Dendritic Cell Neoplasms

Neoplasm	Immunophenotype
Histiocytic sarcoma	The tumor cells are positive for CD45, CD4, CD11c, CD68, and lysozyme. Myeloid markers such as CD13, CD33, and MPO will be negative. Markers for follicular dendritic cells such as CD21, CD23, and CD35 are negative. CD1a, which stains Langerhans cells, is negative.
Follicular dendritic cell sarcoma	The tumor cells are positive for CD21, CD23, and CD35. There is variable positivity for S100, CD68, and EMA.
Interdigitating dendritic cell sarcoma	The tumor cells are positive for S100 and vimentin. Markers for FDC sarcoma, CD21, CD23, and CD35 are negative.
Langerhans cell histiocytosis and Langerhans cell sarcoma	Tumor cells are positive for CD1a, langerin, and S100. Cells may also be positive for vimentin and CD68.

Langerhans cell sarcoma is a high-grade neoplasm in which the tumor cells display overt malignant cytologic features and the immunophenotype is that of LCH cells.

KEY POINTS

- The risk of developing lymphoma in individuals with primary immune deficiency is 10- to 200-fold higher than that of individuals without primary immune deficiency, and the risk and frequency of lymphomas depend on the specific type of immune deficiency.
- Overall in primary immune deficiency patients, DLBCL are most common. Burkitt lymphoma, Hodgkin lymphomas, and T cell leukemias and lymphomas are also seen. In ataxia–telangiectasia, T cell leukemias and T cell lymphomas are more common than B cell neoplasms.
- The most common HIV-associated lymphomas are Burkitt lymphoma, DLBCL (often of the CNS), primary effusion lymphoma, and plasmablastic lymphoma. With the use of HAART, the incidence of non-Hodgkin lymphoma has decreased; however, the incidence of Hodgkin lymphoma has increased.
- PTLDs are seen in individuals who have undergone solid organ or bone marrow transplantations. They develop as a consequence of immunosuppression.
- The majority of PTLDs in solid organ transplants are of host origin. However, in most bone marrow transplant cases, the majority of PTLDs are of donor origin.
- As defined by the WHO classification, histiocytic sarcomas are malignant tumors with morphologic and immunophenotypic features of mature tissue histiocytes.
- Histiocytic sarcoma typically has a diffuse proliferation of large cells with a background of reactive cells. The tumor cells are positive for CD45, CD4, CD11c, CD68, and lysozyme.
- FDC sarcoma is a neoplastic proliferation of cells, typically spindle to ovoid cells with immunophenotypic features similar to those of non-neoplastic FDCs. FDCs are mesenchymal in origin and are located within the lymphoid follicle; they are most numerous in the light zone of the germinal center.
- The tumor cells in FDC are spindle to ovoid and form fascicles, whorls, or a storiform pattern. Diffuse sheets may also be seen. The cytoplasm is eosinophilic. The tumor cells are positive for CD21, CD23, and CD35. There is variable positivity for S100, CD68, and EMA. FDC sarcomas are typically indolent tumors.

- IDC sarcoma is a neoplastic proliferation of cells, typically spindle to ovoid cells with immunophenotypic features similar to those of interdigitating dendritic cells. IDCs are myeloid derived. The tumor cells are spindle to ovoid and demonstrate whorls, fascicles, or a storiform pattern. Features are thus very similar to those of FDC sarcoma. The tumor cells are positive for S100 and vimentin.
- LCH is a neoplastic proliferation of Langerhans cells. Langerhans cells are also myeloid derived.
- LCH cells are oval cells distributed among eosinophils, histiocytes, neutrophils, and small lymphocytes. The nuclei of LCH cells are grooved, folded, indented, or lobulated. Electron microscopy demonstrates Birbeck granules. They may also be identified by langerin expression. LCH cells are positive for CD1a, langerin, and S100. Cells may also be positive for vimentin and CD68.
- Langerhans cell sarcoma is a high-grade neoplasm in which the tumor cells display overt malignant cytologic features and the immunophenotype is that of LCH cells.

References

[1] Van Krieken J. Lymphoproliferative disease associated with immune deficiency in children. Am J Clin Pathol 2004;122(Suppl. 1):S122−7.

[2] Said J. Follicular lymphoma and histiocytic/dendritic neoplasm related? Blood 2008;111: 5418−9.

[3] Filipovich AH, Mathur A, Kamat D, et al. Primary immunodeficiencies: genetic risk factors for lymphoma. Cancer Res 1992;52(Suppl.): 5465s−7s.

[4] Taylor AM, Metcalfe JA, Thick J, Mak YF. Leukemia and lymphoma in ataxia telangiectasia. Blood 1996;87:423−38.

[5] Clifford GM, Polesel J, Rickenbach M, et al. Cancer risk in the Swiss HIV cohort study: associations with immunodeficiencies, smoking, and highly active retroviral therapy. J Natl Cancer Inst 2005;97:425−32.

[6] Zutter MM, Martin PJ, Sale GE, et al. Epstein−Barr virus lymphoproliferation after bone marrow transplantation. Blood 1988;72:520−9.

[7] Pileri SA, Grogan TM, Harris NL, et al. Tumors of histiocytes and accessory dendritic cells: an immunohistochemical approach to classification from the International Lymphoma Study Group based on 61 cases. Histopathology 2002;41:1−29.

[8] Vos JA, Abbondanzo SL, Barekman CL, et al. Histiocytic sarcoma: a study of five cases including the histiocytic marker CD163. Mod Pathol 2005;18:693−704.

CHAPTER 15

Essentials of Coagulation

15.1 INTRODUCTION

Blood clotting (coagulation) is initiated within seconds after vascular injury and is considered one of the fastest tissue repair systems in the human body. The main purpose of coagulation is to seal an injured vessel, and this is accomplished by aggregation of platelets at the site of injury. First, a loose platelet plug is formed, which is then stabilized by the formation of a fibrin network. Both events, respectively known as primary and secondary hemostasis, not only prevent blood loss but also trigger wound healing and tissue regeneration. Because a site of bleeding is also a potential entry point of invading microorganisms, coagulation is one of the first humoral regulatory systems to encounter invading microorganisms. Antimicrobial peptides are released from platelet when coagulation is activated. In addition, an intact platelet–fibrinogen plug can provide an active surface that allows the recruitment, attachment, and activation of phagocytosing cells. Moreover, coagulation factors are able to induce pro- and anti-inflammatory reactions by activating protease-activated receptors on immune cells [1].

15.2 NORMAL HEMOSTASIS

Hemostasis consists of three steps:

- Vasoconstriction: This is mediated by reflex neurogenic mechanisms. Vasoconstriction is augmented by endothelin, which is released from damaged endothelial cells. Vasoconstriction reduces flow of blood, thus reducing the extent of blood loss.
- Platelet plug formation (primary hemostasis): Platelets adhere to the subendothelial collagen, and shape change and release of platelet granule contents occur. Additional platelets are recruited, and a platelet plug is formed. The primary platelet plug that is formed is reversible, but with the help of fibrin, a secondary and irreversible platelet plug is formed.

CONTENTS

A. Wahed and A. Dasgupta: Hematology and Coagulation. DOI: http://dx.doi.org/10.1016/B978-0-12-800241-4.00015-2

- Activation of the coagulation cascade (secondary hemostasis): Activation of the clotting cascade results in the formation of fibrin and cross-linking of fibrin with resultant arrest of bleeding.

15.2.1   Platelets and Platelet Events

Platelets are anucleated discoid-shaped blood cells derived from megakaryocytes and have glycoproteins attached to the outer surface that serve as receptors. Morphologically, a platelet has three zones: The peripheral zone, responsible for adhesion and aggregation and also containing the platelet membrane; the sol-gel zone, responsible for contraction and support of the microtubule system; and the organelle zone. Platelet glycoprotein Ib (GpIb) is a disulfide-linked $\alpha\beta$ heterodimer (molecular weight, 16 kDa) that forms a complex with GpIX (22 kDa). GpV (82 kDa), the only major membrane protein that is a substrate for thrombin, forms a noncovalent complex in the platelet membrane with other platelet glycoproteins (GpIb/IX and GpIb/IX/V), which plays a central role in attachment of platelets with von Willebrand factor (vWF) in the subendothelium of the damaged vessel wall [2]. GpIb/IX/V has a central GpV, and to each side of this is one GpIb-α, GpIb-β, and GpIX respectively. GpIb-α binds to vWF, specifically the A1 domain. Binding of vWF to GpIb-α initiates platelet adhesion. GpIIb/IIIa is responsible for attachment with fibrinogen. GpIa/IIa and GpVI also attach to collagen, where GpIa/IIa acts as a receptor for platelet adhesion to collagen. Binding of collagen to GpVI initiates platelet aggregation and platelet degranulation.

Adenosine diphosphate (ADP) is a physiological agonist that plays an important role in normal hemostasis and thrombosis. ADP causes platelets to undergo shape change, release granule content, and aggregate. At that point, platelets also hydrolyze arachidonic acid from phospholipids and convert it into thromboxane A_2 via cyclooxygenase and thromboxane A_2 synthase. It is well-established that ADP activates platelets through three purinergic receptors: P2Y1, P2Y12, and P2X1 [3]. The following are major characteristics of various membrane receptors present in platelets:

- ADP receptors (P2Y1, P2Y12, and P2X1): P2Y1 receptors once activated are involved in platelet shape change. P2Y1 receptors are widely distributed in many tissues. P2Y12 receptors are only found in platelets and once activated are involved in platelet aggregation, secretion, and thrombus stabilization.
- Thrombin receptors (PAR-1 and PAR-4): Thrombin can activate platelets. This is done via the PAR-1 and PAR-4 receptors.
- Thromboxane receptors: Once activated, they stimulate platelet activation.

- Adrenergic receptors.
- The canalicular system in platelets represents a reservoir of membranes connected to the outer surface that allows release of platelet granule contents to the exterior. Chemicals released from platelets are capable of activating other platelets. Peripheral microtubules are responsible for shape change and release of chemicals.

Platelets are capable of secreting granules that are critical to normal platelet function. Among the three types of platelet secretory granules—α granules, dense granules, and lysosomes—α granules are most abundant (50–80 per platelet ranging in size from 200 to 500 nm). α Granules are stained by Wright–Giemsa stain and contain fibrinogen, fibronectin, factor V, vWF, platelet factor-4 (PF-4), platelet-derived growth factors (PDGF), transforming growth factor-β (TGF-β), and thrombospondin. The contents of the α granules are released into the canalicular system. Deficiency of α granules results in platelets that appear pale and gray on peripheral smear; this is referred to as gray platelet syndrome. It is transmitted in an autosomal recessive manner.

Dense granules (also known as dense bodies or δ granules) are electron dense due to the presence of calcium and appear as dark bodies under the electron microscope. Dense granules are present in low numbers (<10 per platelet) and contain adenosine triphosphate (ATP), ADP, ionized calcium, histamine, serotonin (5-hydroxytryptamine (5-HT)), and epinephrine. The dense granule contents are released directly through fusion with the plasma membrane. Lysosomes contain acid hydrolases. The organelle zone of platelets also contains glycogen and mitochondria.

Platelets circulate for approximately 10 days in the circulation. Approximately one-third of platelets are normally sequestered in the spleen. The three main platelet events are as follows:

- Adhesion and shape change: With exposure of subendothelial collagen to vWF, conformational change in high-molecular-weight multimer occurs, followed by interaction between vWF and GpIb/IX/V receptors. This allows adhesion of platelets to subendothelial collagen via interaction with vWF; vWF, GpIb/IX/V, collagen, GpIa/Iia, and GpVI are all involved in this step. In von Willebrand disease as well as Bernard–Soulier syndrome (in which GpIb/IX/V is lacking), abnormal platelet function (thrombocytopathia) is observed.
- Platelet activation: During this process, there is an increase in cytoplasmic calcium concentration with change in shape of platelets, extension of pseudopodia, and release of chemicals (release reaction). Phosphatidylserine is translocated to the external surface. Release of ADP, activation of thrombin receptors, and production of thromboxane A_2 (from arachidonic acid with the aid of cyclooxygenase enzyme) are all

involved in platelet activation. Activation of the ligand binding site on GpIIb/IIIa also occurs at this time. GpIIb/IIIa then interacts with GpIIb/IIIa on other platelets through fibrinogen, causing platelet aggregation. If there is a deficiency of enzymes responsible for synthesis of chemicals normally stored and released from platelets, abnormalities of receptors through which they act, or abnormalities of normal storage of chemicals, dysfunctional platelets result.

■ Aggregation: Fibrinogen-mediated binding of activated GpIIb/IIIa receptors on adjacent platelets is augmented by thrombospondin, a component of α granules. Thus, with lack of GpIIb/IIIa (known as Glanzmann's thrombasthenia or syndrome; transmitted as autosomal recessive) and hypofibrinogenemia, abnormal platelet function is observed.

15.3   THROMBOCYTOPENIA AND THROMBOCYTOPATHIA

Bleeding due to platelet disorders is mainly caused by thrombocytopenia (low platelet count) or thrombocytopathia (dysfunctional platelets) or both. Manifestation of bleeding is most often in the form of purpuras, mucosal bleeding (e.g., epistaxis, gum bleeding, and gastrointestinal bleed), prolonged bleeding from superficial cuts and abrasions, and menorrhagia.

The following are major causes of thrombocytopenia:

■ Decreased production: Generalized bone marrow failure or selective megakaryocyte depression results in decreased platelet formation. Congenital diseases associated with reduced platelet production include hereditary thrombocytopenias, macrothrombocytopenia with neutrophilic inclusions (MYH9 disorders), Wiskott–Aldrich syndrome (X-linked recessive), and Bernard–Soulier disease.

■ Increased breakdown: Idiopathic thrombocytopenic purpura (ITP), heparin-induced thrombocytopenia (HIT), drug-induced thrombocytopenia, and neonatal and post-transfusion purpura.

■ Increased utilization: Disseminated intravascular coagulation (DIC), thrombotic thrombocytopenic purpura (TTP), and hematolytic uremic syndrome (HUS).

■ Increased sequestration: Kasabach–Merritt syndrome (platelets sequestered in a hemangioma).

However, it is important to be aware of the causes of pseudothrombocytopenia, which, as the name implies, is not true thrombocytopenia. Pseudothrombo-cytopenia may be due to the following:

■ Platelet clumps: This phenomenon occurs when blood is collected in ethylenediamine tetraacetic acid (EDTA)-containing blood collection tubes. Hematology analyzers can flag a sample when they detect clumps. Clumps may be seen on peripheral smear examination. Re-collection of blood should be done with heparin or citrate tubes for accurate platelet counts.

- Platelet satellitism: This phenomenon also occurs when blood is collected in EDTA. Here, platelets surround neutrophils, resulting in satellitism. The platelets are not counted, resulting in thrombocytopenia.
- Large platelets: Large platelets may be counted as red blood cells by the hematology analyzer. If these large platelets are numerous, this may falsely lower the actual platelet count.
- Traumatic venipuncture may result in activation of the clotting process, resulting in thrombocytopenia.

Thrombocytopathia can be congenital or acquired. Congenital disorders such as von Willebrand disease, Bernard–Soulier syndrome, Chediak–Higashi syndrome, Hermansky–Pudlak syndrome, and Glanzmann's syndrome are examples of congenital thrombocytopathia. Acquired thrombocytopathia may be due to the use of nonsteroidal anti inflammatory drugs such as aspirin, uremia, or acquired von Willebrand disease. Acquired von Willebrand disease is an acquired bleeding disorder that may suddenly manifest in an individual without any family history of bleeding disorders. Usually, this disease is associated with monoclonal gammopathy; lymphoproliferative, myeloproliferative, or autoimmune disorders; and a pathogenic mechanism involving autoantibodies against vWF, resulting in inactivation of plasma vWF or rapid clearance of this factor [4]. Various causes of thrombocytopathia are listed in Table 15.1.

15.3.1 Hereditary Thrombocytopenias

Hereditary thrombocytopenia includes thrombocytopenia with absent radii (TAR syndrome), congenital amegakaryocytic thrombocytopenia (in this

Table 15.1 Congenital and Acquired Causes of Thrombocytopathia

Congenital Thrombocytopathia	Acquired Thrombocytopathia
<ul><li>Disorders of platelet adhesion: Von Willebrand disease, Bernard–Soulier syndrome</li><li>Disorders of platelet activation: Storage pool disorders, Chediak–Higashi syndrome, Hermansky–Pudlak syndrome</li><li>Disorders of platelet aggregation: Glanzmann's syndrome</li></ul>	<ul><li>Drugs: Aspirin and various other nonsteroidal anti-inflammatory drugs</li><li>Uremia</li><li>Acquired von Willebrand disease</li><li>Myeloproliferative diseases</li><li>Antiplatelet antibodies</li></ul>

disorder, the thrombopoietin receptor is deficient or defective; thrombopoietin is required for maturation of megakaryoblasts to megakaryocytes), familial thrombocytopenia—leukemia (thrombocytopenia combined with thrombocytopathy with an increased incidence of acute myeloid leukemia), and Fanconi's anemia (autosomal recessive transmission).

Macrothrombocytopenia with neutrophilic inclusions (MYH9 disorders) is a group of disorders characterized by mutations in the *MYH9* gene. This gene encodes the nonmuscle myosin heavy-chain class IIA protein. These disorders are characterized by thrombocytopenia, large/giant platelets, and Dohle-like bodies in neutrophils. This group of disorders includes May—Hegglin anomaly (autosomal dominant), Sebastian syndrome, Fechtner syndrome (nephritis, ocular defects, and sensorineural hearing loss), and Epstein syndrome.

Wiskott—Aldrich syndrome is due to a defect in the *WASP* gene, and this syndrome is characterized by immune deficiency, eczema, and thrombocytopenia with small platelets.

Bernard—Soulier syndrome is due to deficiency of the GpIb/IX/V receptor. There is also thrombocytopenia with large/giant platelets. Some cases of Bernard—Soulier syndrome are due to defects of the GpIb-β gene located on chromosome 22. This gene may be affected in velocardiofacial syndrome or DiGeorge syndrome associated with deletion of 22q11.2.

15.3.2   Idiopathic Thrombocytopenic Purpura

ITP is immune-mediated destruction of platelets causing low platelet count. Bone marrow shows increased megakaryocytes, and these megakaryocytes release relatively immature platelets that are larger than normal. It is thought that in ITP, antibodies are formed against pathogens that cross-react with platelet GpIb/IX or IIb/IIIa. Acute ITP is self-limiting. Chronic ITP lasts for more than 1 year, and 10% of individuals with chronic ITP have splenomegaly. ITP may be seen alone or sometimes as part of Evan's syndrome (with autoimmune hemolytic anemia). ITP may also occur in patients with systemic lupus erythematosus, chronic lymphocytic anemia, or HIV infection, and following stem cell transplantation.

15.3.3   Heparin-induced Thrombocytopenia

There are two types of HIT:

- HIT I: A decrease in platelets within the first 2 days of heparin administration is observed. This disorder is non-immune and due to the direct effect of heparin on platelet activation. There is no need to discontinue heparin therapy.

■ HIT II: This type affects 0.2–5% of patients who are receiving heparin for more than 4 days. Clinical features include heparin administration for more than 4 days with reduction of platelet count by 50% or more. Patients at risk for HIT II are females, surgical patients (especially cardiac and orthopedic surgery), patients with previous exposure to heparin, and patients receiving unfractionated heparin. In HIT II, antibodies are formed that bind to PF-4 and result in platelet aggregation causing thrombosis. It is thought that the antibodies are initially formed against antigens mimicking PF-4–heparin complex (e.g., PF-4 complex with polysaccharide on the surface of bacteria).

The following are tests for HIT:

■ Enzyme-linked immunosorbent assay (ELISA)-based test: Heparin–PF-4 complexes are coated on a plate to detect antibody in the patient's plasma/serum. The test is very sensitive but has low specificity. Results are published as positive or negative based on the optical density (OD) value of the patient compared to the control. It is important to appreciate that individuals with positive ELISA do not necessarily have platelet activating antibody. However, individuals with OD greater than 1.0 are more likely to have platelet activating antibody.
■ Heparin-induced platelet aggregation: Heparin-induced platelet aggregation is performed using donor platelets and the patient's serum. The control test has no heparin . Two other tests are performed—one with a low concentration of heparin (0.1–0.3 U/mL) and the other with a high concentration of heparin (10–100 U/mL). A positive test is one in which there is platelet aggregation (>25%) with a low concentration of heparin but not with a high concentration or in the control.
■ Serotonin release assay: Platelets from a normal donor are incubated with radiolabeled serotonin (^{14}C), and then labeled platelets are incubated with the patient's serum in the absence and the presence of a therapeutic (0.1 U/mL) or a high concentration of heparin (100 U/mL). For a test to be positive, there must be at least 20% radioactivity release in the presence of 0.1 U/mL of heparin, but release of radioactivity must be substantially reduced in the presence of high heparin concentration. Serotonin release assay is considered the gold standard. However, this is often a send-out test, and thus results are available only after considerable delay.

15.4   TESTS FOR PLATELET FUNCTION

Platelets are the main regulator of hemostasis and also interact with a large variety of cell types, including monocytes, neutrophils, endothelial cells, and smooth muscle cells. Platelets are sensitive to manipulation and are prone to artifactual *in vitro* activation. Therefore, care must be exercised during the performance of platelet function tests. Platelet function tests are performed in a variety of patients, including those with bleeding disorders. In most cases, a platelet-mediated hemostatic disorder cannot be established by a single function defect; but, rather, a combination of platelet function abnormalities are required [5]. Applications of platelet function tests are listed in Box 15.1.

Bleeding time—the time required for bleeding to stop after a defined incision is made into the skin—was introduced by Duke in 1910. Ivy made the method more reliable by introducing a blood pressure cuff on the upper arm that was inflated to 40 mmHg and by placing the incision into the anterior surface of the forearm. This protocol is still followed, and drops of blood are absorbed with filter paper disks every 30 sec. The time required for bleeding to stop is noted. This test has been used most often to detect qualitative defects of platelets, vascular defects, or von Willebrand disease, but it has poor clinical correlation.

The capillary fragility test (also known as a Rumpel–Leede capillary fragility test or tourniquet test) determines capillary fragility and is a clinical diagnostic method to determine the hemorrhagic tendency of patients. This test assesses the fragility of capillary walls and is used to identify thrombocytopenia or thrombocytopathia. The test is defined by the World Health Organization as one of the necessary requisites for diagnosis of Dengue fever. A blood pressure cuff is applied and inflated to a point between the systolic and diastolic blood pressures for 5 min. The test is positive if there are 10 or more petechiae per square inch. In Dengue hemorrhagic fever, the test usually provides a definite positive result with 20 petechiae or more. This test does not have high specificity.

The PFA-100 system is a platelet function analyzer designed to measure platelet-related primary hemostasis. The instrument uses two disposable cartridges that are coated with platelet agonist. For analysis, whole blood is collected from the patients in a citrate tube, and testing should be performed

BOX 15.1 APPLICATION OF PLATELET FUNCTION TESTS

- Patients with bleeding disorders
- Monitoring response of a patient receiving antiplatelet therapy
- Platelet function tests to screen donors
- Assessment of platelet function in platelet concentrates
- Assessment of platelet function following platelet transfusion in a patient
- Perioperative assessment of platelet function in a patient

within 4 hr of collection. Blood is transferred into a sample cup. Blood is aspirated from the sample cup by the analyzer and passes through an aperture in a membrane that is already coated with platelet agonists. When platelet aggregation occurs, the aperture closes and the blood flow stops. This is the closure time. One membrane is coated with collagen/epinephrine (CEPI), and the other membrane is coated with collagen/adenosine diphosphate (CADP). If the CEPI closure time is prolonged but CADP closure time is normal, this is most likely due to aspirin. If both CEPI and CADP closure times are prolonged or CEPI closure time is normal and CADP closure time is abnormal, this denotes platelet dysfunction or von Willebrand disease. PFA-100 is insensitive to von Willebrand disease type 2N, clopidogrel, ticlopidine, and storage pool disease. However, the test is very sensitive to von Willebrand disease type I and GpIIb/IIIa antagonists. PFA-100 results are affected by thrombocytopenia (low hematocrit level). Results are not affected by heparin or deficiencies of clotting factors other than fibrinogen.

VerifyNow (VFN) is a rapid, turbidimetric whole blood assay capable of evaluating platelet aggregation. This assay is based on the ability of activated platelets to bind with fibrinogen. The VFN GpIIb/IIIa assay uses fibrinogen-coated microparticles and thrombin receptor activating peptide (TRAP) as agonists to maximally stimulate platelets in order to determine platelet function. If GpIIb/IIIa antagonists are present in a patient's blood, then platelet aggregation should be reduced. In the VFN aspirin assay, arachidonic acid is used as the agonist to measure the antiplatelet effect of aspirin. Results are expressed in aspirin reactive units (ARUs). The VFN P2Y12 assay is similarly used to assess the antiplatelet effect of clopidogrel, and results are expressed in plavix reactive units (PRUs). VFN assays may be used to assess the efficacy of the previously mentioned drugs, to check patient compliance, and also to assess the residual effect of these drugs if patients are to undergo surgery or invasive procedures.

Plateletworks is a platelet function test that uses whole blood. Kits for the test as well as an impedance cell counter (ICHOR II analyzer) are commercially available from the Helena Laboratory so that this test can be used as a point-of-care test. This test assesses platelet function by comparing the platelet count before and after exposure with a specific platelet agonist. For this test, blood is collected in EDTA tubes as well as in other platelet agonist (e.g., ADP, arachidonic acid, or collagen)-containing tubes. Functional platelets should aggregate in the agonist tube, and nonfunctional platelets should not aggregate. Then a hematology analyzer such as ICHOR II is used to count the number of platelets in the EDTA tube and also the number of unaggregated platelets in the agonist tube. The unaggregated platelets are dysfunctional. In order to calculate the number of functional platelets, the platelet count in the presence of a platelet agonist should be subtracted from the platelet count obtained using blood collected in the EDTA tube. Campbell *et al.* noted that the point-of-care test platform Plateletworks is useful for monitoring platelet response in patients receiving antiplatelet agents, including aspirin and clopidogrel [6].

The platelet aggregation test using platelet aggregometry is a widely used laboratory test to screen patients with inherited or acquired platelet function defects. Platelet aggregometry measures the increase in light transmission through platelet-rich plasma that occurs when platelets are aggregated due to the addition of an agonist. For this test, blood should be collected in a citrate tube, and the test should be performed within 4 hr of blood collection. Prior to analysis, the specimen should be stored at room temperature. Platelet-rich plasma is obtained from the sample by centrifugation. Ideally, the platelet count of the platelet-rich plasma should be approximately 200,000−250,000; if the platelet count is higher, it can be adjusted by saline. Prior to actual testing, the platelet-rich plasma should be left at room temperature for approximately 30 min. If the original platelet count of the patient is less than 100,000, then the test might be invalid. If the test needs to be performed, then the platelet count of the control should also be lowered. Here, the extent of aggregation is measured by impedance technique. Various agonists used for this test include arachidonic acid, collagen, ristocetin, ADP, and epinephrine. Platelet aggregation can also be performed using whole blood instead of platelet-rich plasma. In general, agonists such as ADP and epinephrine are considered weak agonists. These two weak agonists, in low concentration in a normal person, demonstrate two waves of aggregation. The primary wave of aggregation is due to activation of the GpIIb/IIIa receptor. The secondary wave is due to platelet granule release. Lack of the secondary wave implies a storage pool disorder due to a reduced number of granules or defective release of granule contents. The use of higher concentrations of these agonists (ADP and epinephrine) results in the two waves merging into one wave of aggregation. Collagen characteristically demonstrates an initial shape change before the wave of aggregation. This is seen as a transient increase in turbidity. Effective aggregation is typically considered as 70−80% aggregation. However, values should be compared with the control value. Values of 60% or more are generally considered to be adequate.

The following are patterns of platelet aggregation:

- Normal: There is adequate aggregation with ADP, collagen, epinephrine, arachidonic acid, and ristocetin at high dose but not at low dose ($\leq$6 mg/mL).
- von Willebrand/Bernard−Soulier pattern: There is adequate aggregation with ADP, collagen, epinephrine, arachidonic acid, but not at a higher dose of ristocetin.

- von Willebrand type IIB pattern/pseudo von Willebrand pattern: There is increased aggregation with a low dose of ristocetin.
- Glanzmann's thrombasthenia/hypofibrinogenemia pattern: There is adequate aggregation with ristocetin but impaired aggregation with all other agonists is observed. Uremia and antiplatelet medication can also produce similar results.
- Disorder of activation (storage pool disorder): Loss of secondary wave of aggregation with ADP and epinephrine at lower doses is the characteristic of this disorder. Storage pool disease is the most common inherited platelet function defect. It is subdivided into α-granule and dense granule storage pool diseases.
- Aspirin effect: Significant impairment with aggregation with arachidonic acid is observed in patients taking aspirin. There may be impairment of aggregation (not as much as arachidonic acid) with ADP, collagen, and epinephrine.
- Plavix (clopidogrel) effect/ADP receptor defect: There is significant impairment of aggregation with ADP.
- Chronic myeloproliferative disorder: There may be impairment of aggregation with epinephrine.

15.4.1 Thromboelastography

Whole blood thromboelastography (wbTEG) is a method of assessing global hemostasis and fibrinolytic function that includes interaction of primary and secondary hemostasis; and, subsequently, defect in one component of hemostasis can affect the other to a certain extent. This technique has existed for more than 60 years, but technological improvements have led to increased utilization of this test in clinical practice for monitoring hemostatic and fibrinolytic rearrangements [7]. wbTEG is a visualization of viscoelastic changes that occur during *in vitro* coagulation. It provides a graphical representation of the fibrin polymerization process, but during interpretation of wbTEG data/tracing, it is important to focus on the most significant defect. In classical wbTEG, a small sample of blood (typically 0.36 mL) is placed into a cuvette (cup) that is rotated gently through 4° 45′ (cycle time, 6 min) to imitate sluggish venous flow and activate coagulation. When a sensor shaft is inserted into the sample, a clot forms between the cup and the sensor. The speed and strength of clot formation are measured in various ways, but they are typically measured by computing the speed at which a specimen coagulates depending on various factors, including the activity of the plasmatic coagulation system, platelet function, fibrinolysis, and other factors that can be affected by illness, the environment, and medications.

wbTEG analysis involves four basic parameters: R (reaction time), K value, angle α, and maximum amplitude (MA). Reaction time is measured in seconds and represents initial latency from the start of the test until the initial fibrin formation (usually at an amplitude of 2 mm). The K value is also measured in seconds and indicates the time taken to achieve a certain level of clot strength (usually at an amplitude of 20 mm). α Angle (degree) measures the speed of fibrin buildup and the cross-linking taking place; thus, it assesses the rate of clot formation. MA (measured in millimeters) represents the ultimate strength of the fibrin clot. Possibly the most important information provided by wbTEG is clot strength, which may help to determine whether the bleeding is related to coagulopathy or is a mechanical bleeding. Clot strength is measured by MA value, and a low MA value indicates platelet dysfunction. G is a computer-generated value reflecting the strength of the clot from initial fibrin blast to fibrinolysis:

$$G = (5000 \times amplitude)/100 - amplitude \ (normal, 5.2-12.4)$$

Therefore, MA and G values represent a direct function of the maximum dynamic properties of fibrin and platelet bonding via GpIIb/IIIa and represent the ultimate strength of the fibrin clot. However, G is the best measurement of clot strength [8]. CI (clot index) represents the hemostasis profile and is calculated based on R, K, α angle, and MA. LY30 (clot lysis at 30 min) indicates the percentage decrease in amplitude at 30 min after MA, indicating the stability or degree of fibrinolysis. Various parameters obtained from wbTEG analysis are listed in Table 15.2. They are also illustrated in Figure 15.1.

Table 15.2 Various Parameters of TEG

Parameter	Comments
R: Reaction (measured in seconds)	The value indicates the time until the first evidence of a clot is detected.
K: Clot kinetics (measured in seconds)	K value is the time from the end of R value until the clot reaches 20 mm, and this value represents the speed of clot formation.
Angle α (measured in degrees)	This measures the rapidity of fibrin buildup and cross-linking (clot strengthening) and is the tangent of the curve made as the K is reached.
MA (maximum amplitude; measured in millimeters) and G: Measures of clot strength	This is a direct measure of the highest point of the TEG curve and represents clot strength. G is calculated from MA.
CI: Clotting index	CI is a mathematic equation calculated from R, K, α angle, and MA values.
Ly30 (clot lysis at 30 min)	This provides information on the fibrinolytic activity during the first 30 min after MA and is a calculated value.

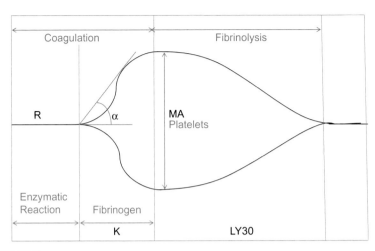

FIGURE 15.1
Different components of wbTEG.

wbTEG may be used to dictate use of blood products and certain medications in patients who are bleeding. Blood products that may be used in bleeding patients include packed red blood cells (PRBCs), fresh frozen plasma (FFP), cryoprecipitate, and platelets. Medications that may be used include protamine, antifibrinolytic agents (e.g., ε amino caproic acid and tranexamic acid). Use of PRBCs is dictated by hemoglobin and hematocrit analysis, and wbTEG is not used for this purpose. However, use of FFP is indicated if the R time is prolonged. In such patients, prothrombin time (PT) and partial thromboplastin time (PTT) should also be prolonged.

Cryoprecipitate is used in bleeding patients with hypofibrinogenemia and uremic thrombocytopathia. Cryoprecipitate is also useful in von Willebrand disease patients who are bleeding. In hypofibrinogenemia, the angle α is low, with a normal MA. These patients also have prolonged thrombin time (TT) and low levels of fibrinogen. Regarding the use of platelets, platelet transfusion is valuable in thrombocytopenia and thrombocytopathia, and in both conditions MA should be low. Complete blood count will also document thrombocytopenia.

Protamine is used to neutralize heparin, and TEG must be performed with and without heparinase (an enzyme that destroys heparin). If the R value in the wbTEG analysis without heparinase is greater than 50% longer than the R with heparinase, protamine is indicated. In these patients, TT and PTT are also prolonged, but PT is not as prolonged as PTT with heparin.

Fibrinolysis may be primary or secondary. Causes of primary fibrinolysis include physiological (due to thrombolytic therapy) or pathological. Secondary fibrinolysis is seen in disseminated intravascular coagulation (DIC). In primary and secondary hyperfibrinolysis, LY30 is prolonged. However, in primary fibrinolysis, the MA value from wbTEG analysis is low, as is the CI value. In secondary fibrinolysis, MA is normal or high with high CI. Antifibrinolytic agents are used only in patients with primary pathologic fibrinolysis.

15.4.2 Platelet Mapping

Platelet mapping is a special wbTEG assay that measures the effects of anti-platelet drug therapy on platelet function. Antiplatelet drugs, whose efficacy can be tested, include the following:

- ADP receptor inhibitors such as clopidogrel and ticlopidine
- Arachidonic acid pathway inhibitors such as aspirin
- GpIIb/IIIa inhibitors such as abciximab, tirofiban, and eptifibatide.

The Platelet Mapping assay specifically determines the MA reduction present with antiplatelet therapy and reports the percentage inhibition and aggregation.

The Platelet Mapping assay measures platelet function in the presence of antiplatelet drugs in a patient's blood sample. The results obtained by the TEG® 5000 analyzer will be reported as percentage inhibition and percentage aggregation. The Platelet Mapping assay measures the presence of platelet-inhibiting drugs using whole blood and four different steps:

- No additive sample to measure total platelet function and the contribution of fibrin to MA yielding $MA_{THROMBIN}$. Thrombin overrides inhibition at other platelet activation pathways and indicates complete activation.
- Activator F is added to measure the contribution of fibrin only to MA and yields MA_{FIBRIN}, which indicates no activation.
- Activator F is added to the sample along with ADP to measure MA due to ADP receptor uninhibited platelets to yield the value of MA_{ADP}, which indicates activation of non-inhibited platelets.
- Activator F is added to the sample along with arachidonic acid (AA) to measure MA due to the TXA_2 pathway, to yield the value of MA_{AA}, which also indicates activation of non-inhibited platelets.

The presence of platelet-inhibiting drugs is reflected in a reduction in the MA values. The percentage inhibition is derived by the following equation:

$$\% \ MA \ reduction = 100 - [\{(MA_P - MA_F)/(MA_T - MA_F)\} \times 100]$$

where MA$_P$ represents MA$_{ADP}$ or MA$_{AA}$, MA$_F$ represents MA$_{FIBRIN}$, and MA$_T$ represents MA$_{THROMBIN}$. The optimum time, from the time the blood enters the syringe to the time it is placed in the TEG instrument, is 4 min.

15.5   SECONDARY HEMOSTASIS

Secondary hemostasis takes place with formation of fibrin clot. The clotting pathways consist of the extrinsic pathway, the intrinsic pathway, and the common pathway. The concept of the coagulation process as a series of stepwise enzymatic conversions was first proposed in 1964 and was described under the categories of the intrinsic pathway (dependent on contact activation by a negatively charged surface and involving coagulation factors V, VIII, IX, XI, and XII) and the extrinsic pathway (dependent on tissue factor being exposed to circulation and involving tissue factor and factor VIII), converging to a common pathway to activate factor X. The intrinsic pathway starts with activation of factor XII, which in turn activates factor XI. Then, factor XI activates factor IX. Activated factor IX and activated factor VIII activate factor X in the common pathway, which in turn—with the help of factor V—converts prothrombin (factor II) to thrombin (factor IIA). Thrombin converts fibrinogen (factor I) to fibrin. Fibrin is stabilized by factor XII. Activation of factor XII requires kallikrein and high-molecular-weight kininogen (HMWK) for activation. Kallikrein is derived from prekallikrein. The following are important points regarding the clotting pathway:

- PT is used to assess the integrity of the extrinsic and common pathways.
- PTT is used to assess the integrity of the intrinsic and common pathways.
- All clotting factors are produced by the liver with the exception of factor VIII, which is produced by endothelial cells.
- Factor VII has the shortest half-life (4−6 hr). With acute liver dysfunction, levels of factor VII are reduced early. Factor VII is required for the extrinsic pathway; therefore, PT is a very good test to assess liver function in the acute setting.
- Individuals with factor XII deficiency or HMWK or kallikrein deficiency do not bleed, but PTT may be prolonged.

All deficiencies of clotting factors are transmitted as autosomal recessive with the exceptions of factor VIII deficiency (hemophilia A) and factor IX deficiency (hemophilia B), which are transmitted as X-linked recessive. Dysfibrinogenemia (which means fibrogen is dysfunctional) is transmitted as autosomal dominant.

15.6   TESTS FOR SECONDARY HEMOSTASIS

PT is a widely used test to evaluate secondary hemostasis. In this test, platelet-poor plasma from a patient (collected in a blood collection tube containing sodium citrate) is mixed with thromboplastin and calcium, and then clotting time is determined at 37°C using a variety of methods, including photo-optical and electromechanical. Automated coagulation analyzers are commercially available for measuring PT along with other coagulation parameters. PT is a functional measure of the extrinsic pathway and the common pathway, and the reference range is 8.8−11.6 sec. Therefore, PT is a useful test to detect inherited or acquired defects in coagulation related to the extrinsic pathway. However, often PT is reported in the form of the international normalized ratio (INR). The thromboplastin used may vary from laboratory to laboratory and from country to country. However, reporting results as INR ensures results are comparable between different laboratories. INR is calculated as follows:

$$INR = [\text{patient PT}/\text{mean normal PT}]^{ISI}$$

where ISI is the international standardized index, which is available from the reagent package insert. The normal value of INR is 0.8−1.2. Causes of prolonged PT include the following:

- Coumarin (warfarin) administration
- Vitamin K deficiency (dietary deficiency)
- Failure of absorption of vitamin K (e.g., cholestasis)
- Liver disease
- Factor deficiency of the extrinsic pathway and the common pathway
- Presence of factor inhibitor.

Vitamin K is responsible for carboxylation of glutamic acid residues of factors II, VII, IX, and X. Dietary vitamin K is vitamin K_1 (phytonadione), and vitamin K that is absorbed from the gut by bacterial activity is vitamin K_2 (menaquinone). Vitamin K_2 has 60% of the activity of vitamin K_1. Vitamin K is converted to the epoxide form by the enzyme epoxidase. The epoxide form is inactive and is converted back to active vitamin K by the reductase enzyme. The full name of the enzyme complex is vitamin K epoxide reductase complex 1 (VKORC1). Warfarin blocks the reductase enzyme. There is genetic variation in the gene for VKORC1 among individuals. Warfarin is metabolized by the P450 CYP system, including CYP2C9. Polymorphism of *CYP2C9* affects CYP2C9 activity. Variation in activity may result in variation in the half-life of warfarin among different patients. Drugs may also affect the activity of CYP2C9. Thus, genetic variation of the *VKORC1* gene, polymorphism of *CYP2C9*, and the effects of other drugs all cause variation in warfarin activity among individuals. The other vitamin K-dependent proteins are protein C and protein S.

PTT (also known as activated partial prothrombin time: (APTT or aPTT) is another useful test for evaluation of secondary homeostasis. In this test, a patient's platelet-poor plasma (citrated plasma, but oxalate can also be used), surface activating agent (silica, kaolin, celite, or ellagic acid), calcium, and platelet substitute (crude phospholipid) are mixed, and clotting time is usually determined using an automated coagulation analyzer. It is a functional measure of the intrinsic pathway as well as the common pathway and can detect hereditary or acquired defects of the coagulation factors XII, XI, X, IX, VIII, V, prothrombin, and fibrinogen. PTT or APTT derives its name from the absence of tissue factor (thromboplastin) in the tests. The normal value varies from laboratory to laboratory, but it is usually between 25 and 39 sec. Causes of prolonged PTT include the following:

- Heparin, direct thrombin inhibitors
- Factor deficiency of the intrinsic and common pathways
- Inhibitors: VIII and IX inhibitors, lupus anticoagulant, or lupus antibody (LA)
- Von Willebrand disease
- HMWK (Fitzgerald factor) deficiency
- Pre-kallikrein (Fletcher factor) deficiency
- Spurious causes.

Important causes of factor deficiencies that lead to isolated (normal PT) but prolongation of PTT include the following:

- Von Willebrand factor deficiency, factor VIII deficiency (von Willebrand factor binds to factor VIII), factor IX deficiency, and factor XI deficiency—may cause normal PT but prolonged PTT. Deficiency of factor VIII causes hemophilia A, deficiency of factor IX causes hemophilia B, and deficiency of factor XI results in hemophilia C. Hemophilia A and B are inherited in an X-linked recessive manner, and hemophilia C is inherited in an autosomal recessive manner.
- Deficiency of factor XII, HMWK, and pre-kallikrein—may result in prolongation of PTT, but there is no bleeding in these individuals.
- Factor VIII and factor IX inhibitors and LA—may cause normal PT but prolonged PTT.
- Heparin and direct thrombin inhibitors (however, both may cause some prolongation of PT). Currently, four parenteral direct thrombin inhibitors (DTIs)—lepirudin, desirudin, bivalirudin, and argatroban— are approved by the US Food and Drug Administration (FDA) [9].

Various scenarios in which PT, PTT, or both may be prolonged are listed in Table 15.3.

Hemophilia A occurs in approximately 1 in 10,000 individuals, and up to 30% of cases may be due to spontaneous mutation. More than 100 mutations

Table 15.3 Interpretation of PT and PTT Results in Various Clinical Scenarios

PT Result	PTT Result	Clinical Scenario
Normal	Prolonged	Factor deficiency of the intrinsic pathway, above the common pathway (e.g., factor VIII and factor IX deficiency)
		Heparin therapy, direct thrombin inhibitor therapy
		Von Willebrand disease
		Inhibitors (e.g., lupus anticoagulant and factor VIII or factor IX inhibitor)
Prolonged	Normal	Liver disease
		Vitamin K deficiency
		Warfarin therapy
		Chronic disseminated intravascular coagulation
Prolonged	Prolonged	Factor deficiency of the common pathway
		Heparin therapy, direct thrombin inhibitor therapy
		Warfarin therapy
		Lupus anticoagulant

in the factor VIII gene have been described. Hemophilia A is clinically divided into mild (>10% factor VIII activity), moderate (2−10% factor VIII activity), and severe (<1% factor VIII activity). Treatment includes use of desmopressin acetate (DDAVP) for mild cases (causes a 2- to 10-fold increase in factor VIII levels) and factor VIII replacement (1 unit/kg raises factor VIII levels by 2%). Factor VIII has a half-life of 8 hr. Typical target values of factor VIII activity in various situations include 100% for surgery, central nervous system bleeding, and gastrointestinal and genitourinary bleeds and 40−80% for bleeding into joints and muscle. Factor VIII levels in female carriers are approximately 50% of normal, but levels of vWF are normal. Thus, most hemophilia carriers have a factor VIII-to-von Willebrand antigen ratio of less than 0.5.

Hemophilia B is less common than hemophilia A, occurring in approximately 1 in 25,000 males. A variation of hemophilia B is known as Leyden phenotype, in which features of hemophilia in childhood improve after puberty. In these patients, there is a mutation in the promoter region of the gene, which disrupts hepatocyte nuclear factor 4 (HNF-4), but the site for the androgen response element is intact. Hemophilia C occurs in approximately 1 in 1 million individuals. However, hemophilia C is the second most common inherited bleeding disorder in women. It also has a high prevalence in Ashkenazi Jews. Unlike factor VIII and factor IX, factor XI deficiency must be treated with fresh frozen plasma. Interestingly, hemophiliacs are protected against coronary artery disease, and co-inheritance of factor V Leiden improves symptoms.

Inhibitors to factor VIII and factor IX may develop spontaneously (autoanti-bodies) or post treatment (alloantibodies) with factor replacement. The inci-dence of alloantibody formation in hemophiliacs is approximately 1%. Autoantibodies may be associated with pregnancy, autoimmune diseases, malignancy, or allergy. Autoantibodies may respond to immune suppression medication. Tests to detect the presence of these antibodies are factor VIII and factor IX inhibitor screen, and when positive, these are followed by fac-tor VIII or factor IX inhibitor assay. The amount of inhibitor quantified by the assay is expressed as Bethesda units (BU) for both, where 1 BU of inhibi-tor is the amount of inhibitor that destroys half the factor VIII or factor IX:C activity in an equal mixture of patient and normal plasma. If the result is 5 or less, it is considered that the antibody is in low titers. In such cases, increasing the doses of factor VIII or factor IX may be useful. If the result is greater than 5, then the individual has high titers. In such cases, factor VIII inhibitor bypass activity (FEIBA) is required. This may be done by use of pro-thrombin complex concentrate or activated factor VII.

An indirect way to document the presence of an inhibitor (e.g., factor VIII inhibitor) is when, for example, PTT is prolonged and mixing study does not show correction. This implies an inhibitor. When factor VIII or factor IX assay are performed, the assay result of one will be low. If the assay result is low due to actual low levels, then mixing study should show correction. If the assay result is low due to inhibitor, then mixing study should not show correction and the factor assay (which is a functional assay) should also be low.

Spurious causes of prolonged PTT include high hematocrit, underfilling of citrate tube, and EDTA contamination, which may occur if the purple-top tube is collected before the blue-top tube. Delay in transport and processing (testing should be done within 4 hr of collection, and the sample should be stored at room temperature) may also give rise to a spurious result.

Thrombin time (TT) is a test in which the patient's plasma is mixed with thrombin and clotting time is determined. It is a measure of functional fibrinogen. Heparin produces prolonged TT. Patients on heparin, however, have a normal reptilase time. Causes of prolonged TT include the following:

- Heparin
- Hypofibrinogenemia
- Dysfibrinogenemia
- Thrombolytic therapy.

Individuals with prolonged PT or prolonged PTT or both may undergo PT/PTT mixing study. The objective of this test is to determine if the cause of pro-longed PT or PTT is a factor deficiency or the presence of inhibitor. In this test, the patient's plasma is mixed with an equal volume of normal plasma, and PTT is measured at 0 and 1 or 2 hr. Failure of correction of prolonged PTT indicates the presence of inhibitors. If results at 0 and 1 or 2 hr are similarly prolonged, this implies lupus anticoagulant. If results show time-dependent prolongation, this implies coagulation factor antibody (e.g., factor VIII or factor IX inhibitor). If PT and PTT are both prolonged and mixing study shows correction, then most likely there is deficiency of factor in the common pathway. Hemophiliacs typically have prolonged PTT with normal PT. Mixing study should show correction.

Causes of prolonged PT and PTT, not showing correction of either, include the following:

- Heparin, DTIs
- Lupus anticoagulant
- Factor inhibitor (factor II, factor V, and factor X inhibitor; factor V is the most frequently found inhibitor of the common pathway).

Causes of acquired clotting factor deficiency include the following:

- Anticoagulants
- Fibrinolytic therapy
- DIC, TTP, or HUS
- Liver disease
- Cardiopulmonary bypass.

Causes of bleeding with normal PT/PTT include dysfibrinogenemia (fibrinogen level is normal; TT is prolonged: autosomal dominant), factor XIII deficiency (clot is not soluble in 5 M urea solution for 24 hr; Rx cryo), and α_2-antiplasmin deficiency (treated with ε-aminocaproic acid).

15.6.1 Factor Assays

PTT-based factor assay is performed for factors VIII, IX, XI, and XII. PT-based factor assay is performed for factors II, V, VII, and X. For these tests, commercially available normal plasma is used as reference plasma; factor-deficient plasmas are also available that are completely deficient in the factor of interest (e.g., factor VIII-deficient plasma for factor VIII assay). The following are the main steps in the procedure:

- Serial dilution of the reference (normal plasma) is done, and each dilution is mixed with an equal volume of factor-deficient plasma. PTT is measured.
- The clotting times (y axis) are plotted against dilution (x axis) on a log-lin graph. A best fit line is drawn. This is the reference line/curve (first line).

- The patient plasma is treated the same way as the reference plasma (i.e., serial dilution and mixed with an equal volume of factor-deficient plasma), and PTT is measured.
- The clotting times (y axis) are plotted against dilution (x axis) on a log-lin graph. A best fit line is drawn. This is the reference line/curve (second line).

The second line should be parallel to the first line or be superimposed with the first line. If the second line is not parallel, then an inhibitor may be present. Often with inhibitor, the clotting time shortens with increasing dilution as the inhibitor is diluted out. If the clotting time is grossly prolonged and does not change with dilution, the clotting factor is likely less than 1%. The two lines are used to calculate factor activity in patient plasma. For example, if a 1:20 dilution of normal plasma shortens the clotting time of deficient plasma to the same extent as a 1:5 dilution of the patient's plasma, the latter sample has 25% of normal activity. Assay results are expressed as percentages. Fibrinogen level is expressed as milligrams per deciliter.

15.6.2 Von Willebrand Disease

Von Willebrand disease is the most commonly inherited bleeding disorder in the general population. An estimated 1 or 2% of the population is affected by this disease ($\sim 60-120$ million people worldwide). This disease is 150 times more prevalent than hemophilia, and it affects men and women equally, including all racial, ethnic, and socioeconomic groups.

Von Willebrand disease is a group of genetically heterogeneous disorders resulting in abnormal function of vWF, and more than 100 mutations have been described. It is predominantly transmitted as an autosomal dominant condition. However, only a small fraction of individuals who inherit the gene suffer from a clinically significant diathesis.

vWF is an unusual, extremely large multimeric glycoprotein composed of repeating units that are polymerized from dimer subunits by disulfide bonds. Pro-peptides are synthesized in the endothelial cells and megakaryocytes and processed. Then vWF is released as a series of multimers, with larger multimeric compounds being more functional. vWF acts as a carrier protein for factor VIII and prevents its inactivation by protein C. It also serves as the ligand that binds glycoprotein Ib receptor on platelets to initiate platelet adhesion to damaged blood vessels. Factor VIII is brought into storage granules (Weibel−Palade bodies in endothelial cells and platelet α granules) by vWF. In the absence of this factor, factor VIII survival is drastically reduced, resulting in secondary factor VIII deficiency. There are three main types of von Willebrand disease:

- Type 1: This is the most common type, accounting for 75% of all cases of van Willebrand disease. These patients have partial quantitative deficiency with 20–50% reduction of vWF, but the factor is structurally normal. This disorder is transmitted as autosomal dominant with reduced penetrance and variable expressivity. Clinical symptoms are usually mild.
- Type 2: These patients have qualitative defect of vWF, and this type accounts for 20–25% of all cases. The type 2 disorder can be classified into four subtypes: 2A, 2B, 2M, and 2N. Type 2N is transmitted as autosomal recessive. All others are transmitted as autosomal dominant.
- Type 3: This is the rarest type, in which patients have total deficiency of vWF resulting in a severe form of disease. This disorder is transmitted as autosomal recessive.
- Pseudo (platelet-type) von Willebrand disease.
- Acquired von Willebrand disease.

Testing for von Willebrand disease includes the following:

- Complete blood count (CBC): In certain subtypes, type 2B and platelet thrombocytopenia may be present.
- Bleeding time should be prolonged because vWF is required for platelet adhesion.
- PTT: PTT should be prolonged due to low levels of factor VIII. Mixing study should show correction.
- von Willebrand panel: This panel of tests consists of those for factor VIII levels, von Willebrand antigen (this assay measures the quantity of vWF), and ristocetin factor activity (vWF:RcoF), which involves a functional test (ability to aggregate normal platelets in the presence of ristocetin). Typically, values for all three should be low.
- Ristocetin-induced platelet aggregation: Ristocetin induces von Willebrand and Gp1b interaction, causing platelet aggregation. Normal healthy individuals should show adequate aggregation with a higher dose of ristocetin. These normal individuals should not show aggregation with low-dose ristocetin (≤ 0.6 mg/mL).
- Multimer analysis (by electrophoresis): This allows analysis of multimer according to size.
- Factor VIII to vWF binding assay: This assesses the binding ability of vWF and factor VIII.

15.6.3 Diagnosis of Various Types of Von Willebrand Disease

It is important to establish the diagnosis of von Willebrand disease by classifying the subtypes. The following approach is usually helpful:

- Type I: CBC does not show thrombocytopenia, but bleeding time may be prolonged. PTT should be prolonged. Abnormal tests on the von Willebrand panel (all decreased) are another characteristic of this type. No aggregation with low-dose ristocetin (as in normal individuals) is observed, but ristocetin at higher dose also fails to induce platelet aggregation. Multimer analysis will demonstrate the presence of all of the various multimers; however, the quantity will be reduced. Type I responds to desmopressin, induces the release of stored vWF and improves clinical condition.

- Type III: CBC does not show thrombocytopenia, but bleeding time may be prolonged. PTT should also be prolonged. There are abnormal tests on the von Willebrand panel (all severely decreased), and ristocetin at higher dose fails to induce platelet aggregation. No aggregation with low dose ristocetin (as in normal individuals) is observed. Multimer analysis should demonstrate the absence of all mutimers.

- Type IIA: In this disorder, reduced levels of large and intermediate-sized multimers are observed. CBC does not show thrombocytopenia, but bleeding time may be prolonged. PTT should be prolonged along with an abnormal von Willebrand panel (all decreased). Ristocetin at higher dose fails to induce platelet aggregation. There is no aggregation with low-dose ristocetin (as in normal individuals). There are two subtypes of type 2A: 2A-1 (defect in intracellular transport) and 2A-2 (vWF more susceptible to proteolysis by the vWF cleaving protease ADAMTS13—a disintegrin and metalloprotease with a thrombospondin type 1, motif 13).

- Type IIB: In this disorder, there is increased affinity of the large vWF multimers for platelet binding (to GpIb). vWF binds to platelets in the bloodstream, and these large bundles of platelets are removed from circulation with resultant thrombocytopenia. Thus, CBC does show thrombocytopenia, but bleeding time may be prolonged. PTT may be normal or prolonged. The von Willebrand panel may be normal or decreased. Ristocetin induces aggregation with low dose, which is a reflection of increased affinity of vWF to platelet GpIb. Multimer analysis should demonstrate reduction of large multimers because large multimers, along with the platelets, are consumed.
  - An identical clinical and laboratory phenotype is seen in the platelet GpIb receptor abnormally, in which there is increased affinity of

GpIb to bind to normal vWF. This is called "pseudo" or platelet-type von Willebrand disease. To distinguish these two disorders, repeating the test using normal (donor) fixed washed platelets and incubating them with the patient's plasma and with a lower or higher dose of ristocetin will not result in aggregation. This is because the donor (normal) platelets are normal, with no increased affinity for vWF.

- Type 2M (M for multimer): In this disorder, there is mutation in the A1 region, resulting in decreased platelet-dependent function. The multimers are present but dysfunctional. Therefore, the von Willebrand disease panel will show normal vWF antigen assay but low functional assay. Multimer analysis should demonstrate a normal multimer pattern.

- Type 2N (N for Normandy): This disorder is due to mutation in the factor VIII binding region of vWF, resulting in rapid turnover of unbound factor VIII causing a reduced plasma level of factor VIII. It is transmitted as an autosomal recessive pattern. Factor VIII levels are quite low (<10%), and clinically it mimics hemophilia A (autosomal hemophilia). In cases of apparent hemophilia A with poor factor VIII recovery and survival *in vivo* and normal mixing studies, von Willebrand disease type 2N should be considered in the differential diagnosis. vWF-to-factor VIII binding assay is necessary to confirm the diagnosis.

- Acquired von Willebrand disease: This condition typically presents as a sudden onset of mucocutaneous bleeding in a previously asymptomatic patient. Mechanisms include antibody formation against vWF with resultant increased clearance of the factor from circulation or inhibition of function of vWF. Another mechanism may be adsorption of vWF by tumor cells. Tumor cells may have aberrant GP Ibα receptor expression.

Management of von Willebrand disease includes use of the following:

- Desmopressin: There is a 2- to 10-fold increase in plasma vWF levels, but this therapy is contraindicated in type 2B and platelet-type von Willebrand disease because release of abnormal vWF may induce thrombocytopenia.
- Cryoprecipitate.
- Factor VIII concentrates.

15.7 ANTIPLATELETS AND ANTICOAGULANTS

The following antiplatelets are used in clinical practice:

- Aspirin: This drug inhibits cyclooxygenase enzyme. The efficacy of aspirin can be assessed by the VFN test, in which effective inhibition is indicated by values less than 550 ARU. No antidotes are available. Bleeding due to aspirin requires platelet transfusion.

- Clopidogrel (Plavix): This drug inhibits P2Y12 ADP receptors. Efficacy can be assessed by the VFN test, in which effective inhibition is indicated by values less than 210 PRU. No antidote is available, and bleeding due to clopidogrel requires platelet transfusion. This drug should be stopped typically 5 days prior to surgery.
- Ticagrelor (Brilinta): This drug also inhibits P2Y12 ADP receptors. The literature suggests that efficacy can be assessed by the VFN test. No antidote is available, and bleeding due to ticagrelor requires platelet transfusion. This drug is typically stopped 3−5 days prior to surgery. Other P2Y12 inhibitors that are FDA approved include prasugrel and ticlopidine.
- Eptifibatide (Integrilin): This drug inhibits GpIIb/IIIa receptors. No antidotes are available. Bleeding due to eptifibatide requires platelet transfusion, and this drug is typically stopped 3−6 hr prior to surgery. Other GpIIb/IIa inhibitors include abciximab and tirofiban.
- Phosphodiesterase inhibitors: These agents inhibit the enzyme phosphodiesterase and cause accumulation of cAMP, resulting in impaired platelet function. In this group, dipyridamole and cilostazol are commercially available for use.

Various anticoagulants are also used in therapy, including the following:

- Warfarin: This drug is a vitamin K antagonist, and efficacy is assessed by measuring PT (INR). If the patient is bleeding while on warfarin, effective management includes fresh frozen plasma infusion. If there is no bleeding but PT is prolonged and requires reversal, vitamin K administration at a dosage of 1−10 mg oral or intravenous is recommended. The goal is to reduce the INR to less than 1.5.
- Unfractionated heparin: This agent inhibits factors IX, X, XI, and XII. Antithrombin III (AIII) is required for effective function, and a low level of ATIII is a cause of heparin resistance. Efficacy is measured by PTT (ideally after 6 hr of dose adjustment). The antidote of heparin is protamine. Patients on heparin should have prolonged thrombin time. During surgery using heparin (e.g., cardiopulmonary bypass), the heparin effect is monitored by activated clotting time. Heparin efficacy can also be monitored by anti-Xa assay. This is applicable when PTT measurement can be relied on—for example, when a patient has a lupus anticoagulant (prolongs PTT) and is on heparin.
- Low-molecular-weight heparin: This agent inhibits factor Xa, and efficacy is monitored by anti-factor Xa assay (ideally 4 hr after dose adjustment). Antidote is protamine (66% efficient compared to neutralization of unfractionated heparin).
- DTIs: Examples of these agents in use are bivalirudin (Angiomax), dabigatran (Pradaxa), argatroban, and lepirudin. None of these agents

has an antidote. Bivalirudin has the shortest half-life (~ 25 min). Impaired renal function results in impaired excretion in all except argatroban (hepatic clearance). Efficacy can be monitored by PTT. The ecarin clotting time can also be used to monitor DTIs. Venom extract of the viper *Echis carinatus* converts prothrombin to meizothrombin. Meizothrombin is an intermediate product in the conversion of prothrombin to thrombin. DTIs inhibit the conversion of prothrombin to meizothrombin. DTIs may be immunogenic and cause formation of antibodies. This will reduce efficacy of the drugs. This is least likely with bivalirudin. DTIs are used for HIT.

■ Direct factor Xa inhibitors: Examples include apixaban, rivaroxaban, idraparinux, fondaparinux, dalteparin, and enoxaparin. All these drugs are excreted mainly by the kidneys except apixaban, which is excreted by the kidneys (25%) and the liver (75%). Enoxaparin and dalteparin may be countered partially with protamine. Others do not have any antidote.

KEY POINTS

■ Hemostasis consists of three steps: vasoconstriction, which is mediated by reflex neurogenic mechanisms; platelet plug (primary hemostasis); and activation of the coagulation cascade (secondary hemostasis).

■ Platelet glycoproteins: GpIb/IX/V attaches to vWF. GpIIb/IIIa attaches to fibrinogen, and GpIa/IIa and GpVI attach to collagen. GpIa/IIa acts as a receptor for platelet adhesion to collagen. Binding of collagen to GpVI initiates platelet aggregation and platelet degranulation.

■ Platelet α granules are granules stained by Wright–Giemsa stain and contain fibrinogen, fibronectin, factor V, vWF, PF-4, PDGF, TGF-β, and thrombospondin. Deficiency of α granules may result in platelets that appear pale and gray on peripheral smear; this is referred to as gray platelet syndrome. It is transmitted as autosomal recessive.

■ Platelet-dense bodies/δ granules are electron dense due to the presence of calcium, which appears as dark bodies under the electron microscope. These dense bodies are present in low numbers (less than 10 per platelet) and contain ATP, ADP, ionized calcium, histamine, 5-HT, and epinephrine.

■ There are three main platelet events. First is adhesion and shape change, which occurs due to interaction between vWF and GP Ib/IX/V receptors. In von Willebrand disease as well as Bernard–Soulier syndrome (in which GpIb/IX/V is lacking), abnormal platelet function (thrombocytopathia) is observed. Second is platelet activation, in which there is an increase in cytoplasmic calcium with change in shape of platelets, extension of pseudopodia, and release of chemicals (release reaction). Third is platelet aggregation, which is fibrinogen-mediated binding of activated GpIIb/IIIa receptors on adjacent platelets that is augmented by thrombospondin, a component of α granules. Therefore, with lack of GpIIb/IIIa (known as Glanzmann's thrombasthenia, transmitted as autosomal recessive) and hypofibrinogenemia, abnormal platelet function is observed.

- Pseudothrombocytopenia may be due to platelet clumps that occur when blood is collected in EDTA, platelet satellitism, large platelets, or traumatic venipuncture.
- Macrothrombocytopenia with neutrophilic inclusions (MYH9 disorders): This group of disorders is characterized by mutations in the *MYH9* gene that encodes the nonmuscle myosin heavy-chain class IIA protein. These are characterized by thrombocytopenia, large/giant platelets, and Dohle-like bodies in neutrophils. This group of disorders includes May−Hegglin anomaly (autosomal dominant), Sebastian syndrome, Fechtner syndrome (nephritis, ocular defects, and sensorineural hearing loss), and Epstein syndrome.
- Wiskott−Aldrich syndrome: This disorder is due to a defect in the *WASP* gene. It is characterized by immune deficiency, eczema, and thrombocytopenia with small platelets.
- Bernard−Soulier syndrome: This disorder is due to deficiency of the GpIb/IX/V receptor. There is also thrombocytopenia with large/giant platelets. Some cases of Bernard−Soulier syndrome are due to defects of the GpIb-β gene located on chromosome 22. This gene may be affected in velocardiofacial syndrome or DiGeorge syndrome associated with deletion of 22q11.2.
- Heparin-induced thrombocytopenia (HIT II): This type of thrombocytopenia is experienced by 0.2−5% of patients on heparin for more than 4 days. Clinical clues for HIT II include heparin administration for more than 4 days with reduction of platelet count by 50% or more. Patients at risk for HIT II are females, surgical patients (especially cardiac and orthopedic surgery), patients with previous exposure to heparin, and patients receiving unfractionated heparin.
- Tests for HIT are ELISA, heparin-induced platelet aggregation, and serotonin release assay. Serotonin release assay is considered the gold standard.
- PFA-100: If the collagen/epinephrine (CEPI) membrane closure time is prolonged but ADP closure time is normal, this is most likely due to aspirin. If both CEPI and ADP closure times are prolonged or CEPI closure time is normal and ADP closure time is abnormal, this denotes platelet dysfunction or von Willebrand disease. However, PFA-100 is insensitive to von Willebrand disease type 2N, patients receiving

clopidogrel and ticlopidine, and storage pool disease. It is very sensitive to von Willebrand disease type I and therapy with GpIIb/IIIa antagonists.

- VerifyNow: VFN is a rapid, turbidimetric whole blood assay that assesses platelet aggregation. It is based on the ability of activated platelets to bind with fibrinogen.
- Normal platelet aggregation: This indicates adequate aggregation with ADP, collagen, epinephrine, arachidonic acid, and ristocetin at high dose but not at low dose ($\leq$6 mg/mL).
- Platelet aggregation in von Willebrand disease/Bernard–Soulier pattern: There is adequate aggregation with ADP, collagen, epinephrine, and arachidonic acid but not at higher doses of ristocetin.
- Platelet aggregation in von Willebrand disease type IIB pattern/pseudo vWD pattern: There is increased aggregation with a low dose of ristocetin.
- Platelet aggregation in Glanzmann's thrombasthenia/ hypofibrinogenemia pattern: There is adequate aggregation with ristocetin but impaired aggregation with all other agonists. Uremia and antiplatelet medication can also produce similar results.
- Platelet aggregation in disorder of activation (storage pool disorder): There is loss of the secondary wave of aggregation with ADP and epinephrine at lower doses. Storage pool disease is the most common inherited platelet function defect. It is divided into α granule and dense granule storage pool diseases.
- Aspirin effect on platelet aggregation: There is significant impairment with aggregation with arachidonic acid. There may be impairment of aggregation (not as much as with arachidonic acid) with ADP, collagen, and epinephrine.
- Plavix effect/ADP receptor defect: There is significant impairment with aggregation with ADP.
- Chronic myeloproliferative disorder: There may be impairment with aggregation with epinephrine.
- Thromboelastography (TEG): TEG is a global test for hemostasis that includes interaction of primary and secondary hemostasis. Parameters of TEG include R (reaction time measured in seconds, representing initial latency from start of the test until the initial clot formation), K (value measured in seconds, representing the time taken to achieve a ceratin clot strength of usually 20 mm), α angle (measures the speed of fibrin buildup), and MA (maximum amplitude, measured in millimeters and represents clot strength). TEG also produces certain other calculated parameters, such as G (computer generated value representing clot strength), CI (clot index, a value calculated based on R, K, α angle, and MA) and LY30 (clot lysis; 30 min indicates the

percentage of decrease in amplitude 30 min after MA and indicates stability of fibrinolysis).

- Platelet mapping is a special TEG assay to measure the effects of antiplatelet drug therapy on platelet function.
- Regarding the clotting pathway, PT is used to assess the integrity of the extrinsic and common pathways; PTT is used to assess the integrity of the intrinsic and common pathways.
- All clotting factors are produced by the liver with the exception of factor VIII, which is produced by endothelial cells.
- Factor VII has the shortest half-life (4−6 hr). With acute liver dysfunction, levels of factor VII are reduced early. Factor VII is required for the extrinsic pathway. Thus, PT is a very good test to assess liver function in the acute setting.
- Individuals with factor XII deficiency or HMWK or kallikrein deficiency do not bleed. PTT may be prolonged.
- All deficiencies of clotting factors are transmitted as autosomal recessive with the exception of factor VIII deficiency (hemophilia A) and factor IX deficiency (hemophilia B), which are transmitted as X-linked recessive. Dysfibrinogenemia, which means fibrogen is dysfunctional, is transmitted as autosomal dominant.
- INR (international normalized ratio) = [patient PT/mean normal PT]^{ISI}.
- Causes of prolonged PT include coumarin (warfarin), vitamin K deficiency (dietary deficiency), failure of absorption of vitamin K (e.g., cholestasis), liver disease, factor deficiency of the extrinsic and common pathways, and factor inhibitor.
- Vitamin K is converted to the epoxide form by the enzyme epoxidase. The epoxide form is inactive; it is converted back to active vitamin K by the reductase enzyme. The full name of the enzyme complex is vitamin K epoxide reductase complex 1 (VKORC1). Warfarin blocks the reductase enzyme. There is genetic variation in the gene for VKORC1 among individuals. Warfarin is metabolized by the P450 CYP system, including CYP2C9. Polymorphism of *CYP2C9* affects CYP2C9 activity. Variation in activity will result in variation in the half-life of warfarin among individuals. Drugs may also affect the activity of CYP2C9. Thus, genetic variation of the *VKORC1* gene, polymorphism of *CYP2C9*, and the effect of other drugs all result in variation in warfarin activity among individuals.
- Causes of prolonged PTT include heparin, direct thrombin inhibitors, factor deficiency of the intrinsic and common pathways, inhibitors (VIII and IX inhibitors, LA), von Willebrand disease, HMWK (Fitzgerald factor) deficiency, pre-kallikrein (Fletcher factor) deficiency, and spurious causes.

- Inhibitors to factors VIII and IX may develop spontaneously (autoantibodies) or post treatment (alloantibodies) with factor replacement.
- One Bethesda unit (BU) of inhibitor is the amount of inhibitor that destroys half the factor VIII or factor IX:C activity in an equal mixture of patient and normal plasma. If the result is 5 or less, it is considered that the antibody is in low titers.
- Causes of prolonged TT include heparin, hypofibrinogenemia, dysfibrinogenemia, and thrombolytic therapy.
- Mixing study: Individuals with prolonged PT or prolonged PTT, or both, may undergo PT/PTT mixing study. The objective of this test is to determine if prolonged PT or PTT is due to a factor deficiency or to the presence of inhibitor. Here, the patient's plasma is mixed with an equal volume of normal plasma. PTT is measured at 0 and 1 or 2 hr. Failure of correction of prolonged PTT indicates the presence of inhibitors. If results at 0 and 1 or 2 hr are similarly prolonged, this implies lupus anticoagulant. If results show time-dependent prolongation, this implies coagulation factor antibody (e.g., factor VIII or factor IX inhibitor). If PT and PTT are both prolonged and mixing study shows correction, then most likely there is deficiency of factor in the common pathway.
- Causes of bleeding with normal PT/PTT include dysfibrinogenemia, factor XIII deficiency, and α_2-antiplasmin deficiency
- von Willebrand disease is the most commonly inherited bleeding disorder in the general population. Type 1 von Willebrand disease is the most common type; 75% of all people with this disease are type 1. In this disorder, partial quantitative deficiency (20–50% reduction) of vWF is observed, but the factor is structurally normal. This disease is transmitted as autosomal dominant with reduced penetrance and variable expressivity. Clinical symptoms are usually mild. In type 2 disorder, there is a qualitative defect of vWF; this type accounts for 20–25% of all cases. There are four subtypes: 2A, 2B, 2M, and 2N. Type 2N is transmitted as autosomal recessive, but the others are transmitted as autosomal dominant. Type 3 is the rarest type, with total deficiency of vWF, and it manifests as the most severe form of von Willebrand disease. This disorder is transmitted as autosomal recessive. Pseudo (platelet-type) von Willebrand disease and acquired von Willebrand disease are also observed.

References

[1] Van der Poll T, Herwald H. The coagulation system and its function in early immune defense. Thromb Haemost 2014;112 [Epub ahead of print].

[2] Modderman PW, Admiral LG, Sonnenberg A, von dem Borne AE. Glycoprotein V and Ib-IX form a noncovalent complex in platelet. J Biol Chem 1992;267:364–9.

[3] Murugappa S, Kunapuli SP. The role of ADP receptors in platelet function. Front Biosci 2006;11:1977−86.

[4] Van Genderen PJ, Michiels JJ. Acquired von Willebrand disease. Baillieres Clin Haematol 1998;11:319−30.

[5] Kehrel BE, Brodde MF. State of the art in platelet function testing. Transfus Med Hemother 2013;40:73−86.

[6] Campbell J, Ridgway H, Carville D. Plateletworks: a novel point of care platelet function screen. Mol Diagn Ther 2008;12:253−8.

[7] Chen A, Teruya J. Global hemostasis testing thromboelastography: old technology, new application. Clin Lab Med 2009;29:391−407.

[8] Da Luz L, Nascrimento B, Rizoli S. Thrombelastography (TEG): practical considerations and its clinical use in trauma resuscitation. Scand J Trauma Resusc Emerg Med 2013;21:29.

[9] Lee CJ, Ansell JE. Direct thrombin inhibitors. Br J Clin Pharmacol 2011;72:581−92.

Thrombophilias and Their Detection

16.1 INTRODUCTION

Thrombophilia is defined as an abnormality of the coagulation or fibrinolytic system that results in a hypercoagulable state and increases the risk of an individual for a thrombotic event in which intravascular thrombus formation may be arterial or venous. The predisposition to form such a blood clot may arise from genetic or acquired factors. Venous thrombosis has an overall incidence of less than 1 per 1000 people, and in the pediatric population it is approximately 1 in 100,000. However, the frequency increases with advancing age [1]. Certain populations, such as individuals with varicose veins and venous ulcers, have a higher prevalence of thrombophilia than that found in controls [2]. Although studies have suggested that coagulation disorders are the main cause of ischemic stroke, only in 1−4% of cases is the prevalence of thrombophilic markers increased in children with stroke compared to control subjects; specifically, factor V Leiden and the presence of antiphospholipid antibodies contribute significantly to stroke occurrence [3].

16.2 THROMBOPHILIA: INHERITED VERSUS ACQUIRED

Thrombophilia may be inherited or acquired, and the hypercoagulability state may arise from an excess or hyperfunction of a procoagulant or a deficiency of an anticoagulant moiety. The risk factors for thrombophilia may be broadly classified into inherited and acquired categories, which are summarized in Table 16.1. Established genetic factors associated with thrombophilia include factor V Leiden, prothrombin gene mutation, protein C or S deficiency, and antithrombin III (AT III) deficiency, whereas rare genetic defects such as hyperhomocysteinemia and dysfibrinogenemia are also established causes of thrombophilia. Intermediate genetic factors related to thrombophilia include elevated coagulation factors such as elevated factor VIII activity. Elevated activities of factors IX and XI may also be associated with

CONTENTS

263

A. Wahed and A. Dasgupta: Hematology and Coagulation. DOI: http://dx.doi.org/10.1016/B978-0-12-800241-4.00016-4

Table 16.1 Common Inherited and Acquired Causes of Thrombophilia

Common Inherited Causes	Common Acquired Causes
Factor V Leiden	Trauma
Prothrombin gene mutation	Surgery
Protein C deficiency	Immobilization
Protein S deficiency	Pregnancy
Antithrombin III (AT III) deficiency	Use of oral contraceptive pill
Hyperhomocysteinemia	Hormone replacement therapy
(e.g., methylenetetrahydrofolate	Paroxysmal nocturnal hemoglobinuria
reductase mutation)	Heparin-induced thrombocytopenia
Increased factor VIII activity	Lupus anticoagulant (LA)
Dysfibrinogenemia	Anticardiolipin antibodies (ACAs)

thrombophilia. Elevated lipoprotein (a) (Lp (a)) is an inherited risk factor for thromboembolism. Because they are critical to the clot formation process, platelet and platelet glycoprotein gene polymorphisms have received increased attention as possible inherited determinants of prothrombotic tendency. However, their role in genetic susceptibility to thrombotic disease is still controversial. A deficiency in tissue-type plasminogen activator or plasminogen could reduce the capacity to remove excessive clot and contribute to thromboembolic disease [4]. Acquired causes of thrombophilia include trauma; surgery; pregnancy; and use of various medications, such as oral contraceptives, lupus anticoagulant, and anticardiolipin antibodies.

16.3   FACTOR V LEIDEN

In 1993, Dahlback *et al.* reported that a poor anticoagulant response to activated protein C was associated with the risk of thrombosis [5]. Further studies demonstrated that a single base pair substitution at nucleotide location 1691 (guanine to adenine) in the factor V gene results in arginine being substituted by glutamine at amino acid location 506 (R506Q) in the factor V protein—the cleavage site of factor V by activated protein C, a natural anticoagulant. Factor V cleaved at position 506 also functions as a cofactor (along with protein S) in activated protein C-mediated inactivation of factor VIIIa. Loss of anticoagulant activity of factor V may contribute to fibrin generation. The mutant factor V is referred to as factor V Leiden because Dutch investigators from the city of Leiden first described this mutation. Factor V Leiden is seen in 4–10% of the Caucasian population, and the risk for venous thromboembolism increases 5- to 10-fold for heterozygous states and 50- to 100-fold for homozygous states. The carrier state of factor V Leiden is highest in individuals from Greece, Sweden, and Lebanon. It is seen significantly less often in African Americans and Asians (<1%). Other variants of factor V are factor V Cambridge, factor V Liverpool, factor V Hong Kong, and HR2 haplotype. The thrombotic risks for these mutations are not clear. Routine clinical testing for factor V Leiden does not detect these variants.

If an individual inherits factor V Leiden (heterozygous state) and is also heterozygous for deficiency of factor V, this is referred to as the pseudo-homozygous state. The activated protein C resistance (APCr) test is the recommended screening test for detection of factor V Leiden, followed by a confirmatory test such as DNA analysis of the factor V gene, which encodes the factor V protein. Polymerase chain reaction (PCR)-based assay is available for identifying factor V Leiden mutation. The first acute thrombosis in patients carrying factor V Leiden is treated according to standard protocol, and the duration of anticoagulation therapy is based on clinical assessment. In the absence of a history of thrombosis, long anticoagulation therapy is not recommend in asymptomatic patients who are factor V Leiden heterozygotes [6].

16.3.1 Activated Protein C Resistance Test

In the original test, activated partial thromboplastin time (aPTT) in the plasma of a patient is measured with, and again without, addition of activated protein C (APC). In normal individuals, the APC capable of cleaving factor V should inactivate it, and then the cleaved factor V (along with protein C and S) should also inactivate factor VIII. As a result, aPTT should be prolonged. However, if the patient has factor V Leiden, then it is not inactivated by APC, resulting in a clotting time that is shorter than that of a normal patient. The APC ratio is then determined using the following formula:

APC ratio = [aPTT in the presence of APC]/[aPTT]

If the ratio is less than 2.2, then the individual is likely to have factor V Leiden (but note that in some laboratories, the normal ratio may be 2.0 or 2.3; values of 1.5−2.0 may be considered borderline). However, there is considerable overlap between healthy individuals and heterozygotes. A limitation of this test is that it requires a normal aPTT in the patient and thus cannot be used in cases in which there is a prolongation of the aPTT (e.g., patients on oral anticoagulants or with a lupus anticoagulant).

Second-generation APCr screens use a protocol in which a patient's plasma is diluted with commercially available factor V-deficient plasma. This modified assay reduces the number of exogenous confounding factors that may affect the aPTT, such as high factor VIII levels, and it makes the test specific for mutations within factor V. However, if lupus anticoagulant (lupus antibody) is present in a patient's plasma, it may compete for phospholipid and can prolong PTT measurements. Therefore, the presence of lupus antibody is a major source of false-positive results in this assay. This modified assay is specific only for mutations within factor V, whereas the original aPTT assay without factor V-deficient plasma predilution protocol measures APC resistance from any cause. Polybrene (hexadimethrine bromide) may be added, which makes it insensitive to unfractionated heparin and low-molecular-weight heparin. The confirmatory test for factor V Leiden is PCR, which is positive in 90% of APC-resistance cases.

16.4   PROTHROMBIN GENE MUTATION

The prothrombin gene *G20210A* mutation (also called Factor II mutation) is due to a single base pair substitution at nucleotide position 20210 (guanine-to-adenine substitution; chromosome 11p−q12), resulting in an increased prothrombin level with increased risk for venous thrombosis because prothrombin is the precursor of thrombin (a serine protease and a key enzyme in the process of hemostasis) and thrombosis. This disorder is an autosomal dominant inherited defect. This prothrombin gene mutation abnormality is the second most frequent factor after factor V Leiden mutation found in patients with thromboembolism (prevalence varies between 5 and 19%). The prevalence of this mutation is between 1 and 5% for the Caucasian population, in which the risk of venous thrombosis is typically increased three- to fivefold (heterozygotes). However, like factor V Leiden, this mutation is rarely found in individuals from Asia or Africa. Jebeleanu and Procopciuc reported that out of 20 patients with venous leg ulcers, 2 patients showed heterozygous *G20210A* mutation [7]. Laboratory testing includes factor II assay and PCR testing for *G20210A*.

16.5   PROTEIN C DEFICIENCY

Protein C is a vitamin K-dependent coagulation inhibitor that is synthesized by the liver and responsible for inactivating factor Va and factor VIIIa after it is activated by thrombin. Protein C deficiency is less common than factor V Leiden or mutation in the prothrombin gene (*G20210A*). This disorder is transmitted as autosomal dominant and is seen in 0.2−0.5% of the population, which also has a 6.5- to 8-fold increased risk of venous thrombosis. Protein C deficiency is found in 6−10% of hypercoagulation cases. Protein C deficiency may be inherited or acquired (Table 16.2). Inherited protein C deficiency may be due to decreased production of protein C, which represents most of the cases of inherited protein C deficiency (type I; ∼85% of cases). A less common inherited disorder is due to dysfunctional protein (type II). The gene for protein C is located on chromosome 2, and more than 160 mutations have been described. The type I deficiency is more common; most affected patients are heterozygous and usually have a 50%

Table 16.2 Inherited and Acquired Protein C Deficiency

Inherited Protein C Deficiency	Acquired Protein C Deficiency (More Common)
Type I: Decreased production of functional protein C (~85% of inherited cases) Type II: Dysfunctional protein with reduced activity	Decreased synthesis: Liver disease, vitamin K deficiency or the presence of vitamin K antagonist (e.g., coumadin), L-asparaginase therapy Increased clearance: DIC, acute thrombosis, etc. Dilution: Post plasmapheresis, i.v. administration of crystalloids, etc.

reduction in plasma protein C concentration compared to controls. In patients with type II disorder, plasma protein C concentration is normal, but the protein has decreased functional activity.

Protein C deficiency can also be acquired and can be induced by certain conditions such as a particular malignancy. Because protein C is a vitamin K-dependent protein (molecular weight, 62 kDa) produced in the liver, liver disease may cause reduced levels. In addition, patients receiving warfarin therapy or patients who are deficient in vitamin K may also show protein C deficiency. Antineoplastic drugs may impair either the absorption or the metabolism of vitamin K, which may indirectly cause protein C deficiency. For example, L-asparaginase, which is used to treat patients with acute lymphoblastic leukemia, is known to cause protein C deficiency. Increased clearance of protein C may occur during acute thrombosis or disseminated intravascular coagulation (DIC). Protein C deficiency may also occur post plasmapheresis due to dilution effect. Acquired protein C deficiency is more common than inherited deficiency.

Clinical manifestations of protein C deficiency are recurrent deep vein thrombosis, pulmonary embolism, neonatal purpura fulminans (in homozygotes), and warfarin (coumadin)-induced necrosis of the skin. The latter, a rare complication of warfarin therapy, is observed mainly in obese perimenopausal middle-aged women being treated with warfarin for deep vein thrombosis or pulmonary embolism. Most cases of warfarin-induced necrosis have been reported in patients with hereditary protein C deficiency, but this may also be due to secondarily acquired protein C deficiency. This complication of warfarin therapy is usually observed during the first few days of treatment (onset within the first 1–10 days but most commonly within the first 5 days) and is due to the shorter half-life of protein C compared to that of other vitamin K-dependent factors, such as factor IX, factor X, and prothrombin, which may cause temporary exaggeration of the imbalance between the pro-coagulation and anticoagulation pathways. However, protein S and antithrombin II deficiency may also cause this episode. This phenomenon can be prevented by carefully adjusting the initial dosage of warfarin (progressively increasing warfarin dosage oven an extended period of time) and use of fresh frozen plasma when starting warfarin therapy [8]. Laboratory testing to diagnose this disorder includes function assay for protein C as well as immunological testings.

16.6 PROTEIN S DEFICIENCY

Protein S (originally discovered and purified in Seattle, Washington; hence the name), a vitamin K-dependent protein, is synthesized in the liver and also in megakaryocytes. Protein S serves as a cofactor of protein C in inactivating factor Va and factor VIIIa. In plasma, 60% of protein S is bound to C4b binding protein (an acute phase reactant) and is functionally inactive. Protein S deficiency is transmitted as autosomal dominant and is observed in 0.7% of the population. This deficiency increases the risk of venous thrombosis by 1.6- to 11.5-fold compared to that of the normal population. Protein S deficiency is found in 5–10% of hypercoagulation cases. Protein S deficiency may be inherited or acquired (Table 16.3).

Inherited protein S deficiency can be type I, II, or III. There are two homologous genes for protein S—PROS1 and PROS2—both of which are located on chromosome 3. Type I is the most common type of inherited disorder of protein S, accounting for an approximately 50% reduction in protein S level in plasma with marked reduction in free protein S level, which is the active form. Type II disorder is characterized by normal total and free protein S levels but diminished protein S functional activity. Type II (also known as type IIa) is characterized by a normal total S level but selectively reduced free protein S levels and protein S functional activity (approximately 40% reduction) [4]. This phenomenon is due to increased binding of protein S to C4b binding protein. However, acquired cases of protein S deficiency are more common than inherited cases of protein S deficiency. Because protein S, similarly to protein C, is a vitamin K-dependent protein, acquired causes of protein S deficiency are similar to acquired causes of protein C deficiency.

Table 16.3 Inherited and Acquired Protein S Deficiency

Inherited Protein S Deficiency	Acquired Protein S Deficiency (More Common)
Type I: Decreased production of functional protein S Type II: Dysfunctional protein Type III: Increased binding to C4b binding protein	Decreased synthesis: Liver disease, vitamin K deficiency or the presence of vitamin K antagonist (e.g., coumadin), L-asparaginase therapy Increased clearance: DIC, acute thrombosis Dilution: Post plasmapheresis, i.v. administration of crystalloids

Common causes of acquired protein S deficiency include liver disease, vitamin K deficiency or therapy with warfarin (vitamin K agonist), or L-asparaginase therapy. Increase clearance of protein kinase S as observed in DIC or acute thrombosis may also cause protein S deficiency. Clinical manifestations of protein S deficiency include recurrent deep vein thrombosis, pulmonary embolism, and neonatal purpura fulminans. Laboratory testings are immunological assays (free and total) as well as functional assay.

16.6.1 Assays for Protein C and Protein S

Available assays for protein C include quantifying protein C antigen (estimate quantity of protein C in plasma but not functional protein C) using an immunoassay (e.g., enzyme-linked immunosorbent assay (ELISA); immunodiffusion or immunoelectrophoresis method), in which a monoclonal antibody against protein C is used for antigen binding. Functional assays for protein C (determine protein C activity) may be clot based or chromogenic (spectrophotometric detection). Multiple immunoassays to determine protein C concentration as well as both clot-based and chromogenic-based functional assays for protein C are commercially available.

Protein C functional assay is useful in determining protein C deficiency and should be the first test performed in a patient with suspected protein C deficiency [9]. If the value is abnormal, then the quantity of protein C in serum may be estimated using an immunoassay. In the clot-based functional protein C assay, the patient's plasma is incubated with Protac, derived from venom of the southern copperhead snake *Agkistrodon contortrix*, which activates protein C. Then an activator (e.g., aPTT reagent or Russell viper venom reagent, which induces clot formation) is added. If protein C is present and is activated, then factor Va and factor VIIIa are inactivated; thus, clotting time is longer, which can be measured by aPTT (if aPTT reagent is used) or Russell venom clotting time. For chromogen-based functional assay, activated protein C is capable of cleaving a chromogen that produces a color that can be measured photometrically. Conditions that may shorten clotting time may cause falsely low levels of protein C. These include high levels of factor VIII, low levels of protein S, and factor V Leiden. Also, vitamin K deficiency and warfarin may also result in falsely low levels because in both cases, protein C is not adequately carboxylated and thus not functional. Situations in which the clotting time is prolonged may result in falsely high protein C levels, such as with heparin therapy, direct thrombin inhibitor therapy, and lupus anticoagulant.

The protein S functional assay, which measures the biological activity of free protein S, is based on the principle of adding protein S-depleted plasma and activated protein C to test plasma (the patient's plasma). Then an activator (Russell viper venom reagent) is added, and aPTT is measured. As with protein C assay, clotting time is longer with adequate levels of protein S. Causes of false-low or -high levels are similar to those with the protein C functional assay. However, immunoassays are commercially available for determining total concentration of protein S or free concentration of protein S in plasma. For determining total concentration of protein S (antigenic assay), ELISA or latex particle-based assays are commercially available. For measuring the free concentration of protein S, C4b-bound protein S must be precipitated using a precipitation agent such as polyethylene glycol; after centrifuge, protein S concentration in the supernatant, which represents free protein S, can be measured. Alternatively, monoclonal antibody directed to protein S epitopes not accessible in the bound form can be used in an immunological assay for direct determination of free protein S.

16.7 ANTITHROMBIN III DEFICIENCY

AT III inactivates thrombin and other factors (Xa, IXa, XIa, XIIa, and kallikrein), and this process is accelerated by heparin. AT III deficiency is transmitted as autosomal dominant and is seen in 0.17% of the population. The risk of thrombosis is increased 5- to 8.1-fold and accounts for 5−10% of hypercoagulation cases. AT III deficiency may be inherited or acquired. Inherited AT III deficiency is of two types:

- Type I: Reduced production of functionally normal AT III, and values are reduced by approximately 50% in heterozygotes. More than 80 mutations have been reported in patients with type I deficiency.
- Type II: Dysfunctional AT III. There are three subtypes. In the reactive site (RS) subtype, there is altered AT interaction with Xa and thrombin active sites. In the heparin binding site (HBS) subtype, there is defective heparin binding to antithrombin. In the pleomorphic subtype (PE), there are multiple defects.

Acquired AT III deficiency may be seen in patients with liver disease (reduced synthesis), nephrotic syndrome (renal loss of AT III as well as reduced synthesis by the liver), DIC, and therapy with L-asparaginase.

Clinical manifestations of AT III deficiency include recurrent deep vein thrombosis and pulmonary embolism. Laboratory testing involves functional assay (chromogenic) and immunologic (antigenic) assay.

16.8 HYPERHOMOCYSTEINEMIA

Homocysteine is a sulfhydryl group-containing amino acid produced from the metabolism of essential amino acid methionine. Hyperhomocysteinemia may be inherited or acquired, and it is a risk factor for cardiovascular diseases as well as thrombophilia. Inherited causes include deficiencies of 5,10-methyle-netetrahydrofolate reductase (MTHFR) or cystathionine-β-synthase enzymes due to defective genes that encode such enzymes. MTHFR is an enzyme in folate-dependent homocysteine remethylation, catalyzing the reduction of 5,10-methylenetetrahydrofolate into 5-methyltetrahydrofolate. Single base pair substitution at nucleotide location 677 (thymine replaces cytosine: 677 C > T) in the *MTHFR* gene results in substitution of alanine by valine at location 223 (A223V) in the 5,10-methylenetetrahydrofolate reductase enzyme with decreased activity causing hyperhomocystinemia. As a result, these individuals are at higher thrombotic risk (mechanisms: blood vessel injury, coagulation activation, fibrinolysis inhibition, and platelet activation). Approximately 11% of the Caucasian population is homozygous for this mutation, with a threefold increased risk of thrombosis. Although this is the most common mutation, at least 40 different mutations of the *MTHFR* gene have been described in individuals with homocystinuria.

Acquired causes of hyperhomocystinemia include vitamin B_6, vitamin B_{12}, and folate deficiencies as well as renal failure. Combined folic acid–vitamin B therapy can reduce elevated homocysteine levels in blood [10]. Laboratory testing includes plasma homocysteine assay using an automated analyzer, but PCR testing to detect genetic mutation is currently not fully accepted as part of the routine battery.

16.9 INCREASED FACTOR VIII ACTIVITY

Factor VIII is a plasma sialoglycoprotein that plays an important role in hemostasis. For a long time, it has been recognized that factor VIII deficiency in patients with hemophilia A results in bleeding episodes. In recent years, research—especially epidemiological studies—has demonstrated that the converse is also true: Elevated plasma factor VIII coagulant activity is now accepted as an independent marker of increased thrombotic risk. The Leiden Thrombophilia Study (LETS) was the first to report the association between high plasma levels of factor VIII and venous thromboembolism. Factor VIII activity is higher in non-O blood group individuals. It is also well recognized that levels of factor VIII increase with advancing age, and this may be related to increased risk of thrombotic events with advanced age. Although thrombosis is rare in children, elevated factor VIII levels (>90th percentile) were observed in 19.5% of children with venous thrombosis compared to only 4.4% in the control group. Currently, no polymorphism of the factor 8 gene is associated with increased levels of factor VIII in plasma [11]. Factor VIII activity may also be elevated in pregnancy, malignancies, infection, and inflammation.

16.10 ACQUIRED CAUSES OF THROMBOPHILIA

Acquired causes of thrombophilia are more common that inherited causes. Trauma, surgery, immobilization, pregnancy, hormone replacement therapy, use of oral contraceptives, paroxysmal nocturnal hemoglobinuria, etc. are all acquired causes of thrombophilia. However, lupus anticoagulant (lupus antibody (LA)) and anticardiolipin antibodies (ACAs) are commonly encountered in patients with a higher risk of thrombotic events. The use of oral contraceptives increases the risk of venous thromboembolism as well as arterial thrombosis. Third-generation oral contraceptives appear to increase the risk of venous thromboembolism compared to the second-generation pill, but such medication usually does not increase the risk of arterial thrombosis. The effect of oral contraceptives on increasing the risk of venous thromboembolism is more pronounced within the first year of therapy, and as expected, such risks are higher in patients with an inborn or acquired tendency for thrombophilia [11]. Although hormone replacement therapy can increase the risk of thromboembolism, the risk is lower in users of estrogen-only hormone replacement therapy than in users of estrogen–progestin hormone replacement therapy.

16.10.1 Lupus Anticoagulant and Anticardiolipin Antibodies

Lupus anticoagulant or LA and ACAs are immunoglobulins that prolong *in vitro* phospholipid-dependent clotting times and are examples of antiphospholipid antibodies. Other examples of antiphospholipid antibodies are anti-phosphatidyl serine and anti-β_2 glycoprotein I antibody. These antibodies may causes arterial or venous thrombosis or miscarriage.

Laboratory testing for lupus anticoagulant includes aPTT, mixing study, dilute Russell viper venom time (dRVVT), platelet neutralization procedure (PNP), and hexagonal phase phospholipid neutralization test (Staclot LA). In the Staclot LA assay, test plasma is incubated with and without hexagonal phase phospholipid reagent, and aPTT is measured. It is a confirmatory test for detecting the presence of lupus anticoagulant antibody. The PNP assay is based on the ability of platelets to bypass the effect of lupus anticoagulant antibody by correcting the prolonged clotting time in various phospholipid-dependent assay systems. In this assay, freeze–thawed platelet suspension (platelet lysate), which is added to reaction mixture, can neutralize lupus antibodies by binding with them.

The dilute Russell viper venom test is a commonly used screening test for LA. It is called the dilute Russell viper venom time test because the phospholipid reagent is diluted such that its concentration becomes rate limiting; thus, it becomes a sensitive assay. The principle of the assay is that Russell's venom directly activates factor X, which then activates prothrombin (factor II) in the presence of factor V and phospholipid, thus initiating coagulation (clot formation in 23−27 sec). If lupus coagulation antibodies are present in a patient's sample, these antibodies should bind with phospholipids, thus reducing their availability to participate in coagulation; as a result, clotting time should be prolonged (positive test).

If the screening test is positive, then mixing study must be performed in which the patient's plasma is mixed with normal plasma in a 50:50 ratio. A prolonged bleeding time that is not corrected by adding normal serum indicates the presence of lupus antibody. In the mixing study, no phospholipid is added, and the ratio is calculated as follows:

dRVVT mix ratio = dRVVT mix (50:50 patient plasma to normal)/dRVVT screen

For confirmation assay, phospholipid is added, which should correct the prolonged bleeding time:

dRVVT confirmation ratio = dRVVT screen/dRVVT with added phospholipid

The dRVVT confirmation ratio, PNP, and Staclot LA use the same principle. If excess phospholipid is present in the test media, all antibodies present must bind to added phospholipids and bleeding time should be reduced because a sufficient amount of phospholipids should be present to participate in the coagulation process. In PNP, a platelet lysate is the source of the phospholipid, whereas in Staclot LA, the source of phospholipid is hexagonal phase phospholipid reagent.

For testing anticardiolipin antibodies, ELISA (for IgG, IgM, and IgA types of antibodies) assays are commercially available. Because cardiolipin is the antigen used for screening tests for syphilis (e.g., Venereal Disease Research Laboratory (VDRL) and rapid plasma reagin (RPR) tests), false-positive tests for syphilis may be seen in individuals positive for anticardiolipin antibodies. Individuals with syphilis infection may also have false-positive tests for anticardiolipin antibodies.

KEY POINTS

- Factor V Leiden can be seen in 4−10% of the Caucasian population, and the risk for venous thromboembolism increases 5- to 10-fold for heterozygous states and 50- to 100-fold for homozygous states. Factor V Leiden is poorly inhibited by activated protein C (APC), which can lead to venous thrombosis. However, factor V activity in these patients is normal.
- In the original activated protein C (APC) test, activated partial thromboplastin time (aPTT) in the plasma of a patient is measured

with, and again without, addition of APC. The APC ratio is then determined using the following formula:

APC ratio = [aPTT in the presence of APC]/[aPTT]

If the ratio is less than 2.2, then the individual is likely to have factor V Leiden. A limitation of this test is that it requires a normal aPTT in the patient and so cannot be used in cases in which there is a prolongation of the aPTT (e.g., in patients on oral anticoagulants or a lupus anticoagulant). Second-generation APCr screens use a protocol in which a patient's plasma is diluted with commercially available factor V-deficient plasma. This modified assay reduces the number of exogenous confounding factors that might affect the aPTT, such as high factor VIII levels, and it makes the test specific for mutations within factor V.

- Prothrombin gene mutation: Single base pair substitution at nucleotide position 20210 (guanine-to-adenine substitution: G20210A). This results in increased prothrombin levels with increased risk for venous thrombosis. Laboratory testing includes factor II assay and PCR testing for *G20210A*.
- Inherited protein C and S deficiencies are transmitted as autosomal dominant. These proteins inhibit factor Va and factor VIIIa. Protein C and S deficiencies may be inherited or acquired. Laboratory testing includes immunological assays and functional assays.
- Principle of protein C functional assay: In this test, a patient's plasma is incubated with Protac, derived from southern copperhead snake venom, which activates protein C. Then an activator such as aPTT is added, and clotting time is measured. Alternatively, Russell viper venom reagent is added to activate clotting, and clotting time is measured. If protein C is present and is activated, then factor Va and factor VIIIa are inactivated; thus, clotting time is longer.
- Principle of protein S functional assay: In this test, protein S-depleted plasma and APC are added to test plasma. Then an activator (e.g., aPTT reagent or Russell viper venom reagent) is added, and clotting time is measured.
- Antithrombin III (AT III) inactivates thrombin and other factors (Xa, IXa, XIa, XIIa, and kallikrein) that are accelerated by heparin. AT III deficiency may be inherited or acquired. Inherited AT III deficiency is transmitted as autosomal dominant. Laboratory testing includes functional assay (chromogenic) and immunologic assay.
- Hyperhomocysteinemia may be inherited or acquired. Inherited causes include deficiencies of MTHFR or cystathionine-β-synthase caused by gene mutations that encode these enzymes. Hyperhomocystinemia is a risk factor for cardiovascular diseases as well as thrombophilia. Laboratory testing includes determination of homocysteine

concentration in plasma using an automated analyzer and PCR testing (currently not fully accepted as part of the routine battery).

- Multiple epidemiologic studies have demonstrated that elevated factor VIII activity is a risk factor for thromboembolism.
- Lupus anticoagulant and anticardiolipin antibodies are immunoglobulins that prolong *in vitro* phospholipid-dependent clotting times. These antibodies are examples of antiphospholipid antibodies and may cause arterial or venous thrombosis or miscarriage.
- Laboratory testing for lupus anticoagulant includes aPTT, mixing study, dilute Russell viper venom time (dRVVT), platelet neutralization procedure (PNP), and hexagonal phase phospholipid neutralization test (Staclot LA). The dilute Russell viper venom test is a commonly used screening test for LA. Dilute means that a minimal amount of phospholipid is present, which is rate limiting. If LA is present, it binds to the phospholipid and prolongs the clotting time. The dRVVT confirmation ratio, PNP, and Staclot LA use the same principle. If excess phospholipid is present in the test media, it should bind all LA and the clotting should be reduced.
- Testing for anticardiolipin antibodies is performed using immunological methods (ELISA assays for IgG, IgM, and IgA types of antibodies).

References

[1] Hoppe C, Matsunaga A. Pediatric thrombosis. Pediatric Clin North Am 2002;49:1257–83.

[2] Darvall KA, Sam RC, Adam DJ, Silverman SH, et al. Higher prevalence of thrombophilia in patients with varicose veins and venous ulcers than controls. J Vasc Surg 2009;49:1235–41.

[3] Kenet G, Sadetzki S, Murad H, Martinowitz U, et al. Factor V Leiden and antiphospholipid antibodies are significant risk factors for ischemic stroke in children. Stroke 2000;31: 1283–8.

[4] Khan S, Dickerman JJ. Hereditary thrombophilia. Thromb J 2006;4:15.

[5] Dahlback B, Carlsson M, Svensson PJ. Familial thrombophilia due to a previously unrecognized mechanism characterized by poor anticoagulant response to activated protein C: prediction of a cofactor to activated protein C. Proc Natl Acad Sci USA 1993;90:1004–8.

[6] Kujovich JL. Factor V Leiden thrombophilia. Genet Med 2011;13:1–16.

[7] Jebeleanu G, Procopciuc L. G20210A prothrombin gene mutation identified in patients with venous leg ulcers. J Cell Mol Med 2001;5:397–401.

[8] Chan YC, Valento D, Mansfield AO, Stansby G. Warfarin induced skin necrosis. Br J Surg 2000;87:266–72.

[9] Khor B, Van Cott EM. Laboratory test for protein C deficiency. Am J Hematol 2010;85:440–2.

[10] Maron BA, Loscalzo J. The treatment of hyperhomocystinemia. Annu Rev Med 2009;60:39–54.

[11] Jenkins PV, Rawley O, Smith OP, O'Donnell JS. Elevated factor VIII levels and risk of venous thrombosis. Br J Haematol 2012;157:653–63.

CHAPTER 17

Sources of Errors in Hematology and Coagulation

17.1 INTRODUCTION

Hematology tests such as complete blood count (CBC) and peripheral blood smear analysis are commonly ordered tests in clinical laboratories. Sources of errors in such tests may be pre-analytical, analytical, or post-analytical. Blood for CBC is typically collected in vacuum tubes that contain the anticoagulant dipotassium ethylenediaminetetraacetic acid (K_2-EDTA). The blood collected in vacuum tubes is analyzed on an automated hematology analyzer to obtain CBC results. For coagulation, citrated plasma is the preferred specimen. Although minimization of analytical errors has been the main focus of developments in laboratory medicine, the other steps are more frequent sources of erroneous results. Carraro and Plebani reported that in the laboratory, pre-analytical errors account for 62% of all errors, with post-analytical errors representing 23% and analytical errors 15% of all laboratory errors [1]. Even using disodium EDTA versus dipotassium EDTA tubes may cause errors because values of hemoglobin, hematocrit, mean corpuscular volume (MCV), and lymphocyte count were higher when collected in disodium EDTA tubes compared to dipotassium EDTA tubes. In addition, tubes must be filled with blood because hematocrit and MCV values may be falsely low [2]. Differences may even be observed between different brands of dipotassium EDTA tubes [3]. Therefore, laboratories must validate blood collection tubes prior to use for routine blood collection.

Errors related to sample collection, transport, and storage must be avoided in order to obtain accurate results in hematology testing. The amount and concentration of dipotassium EDTA in collection tubes require that blood be collected up to a specific mark on the tubes. If too little blood is collected, dilution of the sample can become an issue with alteration of parameters. Relative excess EDTA in such cases also affects the morphology of blood cells. Regarding the transport of specimen, high temperatures must be avoided. Red cell fragmentation is a feature of excess heat.

A. Wahed and A. Dasgupta: Hematology and Coagulation. DOI: http://dx.doi.org/10.1016/B978-0-12-800241-4.00017-6

Prolonged storage may result in degenerative changes in white blood cells (WBCs). This is best illustrated in neutrophils, which have a round pyknotic nucleus. Abnormal lobulation of the lymphocyte nuclei is another established phenomenon that occurs with prolonged storage of blood. These cells may be considered as atypical lymphocytes with the incorrect implication of an underlying lymphoproliferative disorder. Box 17.1 summarizes sources of laboratory errors in hematology testing.

17.2 ERRORS IN ROUTINE HEMATOLOGY TESTING

Automated hematology analyzers can rapidly analyze whole blood specimens for CBC and differential count, which includes red blood cell (RBC) count, WBC count, platelet count, hemoglobin concentration, hematocrit, RBC indices, and leukocyte differential. In addition, modern hematology analyzers are capable of analyzing many leukocytes using flow cytometry-based methods, some in combination with cytochemistry or fluorescences or conductivity to count different types of WBCs, including neutrophils, lymphocytes, monocytes, basophils, and eosinophils (five-part differential). Nucleated RBCs are also detected. Sources or errors in measurement of various hematological parameters are addressed in this section.

17.2.1 Errors in Hemoglobin Measurement and RBC Count

Hemoglobin measurement is based on absorption of light at 540 nm. If the sample is turbid, this will produce higher hemoglobin levels. Examples of such a state include hyperlipidemia [4], patients on parenteral nutrition, hypergammaglobulinemia, and cryoglobulinemia. Turbidity from very high WBC count can also falsely elevate hemoglobin levels. Smokers have high carboxyhemoglobin that may falsely elevate the measured hemoglobin level.

Large platelets may be counted by some instruments as RBCs. Also, red cell fragments greater than 40 fL will be counted as whole red cells. In both situations, the RBC count will be falsely high. Cold agglutinins will cause red cell agglutination *in vitro* and result in low RBC counts. If cold agglutinins are suspected, the sample should be warmed to obtain an accurate RBC count.

17.2.2 Errors in MCV and Related Measurements

If there is red cell agglutination, then red cell clumps will be counted as single red cells but the volume of the estimated cell will be much higher. This may result in falsely high MCV values. Large platelets being counted as red cells may also result in falsely low MCV values, as these platelets typically have less volume than normal red cells.

BOX 17.1 SOURCES OF LABORATORY ERRORS IN HEMATOLOGY TESTING

- Falsely high hemoglobin
  - Turbid sample (hyperlipidemia, parenteral nutrition, hypergammaglobulinemia, cryoglobulinemia, marked leukocytosis)
  - Smokers (high caboxyhemoglobin)
- Falsely low hemoglobin
  - Rare
- Falsely high RBC count
  - Large platelets
  - Red cell fragments
- Falsely low RBC count
  - Cold agglutinin
- Falsely high MCV
  - Cold agglutinin
  - Hyperosmolar state (uncontrolled diabetes mellitus)
- Falsely high WBC count
  - Nucleated red cells
  - Non-lysis of red cells (due to target cells in hemoglobinopathy)
  - Giant platelets or platelet clumps (due to EDTA)
  - Cryoglobulins
  - Microorganisms
- Falsely low WBC count
  - Leukoagglutination (due to EDTA)
  - Cold agglutinin
- Falsely increased lymphocyte count
  - Nucleated red cells
  - Non-lysis of red cells (due to target cells in hemoglobinopathy)
  - Giant platelets or platelet clumps (due to EDTA)
  - Malarial parasites
  - Dysplastic neutrophils (hypolobated neutrophils)
  - Basophilia
- Falsely decreased lymphocyte count
  - Rare

- Falsely high neutrophil count
  - Rare
- Falsely low neutrophil count
  - Neutrophil aggregation
  - Neutrophil with hemosiderin granules (counted as eosinophils)
- Falsely increased eosinophil count
  - Neutrophils with hemosiderin granules (counted as eosinophils)
  - Red cells with malarial pigments
- Falsely low eosinophil count
  - Hypogranular eosinophils
- Falsely increased monocyte count
  - Large reactive lymphocytes
  - Lymphoblasts
  - Lymphoma cells
  - Immature granulocytes
- Falsely low monocyte count
  - Rare
- Falsely high platelet count
  - Fragmented red cells (in microangiopathic hemolysis)
  - Fragmented white cells (in leukemia)
  - Microorganisms
  - Cryoglobulin
- Falsely low platelet counts
  - Partial clotting or platelet activation
  - Giant platelets
  - Platelet clumps and platelet satellitism (due to EDTA)
  - GpIIb/IIIa antagonists
  - Platelet agglutination (due to cold agglutinin)
- False-positive fragility test
  - Immunologically mediated hemolytic anemias

If the patient is in a state of high osmolarity, the cytoplasm of the red cells of such patients is also hyperosmolar. When diluents are added to the blood in the analyzer, water will move into the red cells, causing them to swell in size. MCV values may be higher than those in the *in vivo* state. Examples of hyperosmolar states are uncontrolled diabetes mellitus, hypernatremia, and dehydration. The converse will occur in hypo-osmolar states.

Values for hematocrit, mean corpuscular hemoglobin (MCH), and mean corpuscular hemoglobin concentration (MCHC) are obtained by calculation using hemoglobin levels, RBC counts, and MCV values. If there is an error in any of these values, the calculated values will also be inaccurate.

17.2.3 Errors in WBC Counts and WBC Differential Counts

Falsely high WBC counts are more common than falsely low WBC counts. There are several situations in which the WBC count may be falsely elevated. One of the most frequent situations is high WBC count in the presence of a significant number of nucleated red blood cells (NRBCs). If an accurate WBC count is required, then a corrected WBC count needs to be performed. This can be done with some hematology analyzers by running the sample again in the "NRBC mode" or performing a manual count. Platelet aggregates and non-lysis of red cells are other causes of spuriously high WBC counts. If the high WBC count is due to non-lysis of red cells, this may be an indication of hemoglobinopathies. Target cells seen in hemoglobinopathies are typically resistant to lysis. Platelet aggregates may be due to EDTA, and redrawing blood in a citrate tube may be the solution in such cases. Erroneous WBC counts with spurious leukocytosis can be seen with the presence of cryoglobulins and microorganisms. Spurious leukopenia can be seen in cold agglutinins and EDTA-dependent leukoagglutination. Moreover, when there are giant platelets or NRBCs or red cells resistant to lysis, these may be counted as lymphocytes in some instruments, giving rise to a falsely high lymphocyte count. Hemoglobinopathies and target cells are important causes of non-lysis of red cells. The presence of malarial parasites in RBCs has been also known to increase the lymphocyte count.

In myelodysplastic syndrome, if the myeloid series is affected, then hypolobated and hypogranular neutrophils may be present. Automated analyzers may no longer count these dysplastic neutrophils as such; instead, they may be counted as lymphocytes.

Basophilia is typically seen in chronic myelogenous leukemia. Basophils are cells with coarse granules that may even obscure the nucleus. If the analyzer falsely recognizes all the dense granules of basophils as one single nucleus, then these cells may be counted as lymphocytes.

It is thus apparent that there can be multiple situations in which the lymphocyte count is inappropriately elevated. Whereas falsely low lymphocyte counts are rare, falsely low neutrophil counts are encountered more frequently. Neutrophil aggregation is a documented phenomenon that can result in low neutrophil count. Neutrophils have fine granules, whereas the granules of eosinophils are larger. Basophils have quite large granules.

If neutrophils have hemosiderin granules, they may be counted as eosinophils. If eosinophils are hypogranular, these eosinophils may be counted as neutrophils. Red cells infected by malarial parasites may contain malarial pigments. Malaria-infected red cells are resistant to lysis. These red cells with malarial pigments may be counted as eosinophils.

A key difference between lymphocytes and monocytes is their size, with monocytes being significantly larger. Reactive (activated) lymphocytes typically have more abundant cytoplasm compared to nonreactive lymphocytes. Their size approaches that of a monocyte. Thus, these lymphocytes may be counted as monocytes.

Also, it has been reported that abnormal lymphocytes such as those seen in chronic lymphocytic leukemia, lymphoblasts, and leukemic or lymphoma cells can be miscounted as monocytes. When there is left shift in the WBC series, there tend to be slightly more immature cells, such as bands and metamyelocytes. Cells that are more immature are naturally larger and may also be counted as monocytes. Storage of blood at room temperature and delay in running the sample on the analyzer may also contribute to inaccurate WBC differential values.

When differential counts obtained by automated analyzers are compared to differential counts performed manually, there are frequently differences in the results. Often, the differences are clinically inconsequential. However, it is important to correlate significantly abnormal results with a morphological review of the peripheral smear.

17.2.4 Errors in Platelet Count

In certain situations, hematology analyzers are known to provide a falsely low platelet count when, in fact, the actual platelet count is adequate. This may give rise to suboptimal clinical management.

Partial clotting of the specimen or platelet activation during venipuncture may cause platelet aggregation. Either mechanism may lead to low platelet counts. Checking the specimen for any clot and analyzing the histograms, as well as reviewing the smear are all important steps to avoid misleading low platelet counts. Various other mechanisms explain falsely low platelet counts, otherwise referred to as pseudothrombocytopenia (anticoagulant-induced pseudothrombocytopenia), platelet satellitism, giant platelets, and cold agglutinin-induced platelet agglutination.

Falsely elevated platelet counts are much less common than falsely low counts. Fragmented red cells or white cell fragments may be counted as platelets, giving rise to high platelet counts. Fragmented red cells can be seen in states of microangiopathic hemolysis such as disseminated intravascular coagulation (DIC). White cell fragments can be seen in leukemic or lymphoma states. Patients with leukemia, especially acute leukemia, need supportive therapy in the form of blood component transfusions. If the platelet count is falsely elevated in a patient with acute leukemia, the decision to transfuse platelets may be delayed, with undesirable clinical consequences. Falsely high platelet counts may also be seen in the presence of cryoglobulins and microorganisms present in blood.

17.3 ERRORS IN SPECIFIC HEMATOLOGY TESTING

In this section, specific selected test errors that are often seen in the hematology laboratory are addressed. These errors may be related to the presence of cold agglutinins, cryoglobulins in the specimen, or conditions such as pseudothrombocytopenia.

17.3.1 Cold Agglutinins

Cold agglutinins are polyclonal or monoclonal autoantibodies directed against red blood cell i or I antigens and preferentially binding erythrocytes at cold temperatures. These autoantibodies are typically immunoglobulin M subtype, which may be associated with malignant disorders (e.g., B cell neoplasm) or benign disorders (e.g., post-infection and collagen vascular disease), and can be manifested clinically as autoimmune hemolytic anemia. In the hematology laboratory with automated analyzers, cold agglutinins typically present as a discrepancy between the RBC indexes. The agglutinated erythrocytes may be recognized as single cells or may be too large to be counted as erythrocytes; subsequently measured MCV is falsely elevated and the RBC count is disproportionately low. Whereas the measured hemoglobin is correct due to its independence from the cell count, the calculated indexes are incorrect: The hematocrit (RBC count $\times$ MCV) is low, whereas the MCH (hemoglobin/RBC count) and the MCHC (hemoglobin/hematocrit) are elevated. Hemagglutination may be grossly visible to the unaided eye, and microscopic examination of the peripheral blood smear will show erythrocyte clumping. By rewarming the blood sample to 37°C, the erythrocyte agglutination is alleviated and correct values may be obtained. More severe cases of cold agglutinin may require use of the saline replacement technique if rewarming the sample fails to resolve the RBC index discrepancy.

Spurious leukopenia due to cold agglutinin is also occasionally encountered with automated hematology analyzers. The mechanism is postulated to be an IgM autoantibody directed against components of the granulocyte membranes. Cold agglutinin-induced leukopenia should be recognized as a potential cause of pseudogranulocytopenia so that WBC counts can be accurately reported and unnecessary evaluation of patients for leukopenia can be avoided.

17.3.2 Cryoglobulins

Cryoglobulins are typically IgM immunoglobulins that precipitate at temperatures less than 37°C, producing aggregates of high molecular weight. The first clue to a diagnosis of cryoglobulinemia is laboratory artifacts detected in the automated blood cell counts. The precipitated cryoglobulins of various sizes may be falsely identified as leukocytes or platelets causing pseudoleukocytosis and pseudothrombocytosis. At the same time, the RBC indexes are generally unaffected. Correction of the artifacts for automated counts can be obtained by warming the blood to 37°C or by keeping the blood at 37°C from the time of collection to the time of testing. Peripheral blood smear typically shows slightly basophilic extracellular material, and leukocyte cytoplasmic inclusions are occasionally found.

17.3.3 Pseudothrombocytopenia

Pseudothrombocytopenia is caused by various etiologies, including giant platelets, anticoagulant-induced pseudothrombocytopenia, platelet satellitism, and cold agglutinin-induced platelet agglutination. Pseudothrombocytopenia may occur with giant platelets. Due to their large size, the giant platelets are excluded from electronic platelet counting, causing pseudothrombocytopenia. This scenario is of particular clinical importance in patients with rapid consumption of platelets in the peripheral circulation as observed in DIC, acute immune thrombocytopenic purpura, or thrombotic thrombocytopenic purpura. Effective platelet production by bone marrow in these cases should be present with many large platelets in peripheral blood, many of which may not be identified by automated analyzers. An accurate platelet count can be obtained with a manual count using phase contrast microscopy.

Anticoagulant-induced pseudothrombocytopenia is an *in vitro* platelet agglutination phenomenon generally seen in specimens collected in EDTA tubes. It has been reported both in healthy subjects and in patients with various diseases (including collagen vascular disease and neoplasm), and it has an overall incidence of approximately 0.1%. Although the agglutination is most pronounced with EDTA, it may occasionally occur with other anticoagulants, such as heparin, citrate, and oxalate [5]. Because the platelet aggregates are large, the automated hematology counters do not recognize them as platelets, leading to lower platelet counts. In some cases, the aggregates are large enough to be counted as leukocytes by automated instruments, causing a concomitant pseudoleukocytosis. The platelet aggregation in pseudothrombocytopenia is usually temperature sensitive, with maximal activity at room temperature. The EDTA-induced pseudothrombocytopenia is mediated by autoantibodies of IgG, IgM, and IgA subclasses directed at an epitope on glycoprotein IIb. This epitope is normally hidden in the membrane GpIIb/IIIa because ionized calcium maintains the heterodimeric structure of the GpIIb/IIIa complex. Through its calcium chelating effect, EDTA dissociates the GpIIb/IIIa complex with GpIIb epitope exposure. It has been noted that in Glanzmann's thrombasthenia, a disorder characterized by the quantitative and/or qualitative abnormality of GpIIb/IIIa, pseudothrombocytopenia does not occur. Interestingly, in recent years, abciximab (a GpIIb/IIIa antagonist) has been found to be associated with pseudothrombocytopenia [6]. If anticoagulant-induced pseudothrombocytopenia is suspected, a peripheral blood smear should be examined for platelet clumping.

Platelet satellitism has features similar to anticoagulant-induced pseudothrombocytopenia. In the presence of EDTA, platelets bind to leukocytes and form rosettes. The binding is usually to neutrophils, but binding to other leukocytes has also been reported. Automated analyzers do not identify platelets that bind to leukocytes, resulting in pseudothrombocytopenia. Platelet satellitism is mediated by autoantibodies of IgG type directed at GpIIb/IIIa on the platelet membrane and to an Fcγ receptor III on the neutrophil membrane.

Platelet agglutination due to cold agglutinins causing pseudothrombocytopenia is a rare condition. The platelet agglutination is anticoagulant independent, usually occurs at 4°C, and is mediated by IgM autoantibodies directed against GpIIb/IIIa. Because these autoantibodies have little activity at temperatures greater than 30°C, these are not associated with any clinical significance.

17.3.4 Spurious Leukocytosis

The presence of microorganisms in the peripheral blood can result in spuriously high WBC counts or differentials by automated analyzers. Organisms that have been shown to be associated with this artifact include *Histoplasma capsulatum*, *Candida* sp., *Plasmodium* sp., and *Staphylococcus* sp. Spurious leukopenia due to EDTA is sometimes encountered. Leukoagglutination has been reported as a transient phenomenon in neoplasia (especially lymphoma), infections (infectious mononucleosis, acute bacterial infection, etc.), alcoholic liver diseases, and autoimmune diseases (rheumatoid arthritis, etc.). It can also occur in the absence of any obvious underlying disease, even though an inflammatory condition is often found. Other EDTA-dependent counting errors are well-known—platelet clumps and platelet-to-neutrophil satellitism. Association of neutrophil clumping and platelet satellitism has also been observed.

17.3.5 False-positive Osmotic Fragility Test

The osmotic fragility test is useful for diagnosis of hereditary spherocytic hemolytic anemia. Spherocytes are osmotically fragile cells that rupture more easily in a hypotonic solution than do normal RBCs. Because these cells have a low surface area:volume ratio, they lyse at a higher solution osmolarity than do normal RBCs with discoid morphology. After incubation in a hypotonic solution, a further increase in hemolysis is typically seen in hereditary spherocytosis. Cells that have a larger surface area:volume ratio, such as target cells or hypochromic cells, are more resistant to lysing in a hypotonic solution.

Conditions associated with immunologically mediated hemolytic anemias may present with many microspherocytes in peripheral blood. Consequently, the fragility test can be positive in immunologically mediated hemolytic anemias, besides hereditary spherocytosis, but the former would have a positive direct Coombs test and the latter would not.

17.4 ERRORS IN COAGULATION TESTING

Patients with coagulation disorders may either bleed or form thromboses. Hemostasis involves activation of the clotting factors and platelets. Evaluation of platelet events may include a CBC, examination of the peripheral smear, bleeding time (BT), and platelet aggregation test. Evaluation of the clotting factors is typically done by partial prothrombin time (PPT) and activated partial thromboplastin time (aPTT; also abbreviated as APTT) measurements. Abnormal PT or aPTT will usually lead to the performance of mixing studies to determine whether the abnormal result is due to factor deficiency or inhibitors. If there is correction of the prolonged clotting time in the mixing study, then factor assays will be performed to identify the deficient factor(s). Inhibitors include specific clotting factor inhibitors as well as lupus anticoagulants. Inhibitor screen and inhibitor assays or confirmatory tests for lupus anticoagulants will follow. Blood samples for coagulation tests are typically obtained in tubes with sodium citrate buffer as anticoagulant. The various sources of erroneous test results in coagulation testing are addressed in this section.

17.4.1 Errors in PT and aPTT Measurements

PT measures the time required for a fibrin clot to form after addition of tissue thromboplastin and calcium to platelet-poor plasma collected in a citrated tube. PT measures the activity of clotting factors VII, X, V, II, and fibrinogen. If the aPTT is normal, then a prolonged PT is due to factor VII deficiency. PT is relatively insensitive to minor reductions in the clotting factors. aPTT is prolonged with deficiencies of XII, XI, X, IX, VIII, V, II, and fibrinogen. Just like PT, aPTT can be normal in minor deficiencies. In general, the deficient factor must be approximately 20–40% to cause a prolonged aPTT. Most laboratories use automated methods for PT and aPTT measurements. Either optical or mechanical methods are employed to monitor clot formation. If measured with optical methods, shortened times may be seen with turbid plasma (e.g., hyperlipidemia and hyperbilirubinemia). aPTT is a test conventionally used to monitor heparin therapy. It is important to properly separate plasma from platelets as soon as possible. Platelet factor 4 can neutralize heparin, thus spuriously reducing aPTT values. Factor VIII levels are reflected in aPTT measurements. Factor VIII is an acute phase reactant. Again, if aPTT is being used to monitor heparin therapy in a patient who has an underlying cause for acute phase reactants to be elevated, aPTT values may be falsely lower than expected.

17.4.2   Errors in Thrombin Time Measurement

Thrombin time (TT) measures the time to convert fibrinogen to fibrin. Dysfibrinogenemia, elevated levels of fibrin degradation product, and paraproteins can interfere with fibrin polymerization, thus falsely prolonging TT. Amyloidosis can inhibit the conversion of fibrinogen to fibrin, also prolonging TT. In certain malignancies, heparin-like anticoagulants have been known to be the cause of prolonged TT. Next, specific selected test errors that are often seen in coagulation laboratory are discussed in more detail.

17.4.2.1   Incorrectly Filled Tubes

Citrate tubes for coagulation tests are designed for a 9:1 ratio of blood to citrate buffer. Underfilling or overfilling of the citrate tube results in imbalances in this blood:buffer ratio and produces artificially prolonged or shortened clotting times, respectively. Underfilling or overfilling respectively results in too little or too much blood sample for a fixed amount of anticoagulant in the tube. The amount of blood that fills the citrated tubes is controlled by vacuum, which maintains the proper 9:1 ratio of blood to anticoagulant. Underfilling may be caused by air bubbles in the tube, vacuum loss, or not allowing the tube to completely fill during the blood collection process. If the tube stopper is removed, it becomes difficult to obtain the correct amount of blood and attain the proper 9:1 ratio. If the patient's hematocrit is known in advance of blood collection to be greater than 55% (e.g., in patients with polycythemia) or less than 21% (patients with severe anemia), the amount of sodium citrate must be adjusted using the following formula:

$$C = 0.00185 \times (100\ H) \times V$$

where C is the volume of 3.2% sodium citrate in milliliters, H is hematocrit in percentage, and V is the volume of blood in milliliters.

Errors in clotting tests due to hematocrit changes without adjustment in the citrate volume are most significant with elevated hematocrits because even severe anemia does not significantly change PT or aPTT.

17.4.2.2 Dilution or Contamination with Anticoagulants

Blood collection from indwelling lines or catheters is a potential source of testing error. Sample dilution from incomplete flushing or hemolysis caused by improper catheter insertion can alter coagulation test results. Heparinized lines should be avoided if blood must be drawn from an indwelling catheter. If it is necessary to use a heparinized line, adequate line flushing must be achieved before blood collection. The National Committee for Clinical Laboratory Standards (NCCLS) recommends flushing lines with 5 mL of saline. At least 5 mL, or six times the dead space volume of the catheter, should be discarded before blood collection. The *Intravenous Nursing Standards of Practice* recommends that manufacturers' instructions should always be followed with regard to the appropriate discard volume. These guidelines also state that blood should not be acquired from various types of indwelling cannulae, venous administration sets, and indwelling cardiovascular or umbilical lines. Even when the initial volume drawn is discarded before blood collection according to these guidelines, specimens drawn from a heparinized line are still easily contaminated with heparin. Consequently, it is best to draw blood for coagulation tests directly from a peripheral vein, avoiding the arm in which heparin, hirudin, or argatroban is being infused for therapy. Before coagulation testing, heparin may be removed or neutralized with polybrene in the coagulation laboratory; however, residual heparin may continue to cause testing interference.

By enhancing antithrombin activity, heparin inhibits activated factors II (thrombin), X, IX, XI, XII, and kallikrein. In contrast, lepirudin, danaparoid, and argatroban inhibit only activated factor II. These anticoagulants (heparin, lepirudin, and argatroban) prolong aPTT and interfere with coagulation tests such as factor assays and lupus anticoagulant assays. Factor assays may yield falsely low levels, whereas lupus anticoagulant may be falsely positive.

17.4.2.3 Traumatic Phlebotomy

Traumatic phlebotomy can result in artificially shortened coagulation results such as for PT and aPTT. This is due to excessive activation of coagulation factors and platelets by release of tissue thromboplastin from endothelial cells. A proper free-flowing puncture technique will avoid the release of tissue thromboplastin and also avoid this artifact.

17.4.2.4   *Fibrinolysis Products and Rheumatoid Factor*

Fibrinolysis is mediated by plasmin, which degrades fibrin clots into D-dimers and fibrin degradation products. Plasmin also degrades intact fibrinogen, generating fibrinogen degradation products. Fibrin degradation products and fibrinogen degradation products are collectively known as fibrin/fibrinogen degradation products (FDPs) or fibrin/fibrinogen split products (FSPs). Assays for D-dimer and FDPs are semiquantitative or quantitative immunoassays:

> *Latex agglutination:* Patient plasma is mixed with latex particles that are coated with monoclonal anti-FDP antibodies. If FDP is present in the patient plasma, the latex particles agglutinate as FDP binds to the antibodies on the latex particles. These agglutinated clumps are detected visually. Various dilutions of patient plasma can be tested to provide a semiquantitative result known as FDP titer. Latex agglutination assays are also available for D-dimers. Various automated and quantitative versions of this assay are commercially available for D-dimers, in which the agglutination is detected turbidimetrically by a coagulation analyzer rather than visually by a technologist.
>
> *Enzyme-linked immunosorbent assay (ELISA):* Quantitative ELISAs are also available for FDP and D-dimers. The traditional ELISA method is accurate but is not useful due to a long analytical time. An automated, rapid ELISA assay for D-dimers is also available (VIDAS, bioMérieux, Durham, NC).

One of the most important limitations of D-dimer and FDP assays is interference by high rheumatoid factor (RF) levels. This may cause false-positive results with almost all available assays. The most useful clue to detect this interference is evaluation of the DIC panel. If all values in this panel (PT, aPTT, TT, and fibrinogen) are normal except for FDP or D-dimer, the presence of rheumatoid factor is most likely the cause.

17.4.3   Platelet Aggregation Testing with Lipemic, Hemolyzed, or Thrombocytopenic Samples

Platelet aggregation measures the ability of platelets to adhere to one another and form a hemostatic plug, which is the key component of primary hemostasis. It can be performed using either platelet-rich plasma (PRP) or whole blood. Substances such as collagen, ristocetin, arachidonic acid, adenosine 5′-diphosphate (ADP), epinephrine, and thrombin can stimulate platelets and hence induce aggregation. Response to these aggregating agents (known as agonists) provides a diagnostic pattern for different disorders of platelet function. Measurement of aggregation response is typically based on changes in the optical density of the sample.

Platelet aggregation is affected by a number of confounding variables. Lipemic and hemolyzed samples complicate aggregation measurements because they obscure spectral changes due to platelet aggregation. Thrombocytopenia also makes platelet aggregation evaluations difficult to interpret because a low platelet count by itself may yield an abnormal aggregation pattern.

17.4.4 Challenges in Anticoagulants and Lupus Anticoagulant Tests

The International Society on Thrombosis and Haemostasis Subcommittee on Lupus Anticoagulant have recommended two sensitive screening tests for lupus anticoagulants that assess different components of the coagulation pathway [7]. These tests are clotting time-based assays, such as the dilute Russell viper venom time (dRVVT), and aPTT-based assays, such as kaolin clotting time, and dilute prothrombin time (tissue thromboplastin inhibition test). Lupus anticoagulants prolong various phospholipid-dependent clotting times in the laboratory because they bind to phospholipid and thereby interfere with the ability of phospholipid to serve its essential cofactor function in the coagulation cascade. Lupus anticoagulant screening assays usually have a low concentration of phospholipid to enhance sensitivity. Any abnormal (prolonged) screening result typically requires a 1:1 mixing study in which the patient plasma is mixed with one equal volume of normal plasma to demonstrate that the clotting time remains prolonged upon mixing. Confirmatory assays are performed if the screening assay remains abnormal after the 1:1 mixing. Confirmatory assays typically demonstrate that upon addition of excess phospholipid, the clotting time shortens toward normal. The platelet neutralization procedure (PNP) is a confirmatory assay in which the source of the excess phospholipid is freeze–thawed platelets. The hexagonal phospholipid neutralization procedure is also based on the same principle—that is, the clotting time becomes corrected after addition of phospholipid in the hexagonal phase. Note that aPTT may or may not be prolonged, depending on the amount of phospholipid in the reagent.

In many lupus anticoagulant assays, heparin (including subcutaneous low-dose heparin) may cause false-positive lupus anticoagulant results. By enhancing antithrombin activity, heparin inhibits activated factors II (thrombin), X, IX, XI, XII, and kallikrein. Subsequently, clotting times such as PT and aPTT are prolonged and interfere with lupus anticoagulant assays. Lepirudin, danaparoid, and argatroban inhibit activated factor II and also can prolong clotting times. Before coagulation testing, heparin may be removed or neutralized with polybrene in the coagulation laboratory; however, residual heparin may continue to cause testing interference. Results for

BOX 17.2 SOURCES OF LABORATORY ERRORS IN COAGULATION

- Falsely prolonged clotting times
  - Underfilling of citrate tube
  - Polycythemia
  - Sample from indwelling catheters (dilution or contamination with anticoagulant)
- Falsely shortened clotting times
  - Overfilling of citrate tube
  - Traumatic phlebotomy
  - Turbid plasma (e.g., hyperlipidemia and hyperbilirubinemia) in optical instrument
- Falsely shortened aPTT in patients on heparin
  - Delay in separation of plasma from platelets
  - Elevated factor VIII (acute phase reactant)
- Falsely prolonged TT
  - Dysfibrinogenemia
  - Elevated levels of FDP and paraproteins
  - Amyloidosis

- Heparin-like anticoagulants (in malignancy)
- Falsely high FDP and D-dimer
  - Rheumatoid factor
- Falsely abnormal platelet function
  - Lipidemia
  - Hemolysis
  - Thrombocytopenia
- Falsely low factor levels
  - Heparin
  - Lepirudin
  - Danaparoid
  - Argatroban
- False-positive results of lupus anticoagulant tests
  - Heparin
  - Lepirudin
  - Danaparoid
  - Argatroban

lupus anticoagulant assays can be interpreted correctly in patients on coumadin. Box 17.2 summarizes important laboratory errors arising in the coagulation laboratory due to various entities.

KEY POINTS

- Hemoglobin measurement is based on absorption of light at 540 nm. If the sample is turbid, this will produce higher hemoglobin levels. Examples of such a state include hyperlipidemia, patients on parenteral nutrition, hypergammaglobulinemia, and cryoglobulinemia. Turbidity from very high WBC count can also falsely elevate hemoglobin levels.
- Smokers have high carboxyhemoglobin that may falsely elevate the measured hemoglobin level.
- Large platelets may be counted by some instruments as RBCs. Also, red cell fragments greater than 40 fL will be counted as whole red cells. In both situations, the RBC count will be falsely high.
- Cold agglutinins will cause red cell agglutination *in vitro* and result in low RBC counts. If cold agglutinins are suspected, the sample should be warmed to obtain an accurate RBC count.
- If there is red cell agglutination, then red cell clumps will be counted as single red cells but the volume of the estimated cell will be much higher. This may result in falsely high MCV values. If large platelets

are counted as red cells, then these platelets typically have less volume than a normal red cell. This may also result in falsely low MCV values.

- One of the most frequent situations is high WBC count in the presence of a significant number of NRBCs. If an accurate WBC count is required, then a corrected WBC count should be performed.
- Platelet aggregates and non-lysis of red cells are other causes of spuriously high WBC counts. Hemoglobinopathies and target cells are important causes of non-lysis of red cells.
- In myelodysplastic syndrome, if the myeloid series is affected, then hypolobated and hypogranular neutrophils may be present. Automated analyzers may no longer count these dysplastic neutrophils as such; instead, these cells may be counted as lymphocytes.
- A key difference between lymphocytes and monocytes is their size, with monocytes being significantly larger. Reactive (activated) lymphocytes typically have more abundant cytoplasm compared to nonreactive lymphocytes. Their size approaches that of a monocyte. Thus, these lymphocytes may be counted as monocytes.
- Partial clotting of the specimen or platelet activation during venipuncture may cause platelet aggregation. Both mechanisms may lead to low platelet counts. Checking the specimen for any clot and analyzing the histograms, as well as reviewing the smear are all important steps to avoid misleading low platelet counts. Various other mechanisms explain falsely low platelet counts, otherwise referred to as pseudothrombocytopenia (anticoagulant-induced pseudothrombocytopenia), platelet satellitism, giant platelets, and cold agglutinin-induced platelet agglutination.
- Falsely elevated platelet counts are much less common than falsely low counts. Fragmented red cells or white cell fragments may be counted as platelets, giving rise to high platelet counts.
- Cryoglobulins are typically IgM immunoglobulins that precipitate at temperatures less than 37°C, producing aggregates of high molecular weight. The first clue to a diagnosis of cryoglobulinemia is laboratory artifacts detected in the automated blood cell counts. The precipitated cryoglobulins of various sizes may be falsely identified as leukocytes or platelets causing pseudoleukocytosis and pseudothrombocytosis.
- Conditions associated with immunologically mediated hemolytic anemias may present with many microspherocytes in peripheral blood. Consequently, the fragility test can be positive in immunologically mediated hemolytic anemias besides hereditary spherocytosis, but the former would have a positive direct Coombs test and the latter would not.

■ Citrate tubes for coagulation tests are designed for a 9:1 ratio of blood to citrate buffer. Underfilling or overfilling of the citrate tube results in imbalances in this blood:buffer ratio and produces artificially prolonged or shortened clotting times, respectively.

■ aPTT (also abbreviated as APTT) is a test conventionally used to monitor heparin therapy. It is important to properly separate plasma from platelets as soon as possible. Platelet factor 4 can neutralize heparin, thus spuriously reducing aPTT values. Factor VIII levels are reflected in aPTT measurements. Factor VIII is an acute phase reactant. Again, if aPTT is being used to monitor heparin therapy in a patient who has an underlying cause for acute phase reactants to be elevated, aPTT values may be falsely lower than expected.

■ TT measures the time to convert fibrinogen to fibrin. Dysfibrinogenemia, elevated levels of fibrin degradation product, and paraproteins can interfere with fibrin polymerization, thus falsely prolonging TT.

■ Traumatic phlebotomy can result in artificially shortened coagulation results such as for PT and aPTT. This is due to excessive activation of coagulation factors and platelets by release of tissue thromboplastin from endothelial cells.

■ One of the most important limitations of D-dimer and FDP assays is interference by high rheumatoid factor (RF) levels. This may cause false-positive results with almost all available assays. The most useful clue to detect this interference is evaluation of the DIC panel. If all values in this panel (PT, aPTT, TT, and fibrinogen) are normal except for FDP or D-dimer, the presence of rheumatoid factor is most likely to be the cause.

■ Platelet aggregation is affected by a number of confounding variables. Lipemic and hemolyzed samples complicate aggregation measurements because they obscure spectral changes due to platelet aggregation. Thrombocytopenia also makes platelet aggregation evaluations difficult to interpret because a low platelet count by itself may yield an abnormal aggregation pattern.

References

[1] Carraro P, Plebani M. Errors in a stat laboratory: types and frequencies 10 years later. Clin Chem. 2007;53:1338−42.

[2] Chen BH, Fong JF, Chiang CH. Effect of different anticoagulant, underfilling of blood sample and storage stability on selected hemogram. Kaohsiung J Med Sci 1999;15:87−93.

[3] Lima-Oliveira G, Lippi G, Salvagno GL, Montagnana M, et al. Brand of dipotassium EDTA vacuum tube as a new source of pre-analytical variability in routine hematology testing. Br J Biomed Sci 2013;70:6−9.

[4] Nosanchuk JS, Roark MF, Wanser C. Anemia masked by triglyceridemia. Am J Clin Pathol 1977;62:838−9.

[5] Schrezenmeier H, Muller H, Gunsilius E, Heimpel H, et al. Anticoagulant-induced pseudo-thrombocytopenia and pseudoleucocytosis. Thromb Haemost 1995;73:506−13.

[6] Stiegler H, Fischer Y, Steiner S, Strauer BE, et al. Sudden onset of EDTA-dependent pseudo-thrombocytopenia after therapy with the glycoprotein IIb/IIIa antagonist c7E3 Fab. Ann Hematol 2000;79:161−4.

[7] Brandt JT, Triplett DA, Alving B, et al. Criteria for the diagnosis of lupus anticoagulants: an update. Thromb Haemost 1995;74(4):1185−90.

Index

Note: Page numbers followed by *"f"*, *"t"* and *"b"* refers to figures, tables and boxes respectively.